THE PHYSICAL BIOLOGY
OF PLANT CELL WALLS

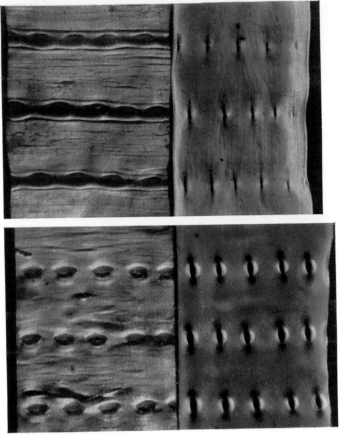

Fig. 9.1 Polarization photomicrographs with a Red I plate of the outer (left) and inner (right) wall layer of the central siphon of some Dasycladaceae. For explanation, see text.

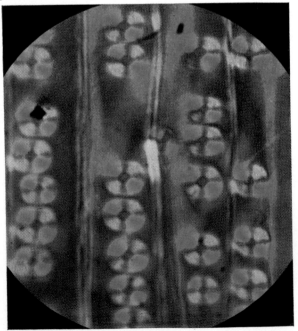

Fig. 10.16 The appearance of bordered pits in conifer tracheids between crossed polaroids. For explanation, see text.

The Physical Biology of Plant Cell Walls

R. D. Preston F.R.S.

Professor Emeritus and former Head of Astbury Department
of Biophysics, University of Leeds

London

Chapman and Hall

First published 1974
by Chapman and Hall Ltd.
11 New Fetter Lane, London ECP 4EE

© *R.D. Preston 1974*

Set by EWC Wilkins Ltd, London N12 0EH
and printed in Great Britain by Fletcher & Son Ltd, Norwich

ISBN 0 412 11600 6

Distributed in the U.S.A.
by Halsted Press, a Division
of John Wiley & Sons, Inc., New York

Library of Congress Catalog Card Number 73/21335

To Eva, my wife and colleague

Contents

Preface   *page*   xii

1   Introduction   1
1.1 Early wall studies   3
1.2 The morphology of cells   7
1.3 Primary *v*. secondary walls   11

2   Intra-atomic and intramolecular bonding
     and molecular models   15
2.1 The electron   16
2.2 Covalent bonds   19
2.3 Ionic bonds   21
2.4 Weak bonds   21
     2.4.1 Permanent dipole-dipole interaction   21
     2.4.2 Permanent dipole-induced dipole   24
     2.4.3 Transient dipole-induced dipole   24
     2.4.4 Hydrogen bond interaction   25
2.5 Molecular models   27
     2.5.1 Skeleton models   27
        (a) The ball and spoke model   27
        (b) The wire model   29
     2.5.2 Space filling models   29

3   The chemical components   32
3.1 Monosaccharides   32

3.2 Disaccharides   *page* 38
3.3 Polysaccharides   39
3.4 Cell wall fractionation   40
3.5 The structure of polysaccharides   43
    3.5.1 Skeletal polysaccharides   44
        (a) Cellulose   44
        (b) Non-cellulosic polysaccharides   49
    3.5.2 Hemicelluloses   53
    3.5.3 Polyuronides   56
    3.5.4 Lignin   57
    3.5.5 Protein   59
        (a) Structural proteins   59
        (b) Enzymes   65

4   Structure determination – optical microscopy   68
4.1 The principles of microscopy   68
4.2 The wave nature of electromagnetic radiation   69
4.3 Resolving power   71
4.4 The polarizing microscope   75
    4.4.1 The extinction directions   77
    4.4.2 The major extinction position (m.e.p.)   79
    4.4.3 Determination of refractive indices and
          birefringence   86
        (a) de Sénarmont compensator   88
        (b) Elliptic or Brace-Koehler compensator   93
    4.4.4 The interpretation of the m.e.p. and the path
          difference   94
        (a) Microfibrils tilted to plane of surface   95
        (b) Superposed wall lamellae differing in m.e.p.   98
    4.4.5 Birefringence and its interpretation   105
4.5 Interference microscopy   108

5   Structure determination – X-ray diffraction   119
5.1 The unit cell – two dimensions   120
5.2 The unit cell – three dimensions   123
5.3 The determination of interplanar spacings   126
5.4 Diffraction by plant fibres   128
5.5 The X-ray-derived structure of cellulose   133

5.6 Regenerated cellulose — cellulose II *page* 140
5.7 The structure of chitin 141
5.8 Crystallite size 144
    5.8.1 Line-broadening 144
    5.8.2 Low-angle scattering 146

6   Structure determination — electron microscopy 148
6.1 Historical development 148
6.2 The instrument 150
6.3 The specimen 153
    6.3.1 Specimen preparation 154
        (a) Manual dissection 154
        (b) Surface replication 155
        (c) Ultra-thin sectioning 159
        (d) Contrast 161

7   General principles of wall architecture 163
7.1 Introduction 163
7.2 Microfibril size and structure in fresh tissue 172
7.3 Elementary microfibrils 174
7.4 Disposition of the wall substances 183

8   Detailed structure — cellulosic algae 192
8.1 *Valonia*-type walls 192
    8.1.1 *Valonia* 193
    8.1.2 The Cladophoraceae 203
        (a) Swarmers 217
    8.1.3 *Glaucocystis* and *Oocystis* 218
8.2 Other algal walls 222
    8.2.1 Green algae 224
    8.2.2 Brown algae 228
        (a) Alginic acids 230
    8.2.3 Red algae 237

9   Detailed structure — non-cellulosic algae 238
9.1 The mannam weeds 239

9.1.1 Chemical constitution   *page* 239
9.1.2 Physical organization   240
    (a) Polarization microscopy   240
    (b) X-ray diffraction   243
    (c) Electron microscopy   251
9.2 The xylan weeds   255
    9.2.1 Chemical constitution   255
    9.2.2 Physical organization   257
        (a) Polarization microscopy   257
        (b) Electron microscopy   258
        (c) X-ray diffraction   262
9.3 Xylan-mannan algae   271
9.4 Taxonomic considerations   273

10   Flowering plants; secondary walls   276
10.1 Wood tracheids and fibres   276
    10.1.1 Determination of microfibril angle   278
        (a) Optical microscopy   278
        (b) X-ray diffraction   282
        (c) Electron microscopy   288
    10.1.2 Wall layering   288
        (a) Chemical differentiation   288
        (b) Physical differentiation   291
    10.1.3 Bordered pits   303
    10.1.4 Pit aspiration   310
    10.1.5 Helical thickening   311
    10.1.6 The warty layer   312
10.2 Other mature cell types   312
    10.2.1 Collenchyma cells   312
    10.2.2 Phloem fibres   315
    10.2.3 Vessel elements   315
    10.2.4 Parenchyma cells   316
10.3 Structural variation in homologous cells   317

11   Viscoelastic properties of secondary cell walls 327
11.1 Tensile properties   327
    11.1.1 Tensile properties of individual cells   331
    11.1.2 Tensile properties of whole tissues   334

11.2 Mechanical failure   *page* 346
    11.2.1 Failure under tension   347
    11.2.2 Compression failure   348
11.3 Growth stresses   352
11.4 Swelling and shrinkage   353
    11.4.1 Single cells   356
    11.4.2 Whole wood   360
11.5 Porosity   367
    11.5.1 Permanent pit membrane pores   369
    11.5.2 Cell wall capillaries   375
        (a) Density considerations   376
        (b) Nitrogen adsorption   378

12   Wall extension and cell growth   383
12.1 Wall extension and microfibrillar orientation   385
    12.1.1 Experimental tests of the multi-net growth
            hypothesis (M.G.H.)   392
        (a) Difficulties with the M.G.H.   393
12.2 Wall extensibility   397
12.3 Spiral growth   409
12.4 Growth and turgor pressure   413
12.5 Cell growth and wall protein   415
12.6 Wall synthesis and auxin   419
12.7 The wall growth process — a summary   421

13   Wall biosynthesis   425
13.1 Matrix substances   429
13.2 Cellulose   433
    13.2.1 Biochemical pathways   433
    13.2.2 Structural considerations   437
        (a) Wall structural factors limiting the synthetic
            mechanism   439
    13.2.3 The microtubule or template hypothesis   442
    13.2.4 The ordered granule hypothesis   444

References   457

Author Index   480

Subject Index   487

Preface

In reviewing during 1959 *The plant cell wall* by the later
P.A. Roelofsen I commented that this could be the last book to be
written on the cell wall of plants as a whole and gave it even then
as my view that subsequent texts would need to be confined to
limited aspects only. The literature was already exploding and even
at that time only an encyclopaedic mind could encompass the data
and only a man like Roelofsen could have written so informatively
about them. Fifteen more years of water under the bridge have
continued the flood and, if some authors' jewels have been jetsam,
there has been nevertheless an alarming amount of flotsam for the
net of the fisherman. It will consequently be understood that no
volume even about 500 pages in length can present a comprehen-
sive account of the totality of plant cell wall investigation now in
process. I have therefore limited myself to a statement of what
seem to me the more important *physical* studies currently
blossoming in this field and this has meant dealing slightly, if at all,
with some otherwise highly significant investigations. Even within
this limitation the reader will find some omissions. There is, for
instance, hardly any mention of fungi and bacteria and no treatment whatever of the relationship which is currently being
increasingly explored between the wall of a host plant and an
invading pathogen; nor is any account taken of the copious information now available relating to the protection of cell walls in, for
instance, timber against microbial attack. Though this is in some
measure a mark of my incompetence and a consequence of my
own special interests, in those researches which I and my colleagues
have been heavily engaged and which are hardly touched upon
here it is especially because of reasons of space.

It is my impression that botanists nowadays are very much more competent in, and willing to deal with, mathematical statements than they were 20 years ago and for this reason I have ventured to deal with the subjects covered in a fashion much more mathematical than seemed advisable in my 1952 book. I hope that my fellow botanists will accept this not as a hurdle unnecessarily put in their way but as a tribute; though those who wish can, I believe, skip the mathematics and still achieve understanding of what the book is about! My fellow physicists, many of whom I trust will read the book, will find, hopefully, sufficient of the biology of cell walls to bring the study to life for them; sufficient at least to represent my deep conviction that, urgent though it is that physicists should enter more and more intimately into this and all other biological disciplines, none should attempt to do so without due regard for the necessary prerequisite for the appropriate biological experience. I can only beg that neither group of scientists should feel that I have pandered too much to the other.

I leave it to the book to demonstrate that, during my working lifetime, the cell wall has passed from a relatively insignificant and inactive outer framework of plant cells, the domain therefore of only a small group of enthusiasts, to become the basis of themes central to many issues in plant science, demanding investigation by scientists of all persuasions, pure and applied, in very considerable numbers. It has brought me into contact with innumerable leading authorities to each of whom I owe a debt. Many of them are still alive and it would be invidious to name them; they know who they are and I hope they will accept this anonymous thank you. It is in the nature of things that still more of them, my early mentors and colleagues, are no longer with us and these I can thank only in spirit. I am bound to record, however, that I still owe an especial debt and one that I never knew how to repay, to my late chief, the great Professor J.H. Priestley, and to my old friend the late Professor W.T. Astbury, F.R.S. Most of the work of my own which appears in this book has been carried out in collaboration with generations of students and colleagues, and my debt to them is of course beyond measure. In particular, though, I owe most of all to my colleague and wife, Dr. Eva Frei.

The mere preparation of a book even of this size puts heavy demands on a number of key people. I would like to thank most warmly all those who accepted these demands: Miss Ann Gudgeon, my Secretary, and through her Mrs. M. Cutter and Mrs. J. Haigh, who coped with my illegible handwriting, with several retypings and with all the correspondence involved; Mr. R.K. White, my electron microscopist and Mr. W.D. Brain and Mr. L. Child, my photographers, for the beautiful electron micrographs for which this department is known; to Drs. B. Sheldrick, W. Mackie, K.D. Parker, B.J. Nelmes, J.E. Lydon and D.B. Sellen on my staff for many helpful discussions; and to my wife for enormous help with proofreading and the indexes. A number of investigators have allowed me the privilege of reproducing some of their published illustrations; these are all acknowledged in the body of the book and my debt to them is hereby acknowledged. I am also grateful to the publishers, for the courteous handling of the manuscript and of my requests, for their careful attention to detail, and for the speed with which they proceeded. In these ways I have had great help in eradicating mistakes from the book; any errors which still remain are entirely my own responsibility.

Astbury Department of Biophysics,   R.D.P.
The University,
Leeds.
1973.

CHAPTER 1

Introduction

Every living cell, plant or animal, is delimited by a membrane completely surrounding it based upon a complex assembly of lipids and proteins. For almost all plant cells, but in few animal cells, this membrane — the plasmalemma — elaborates on its outer surface a layer — the cell wall — which is rigid or semi-rigid and is chemically distinct in that the structurally important components are polysaccharide. Non-polysaccharide components are also incorporated in the wall, such as lignin in xylem and cutin in epidermal walls, and, as will be shown in this book, proteins are also present and important; nevertheless, the essential and sudden change at the plasmalemma surface is from a basically lipo-protein to a basically polysaccharide formulation.

In all plant cell walls one, and usually only one, of these polysaccharides is crystalline. The crystals are submicroscopic in size and have no external crystalline form but the crystallinity can be recognised by all those methods, some of which are detailed in this book, which can detect and measure the molecular ordering of material in the crystalline state. It was for a long time believed that cellulose — a linear polymer of β-D-glucose — was the unique crystalline polysaccharide of the plant cell wall but we now know that this is not the case. Plants have exploited a variety of polysaccharides in combinations which produce walls with clearly similar physical properties in spite of rather profound chemical differences. Nevertheless, cellulose predominates and is perhaps the only crystalline wall polysaccharide of higher plants. Its importance, and the importance of cell walls, even in the carbon economy of the biosphere ought not to be overlooked. It has been estimated that about 10^{11} tons of cellulose are produced and destroyed annually (Hess, 1928) and therefore about 10^{12} tons of total wall material. This overshadows by far the 10^9 tons per annum of its closest rival chitin (Tracey, 1957). The energy equivalent is of the order of 5×10^{15} kWh so that cell walls form a storehouse, both of nutrition and energy which undoubtedly, as pointed out by Colvin (1972), tends to smooth out wide fluctuations in supply.

Temporary withdrawal of cell wall substances from the natural turnover has been practised by man from well beyond historic times, for clothing, instruments, weapons and as building material and these uses have continued down to the present day, and even extended in spite of the increasing use of man-made plastics in our present era. The current world production of timber alone (which is entirely cell wall) may be estimated at about 140×10^9 board ft, enough to make a board walk 1 in thick and 1000 ft wide around the equator. The turnover in Britain alone approaches £300 million annually so that cell walls are big business; and this is not, of course, the end of it. To this must be added the needs of the paper trade, the cotton and other vegetable fibre industries and the industries based on regenerated cellulose and industries using cell wall products as additives (e.g. alginic acids in foods and other commodities). Moreover, cell walls are important in industries in which they are not used; the efficiency of oil extraction from oil seeds depends on cell wall properties, the quality of potatoes and apples depends in some degree on cell walls. All in all there is clear reason why industry should be, and is, supporting research into the physical properties of cell walls which make them uniquely important to the exploiter. These properties are identically those which underly the functions of cell walls in plants, at an understanding of which this book is in principle aimed. Occasion has therefore been taken to deal with some properties in detail which, though significant for their biological implications, are especially of concern to cell wall, including timber, technologists. The fascination of cell wall studies lies, however, in considerations far removed from commercial exploitation.

The question: 'Why do plants develop a cell wall?' is unanswerable in scientific terms as are all questions beginning 'Why?' The effects of the presence of a cell wall are, however, readily defined.

Once a protoplast is surrounded by a cell wall, sudden, quickly reversible, changes of shape are prohibited and it is no longer

2

possible for a tissue to develop into a muscle even if the biochemistry of the constituent cells would allow it. The separation between organisms with and organisms without a wall is therefore a fundamental one. Some plants do, of course, undergo rapid movement, such as the response to touch of the sensitive plant *Mimosa pudica* or of the insectivorous plant Venus flytrap (*Dioneea*), but these are not a display of muscular activity and are not quickly reversible. Growth is nevertheless allowed by a relatively slow yielding of cell walls under stress and it is to be anticipated that the rate of volume increase of a cell — and therefore of a tissue and of a whole organism — should be a function of the properties of the wall as dictated by its structure. Moreover, cessation of growth might well also be dictated by the wall. Both of these anticipations are to a large extent realized. Moreover, since the wall is deposited over the plasmalemma and new lamellae are continuously added during the life of a cell, changes in wall structure in a spatial sequence from outside the cell to the inside express also a time sequence depicting to some extent the history of the cell. Again, the precise architectural details of a wall at any stage of the development of a cell must contain information concerning the mechanism by which it has been elaborated and therefore may reflect the structure of the synthesizing machinery — perhaps of the plasmalemma itself — which can be obtained in no other way. More directly, the wall is important in the life of a plant in keeping many higher plants upright and as the first barrier to entry into the cell of water and nutrients (and of invading organisms) and the last barrier to water loss in land plants. Whereas all other functions of a cell wall depend upon the structure of the solid framework of the wall this last function concerns the porosity of the wall, the interstices between the framework and, as a consequence, no statement of wall structure is complete without consideration of those parts of the wall from which structural components are absent. The growing belief that water moves through the roots and leaves of plants only along the cell walls adds a particular emphasis.

1.1 Early wall studies

The only part of a plant cell immediately obvious under a microscope is the cell wall and it is therefore understandable that this was the first feature to be remarked upon in plants and that it should have continued to be the sole centre of interest in botany for very many decades. The credit for the first observation is usually given to Henshaw who observed the walls of vessels in the wood of walnut trees as early as 1661. Study may be said, however, to have properly begun with the publication of Robert Hook's *Micrographia* in 1667. Hook was not in any sense a botanist and his delineation of cell structure in cork may perhaps be regarded as trivial. The few years which followed, however, saw rapid advances at the hands of Marcello Malpighi in Italy and Nehemiah Grew in England. These two presented their preliminary communications to the Royal Society of London in the same year (1671), though a number of years separated their major publications (Malpighi's *Anatomae Plantarum* in 1675 and Grew's *The Anatomie of Plantes* in 1682). They both described in some detail the solid framework of plants, Malpighi with the tasteful elegance of a masterly sketch and Grew with an eye to the relevance of the chemistry and physics of his day and particularly to Cartesian philosophy. Neither worker had, of course, any clear idea of cell structure — that was to come later — but each showed in his way what an acute observer can make out even with inefficient instruments. Grew in particular became convinced that all the solid framework of what we now call cell walls consisted of exceedingly fine threads woven and twisted together, lying horizontally in parenchyma (a term which he introduced) and vertically in fibres. These threads he considered to be very fine indeed:

'So in the Pith of a Bulrush or the common Thistle, and some other plants: not only the threads of which the Bladders; but also the single Fibres of which the Threads are composed; may sometimes with the help of a good Glass, be distinctly seen. Yet one of these Fibres, may reasonably be computed to be a thousand times smaller than an Horse-Hair'.

Since horse-hairs are about 0.2 mm diameter, the fibres would then be 200 nm in diameter and it must be supposed that here Grew was extrapolating. Nevertheless he had seen threads grading in fineness to the limits then observable. It is always easy to read modern ideas into older and vaguer writings, but Grew expresses himself so unequivocally here that one is bound to conclude that his inspired vision led him closer to the truth than anyone else was to reach for almost the next two hundred years.

The eighteenth century was one of stagnation as far as structure is concerned and such details of anatomy as were needed for the then current upsurge in physiology were all too frequently taken bodily from Grew's work. The concept of a cell was advanced no further than it had been by Grew and Malpighi and even retrograded, as exemplified by the notion introduced by Wolff in 1759, that the young parts of plants consist of a transparent gelatinous substance within which drops of sap are secreted to produce cells. This work, incidentally, constituted a denial of the fibrillar concept of Grew.

Further advances in anatomy had, indeed, to await improvements in instrumentation and these were not forthcoming until the first quarter of the nineteenth century. Between 1812 and 1828 Selligue and Amici introduced achromatic and aplanatic microscope objectives with three double lenses and the effect was dramatic. Moldenhawer (the younger), using maceration techniques, saw for the first time whole cells separated from each other so that the cell wall could in this context for the first time be understood. His reversion to the hitherto despised fibrillar hypothesis of Grew culminated with Meyer who visualized cell construction very much in the way Grew had done.

The greatest figure of this period was, however, undoubtedly von Mohl. He gave what amounts to the modern view both on the primary and the secondary walls of cells and, though he misunderstood the bordered pit, his work enjoyed the security which comes from scrupulous care in observation and a refusal to indulge in unwarranted generalizations and abstractions. He was the first to use the polarizing microscope; while he cannot be said in any sense to have anticipated von Nägeli, his interpretation of wall striations as lines of cleavage in crystals indicates the lines along which he was thinking. von Mohl introduced the concept that the secondary wall is deposited as successive lamellae laid one upon the other from the inside by a process which he therefore called *apposition* and by 1850 had concluded that the intercellular substance (the modern true middle lamellae) is only a cementing substance. Others, including Sanio, insisted that it was part of the primary wall, though chemically modified, and thus set in motion a confusion in terminology which has persisted.

It was, therefore, recognized by now that growing cells were covered with thin expanding walls (known to contain cellulose).

These cells were further acknowledged to be cemented together by a cement which was probably formless though no attempt was made at an understanding of submicroscopic features. This was soon remedied by von Nägeli.

Although von Nägeli was a contemporary of von Mohl, the success of his application of physical principles to wall problems singles him out as the first of the modern wall biophysicists. In his *Stärkekorner* of 1858 he demonstrated what can be done by informed use of the polarizing microscope. His observations of starch grains, and subsequently of cell walls, led him to deduce, correctly, the basic crystallinity and to refute the criticisms of Hofmeister and Strasburger. He deduced, partly from swelling reactions, the existence both in starch grains and cell walls of submicroscopic elongated crystallites which he called *micelles*. These he imagined separated by water and stabilized in aggregates by the opposing attractions of the grains for each other and for water. Perhaps his greatest contribution was the concept that these aggregates grow in the wall by insertion of new micelles between the old, a process which he called *intussusception* as against the apposition of von Mohl, and in so doing creating a controversy which has not completely died down even today.

During this time, parallel investigations of the chemistry of the wall had also been proceeding. As early as 1825 the 'cellulose' extractable from the cell wall was known to be a mixture of the substances *cellulose* and *pectose*, an acknowledgement that extractable cellulose is impure which was lost sight of until fairly recently. The first use of cuprammonium to remove cellulose was made by Frémy in 1859 and subsequent staining and solubility tests soon led to the conclusion that the bulk of the pectic compounds was located in the middle lamella. The real study of the pectic compounds began, however, only in the last decade of the nineteenth century with Mangin in the 1890s. He showed correctly that they consist of two series of compounds — acidic and neutral — differing in chemical constitution from cellulose and often associated with other polysaccharides which are now called hemicelluloses. The next twenty years saw tremendous advances.

This brings us, however, into the twentieth century and to developments, treated in the following chapters, marked by progressive involvement with tools and techniques developed by physicists and applied by biologists — mostly by what have come to be called biophysicists or molecular biologists — not only to cell walls but to the whole spectrum of cytology, biochemistry and physiology. It brings us, therefore, into the body of the book.

1.2 The morphology of cells

All plants, like all animals, originate from individual drops of cytoplasm which are separated from the environment (or, better, connected with it!) by a surface layer called the plasmalemma. The structure of this layer will be looked at briefly in a later chapter; in conventional electron microscope preparations it appears in section as a tri-partite lamella (Fig. 1.1b) some 7.5 nm thick. The plasmalemma, originally thought of as a rather firm organelle, is coming to be thought of as an almost two-dimensional liquid in the form of a lipid bilayer within and upon which large (c. 10 nm) protein and other molecules are present and can migrate (Fig. 1.1c). The fact, however, that some plant cells without a wall (such as swarmers in the algae) may be pear-shaped, argues for an occasional anisotropy hardly consistent with a liquid surface. Such isolated cells are always rounded, however, never angular.

Once a cell wall has been deposited the approximately spherical shape is often lost even with isolated cells. The cell may then become cylindrical, very long and thin, branched, star-shaped and so forth. Such anisotropy of shape begins with the onset of wall deposition opening up the proposition, to be examined in detail in later chapters, that the wall dictates the shape. Some cells growing in isolation do, of course, remain spherical — unicellular algae such as *Chlorella* spring immediately to mind — and their walls must differ materially from those which do not.

When, however, an original single cell develops, by successive divisions, into a tissue of similar cells, as in the meristems of higher plants, then the spherical shape is lost for another reason. The question of the shape which is then acquired has been of concern even in relatively recent times (Naum, 1955; Matzke and Duffy, 1955; Matzke 1956; Meretz, 1962). The problem resolves itself into the mathematical one of defining those polyhedra which can fill space completely when identical polyhedra are placed side by side, and the selection of the one or more most closely fitting the observations which can be made upon the appropriate tissue.

7

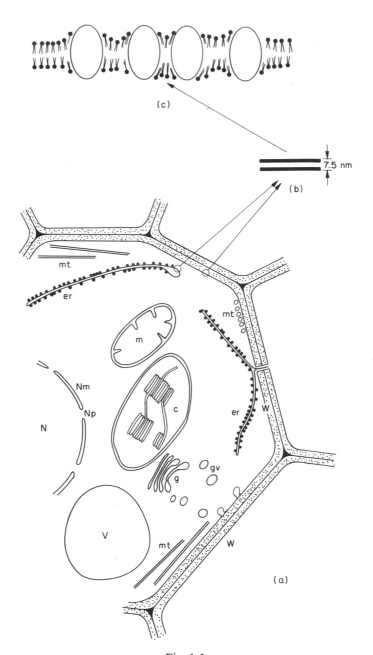

(c)

7.5 nm

(b)

mt

er

mt

m

er

W

Nm

Np

c

N

gv

g

V

mt

W

(a)

Fig. 1.1

Fig. 1.1 **Diagrammatic representation of a section of a 'typical' plant cell and of some organelles**
(a) **W, two primary walls (stippled) and middle lamella, mt microtubules, v vacuole, N nucleus, Nm nuclear membrane, Np nuclear pore, m mitochondrion, g golgi body, gv golgi vesicles, c chloroplast, er endoplasmic reticulum (rough) carrying ribosomes.**
Each membrane shown as a line is a unit membrane which, in conventional electron microscopy, shows in section two dark lines (stained by osmium) with a clear space between as in (b).
(b) **the unit membrane of the plasmalemma.**
(c) **simplified diagrammatic view of current ideas on the structure of cell membranes seen in section; the ellipses represent protein assemblies and otherwise only phospholipids are represented.**

Otherwise an informed guess can be made by analogy with inanimate complexes which appear to develop under similar geometrical conditions e.g. from the froth on soap solutions or from packed and compressed plastic spheres. Finally, the tissue itself may be examined, but this direct approach is not as easy as it seems. All these approaches have, however, yielded the same answer. It is now agreed that cells in such a tissue tend to adopt the form of an orthic tetrakaidecahedron or cubo-octahedron almost in the form adopted by Kieser more than 150 years ago. This was originally suggested as very nearly the form taken by bubbles in a froth by Kelvin and is shown in Fig. 1.2m. Since, however, in froth as in the cells, any three plane faces must meet at 180° and any four edges must meet at the tetrahedral angle, the eight hexagonal faces must be slightly curved, again as pointed out by Kelvin, producing the so-called 'body of Thompson'. Van Iterson and Meeuse (1941) demonstrated this habit in the meristems of higher plants.

Since in a meristem the cells are constantly dividing they are not all of the same size. This means that not all cells will at all times adopt this 14-sided shape. Nevertheless all the cells, and all descendants from them, must adopt a shape derived from the 14-sided figure. Cells derived from the meristems of shoot and root — and in some cases from the equivalent apical cell in lower plants — develop by expansion of some facets of the polyhedra more than others. Some of the factors causing this anisotropic extension are examined in a later chapter. All we need to note here is that this leads to cell elongation. In the elongated cell, such as a procambial cell, such elongation is enormous (100 times or more) but of the six expanded faces, four are still basically

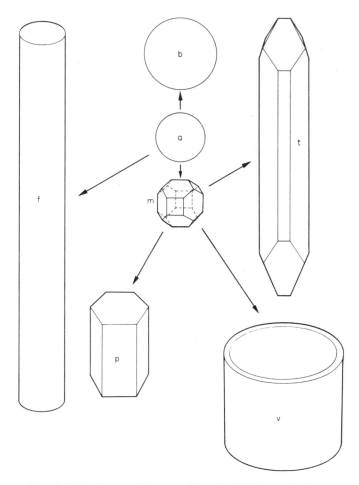

Fig. 1.2 The forms of some cells and their interrelations
- a   idealized spherical cell
- m   isodiametric cell in a meristematic tissue
- p   a parenchyma cell derived from m
- v   a vessel derived from m
- t   a tracheid derived from m
- f   a filamental cell derived from a
- b   a globular cell derived from a

The forms b and f usually develop only in isolation.
In v the end walls have disappeared leaving only a lip.

hexagonal and two tetragonal (Fig. 1.2t). This has been shown to be the case by Lewis (1935). Basically, the cell as a whole has two-fold rotational symmetry about the longitudinal axis, implying that opposite walls should be similar though adjacent walls need not be. As a cell increases in size, it may of course develop new cell contacts and therefore new facets and some of these can be seen in the beautiful plaster casts of Lewis (1935). Moreover, in very elongated cells in which the tips may be ignored, the rotational symmetry becomes sixfold. This may be reduced to twofold if two opposite faces expand laterally or may increase to cylindrical symmetry if intercellular spaces develop to the extent that the cell becomes transversely rounded (Fig. 1.1t and v). The basic twofold symmetry may be lost for any of these reasons.

1.3 Primary v. secondary walls

During this extension the wall remains thin and constitutes the primary wall examined in Chapter 12. The membrane between two cells consists of three layers: the primary wall of each adjoining cell with the cementing middle lamella sandwiched between. Extension has involved the intake of water into the vacuole which has developed as a prominent feature of the cell contents. Otherwise all the cell organelles, nowadays well characterized by electron microscopy, have remained active. They are shown diagramatically in Fig. 1.1, which contains some features to be dealt with later, and electron micrographs of some organelles will be found as Figs. 13.1, 13.2, 13.10 et seq. In most cells a secondary wall layer is now deposited from within which then thickens to an extent which depends upon the type of cell into which the growing cell has developed. In many parenchyma cells (Fig. 1.1p) the layer is thin though the protoplasmic contents remain alive for long periods. In other cells, secondary walls develop to an extent almost to fill the cell, as in some phloem fibres which may then give the appearance of a rod with a fine thread-like cavity running down the centre (which is empty since the contents have died and autolysed). In most elongated cells, however — wood fibres and tracheids, collenchyma cells, many phloem fibres — and some which are not so elongated — vessel elements in dicotyledons, vegetative cells in algae etc. — the wall becomes appreciably thick. If, as in the higher plant cells, secondary wall formation inhibits further extension, the protoplasm then again degenerates, leaving a hollow thread consisting of a thick wall envelope surrounding a fairly large cavity, the *lumen*. If the thickening of the wall does not inhibit further extension, as with some of the algae (p. 192) then the extending cell can come to a dynamic equilibrium in which the cell wall reaches a steady thickness without a reduction in the volume of the living cell contents though cell wall deposition continues. It is with the inter-relationships between growth and primary wall structure and with the properties conferred upon a tissue by the structure of its secondary walls, that this book is largely concerned.

The secondary wall which thus surround the lumen in this condition is usually rather complicated in structure. It can, however, vary from the apparently simple and homogeneous (as in some vessels) to a complex of several concentric lamellae alternating either in chemical make-up or in physical constitution or both. Normally these layers are distinguishable under the ordinary microscope even without pretreatment, but in many cases special procedures have to be adopted, such as staining and the like, in order to make them out at all. Invariably the visible differences arise through a complex of many underlying chemical and physical factors and it will be a large part of the endeavours here to sort out the various factors which impart these differences to the layers and therefore to the cells as a whole. In addition to this complexity in the fundamental building material of the wall layers there are other disturbances of structure which, although perhaps not of such immediate important to an understanding of the fundamentals of wall structure, must nevertheless be taken into account particularly when attempting to define the properties of single walls from those of a whole tissue, however homogeneous. In the majority of cell types the secondary wall is not uniformly thick over the whole surface. Here and there, arranged sometimes at random but more often in some remarkably uniform pattern, there occur thin places in the wall where secondary deposition has never taken place. A further remarkable circumstance is that wherever one cell has such a thin place, the adjacent cell in contact with it has a similar thin place at the corresponding point in the wall. These are spoken of as *pit* pairs. In parenchyma cells this leads to the development of cylindrical canals in the wall which are closed off merely by the two primary walls of each cell and the tenuous middle lamella between them — the *simple pits* (Fig. 10.13d). Such pits have a most profound influence on the structure of the wall area immediately adjacent to them. Of even more consequence are the bordered pits

12

typical of, for instance, the tracheids of conifer stems. Here, as the secondary wall becomes thicker by the continual depositions of new layers within the old, the new layers encroach over the region so that this becomes over-arched and, by the time that secondary deposition ceases, the 'hole' in the wall has become much smaller (Fig. 10.13a). This leaves, scattered over the wall, a series of over-arched regions within which, and to some less extent surrounding which, the wall structure is quite different from elsewhere. This is of particular importance in, for instance, the walls of the vessels in many dicotyledonous trees where the pitting is crowded over almost the whole wall surface and is bordered, narrowly in contact with ray cells and wood parenchyma and more widely in contact with other vessels.

Naturally, it is desirable to begin with the simplest possible case. To those who are interested primarily in the more academic aspects of plant behaviour, this would lead to a study of those cells which grow unimpeded by the obstacles to growth, or the control of growth, consequent on tissue formation, e.g. to the algae where the reactions of cells can be investigated independently. Historically, however, the cells which took precedence in the modern investigations of submicroscopic structure in plant cells were the phloem fibres, and in particular ramie fibres, for several reasons not the least of which is the economic importance of such cells. These are, in fact, rather satisfactory, since they are composed of cells all of one type and approximately of the same size; long thin cells, some one hundred or more times longer than wide, with long, tapering ends and with thick cell walls. These fibres are, moreover, arranged quite parallel to each other, or are so long that cells separated chemically from the tissue can be laid strictly parallel to form a bundle, the properties of which reflect largely the properties of a single cell. This, together with other simplifications in fibre structure, which will appear later, enables the fundamental structural features of the material constituting the walls of plant cells to be determined with some certainty. Investigation of structure in other cell types in more complicated tissues then resolves itself into the application of the knowledge thus obtained to the more complex forms. Even with the fibres, however, there is the complexity that the cells are not strictly the continuous cylinders into which physicists, for the best of reasons, prefer to sublimate their ideas. The long, tapering ends scattered up and down any naturally occurring bundle of fibres introduce some degree of uncertainty into the orientation of the constituent cell walls. As far as the investigation of the fundamental features associated with the crystallinity of the walls is concerned, this does not have any important effect; but it cannot be ignored when attempting completely to delineate the properties of any single cell from the properties of a bundle, or *vice versa*.

CHAPTER 2

Intra-atomic and intramolecular bonding and molecular models

The structures of cell wall components and the inter-relationships between them which confer upon a wall its physical properties can no more be understood without some knowledge of chemical bonds than can those of any other molecular assembly. It is consequently of value to consider first the bonding between atoms and molecules such as are concerned in holding the solid, or semi-solid, cell wall together. The aim here is to give some physical insight into the nature of bonds rather than to attempt any proof, experimental or theoretical, of the statements made. For more detailed treatment the reader is referred to other texts such as Brand and Speakman (1961), Pimentel and McClellan (1960) and the brief but very informative chapter in Setlow and Pollard (1962).

A wide range of substances including cell wall substances consist of finite aggregates of atoms 1–2 Å apart separated by distances usually of the order of, or greater than, 3 Å. This may be considered evidence for the existence of strong, short-range forces — bonds — between atoms and weaker, longer range forces between molecules, using the terms *stronger* and *weaker* in terms of energy relative to the energy of thermal bombardment. Since atoms within molecules come to lie at certain minimal distances, some kind of repulsive forces must also come into play. These obey an inverse power law of high order so that they rise from a negligible to a high value over a very small decrease in separation. Accordingly, it is reasonably satisfactory to think of atoms as bodies with firm boundaries which come sharply into contact when the centres are at a certain critical distance apart. Since this distance is the sum of the effective radii of each atom it is related to the term b in van der Waal's equation ($P(V - b) = RT$) allowing for the volumes of

15

the atoms in a gas at pressure P and volume V; the individual radii are called van der Waal's radii. Approximate values for some relevant radii are as follows:

H, 1.2 Å;   O, 1.40 Å;   N, 1.50 Å; $-CH_3$, 2.0 Å;   S, 1.85 Å.

Although there are some exceptions, it is broadly true that un-bonded atoms cannot come closer to each other than the sum of the van der Waal's radii without severe steric repulsion. This is a most useful outcome in terms of structure determination and model building.

It is impossible to measure the force between atoms directly but the energy required to separate atoms — the bond energy — can be readily determined. As the primary observable parameter it is better to think of energies rather than forces, with the advantage that derivations from energy considerations are the more valuable; and that will be the treatment here. Since, however, the relation of the force F between two atoms spaced x units apart and the energy V in that condition is $F = -dV/dx$, then the force can be calculated if the relation between V and x is known. Energies are commonly expressed in joules per particle, electron-volts per particle or calories per mole and these can be interconverted as follows:

$$1 \text{ J particle}^{-1} = 6.2 \times 10^{18} \text{ eV particle}^{-1}$$

$$1 \text{ eV particle}^{-1} = 1.60 \times 10^{-19} \text{ J particle}^{-1} = 23 \text{ kcal mole}^{-1}$$

$$1 \text{ kcal mole}^{-1} = 0.044 \text{ eV particle}^{-1}$$

It is convenient to note that at room temperature thermal energy amounts to 0.60 kcal mole^{-1} or 1/40 eV particle^{-1}.

2.1 The electron

In attempting to consider the derivations of any of the interaction energies between atoms and molecules it is first necessary to con-sider the nature of the electron since the boundaries of an atom must be related to the distribution of its electrons. The concept introduced by Bohr of electrons circulating round a nucleus in orbits, much as the planets revolve round the sun, could no longer be countenanced upon the introduction by Heisenberg, in 1927, of the *uncertainty principle*. Nevertheless it presented an image which was immediately seized upon and which has persisted. We need to follow it a little in preparation for currently more acceptable formulations.

The Bohr atom gave a ready explanation of observations such as (a) metallic sodium exposed to water catches fire, explodes or both, (b) chlorine dissolved in water is poisonous, (c) sodium chloride shows neither property. The only imaginable concept would be that in sodium chloride the sodium and chloride atoms have transferred an electron, elemental sodium having 'one too many' and chloride 'one too few'. It is now well recognized that some numbers of electrons are more stable than others and that atoms which have these stable numbers are highly inactive chemically. Thus, neon which has eight electrons in the outer orbit is stable; sodium, which has an extra one, is not. Similarly, chlorine, in which the number of extra electrons has reached seven, is not stable. Exchange of an electron leaves a positively charged sodium ion with the electronic configuration of neon and a negatively charged chloride ion with eight outer electrons (equivalent to the stable argon), both stable and with an electrostatic attraction for each other forming an *ionic bond*, one of the strong bonds (c 120 kcal mole^{-1}). On the same principle, bonding between atoms with similar electronic configurations can be understood if atoms *share* electrons; a hydrogen atom, for instance, with only one electron can share this with another hydrogen atom so that each in a sense has two (as in helium), while carbon, with four electrons, can share the electrons of four hydrogen atoms so that each hydrogen atom is related to two electrons (as in helium) and the carbon to eight (as in neon). The electrons which are shared are the valency electrons and the bond in which two electrons are shared is the *covalent bond*, another strong bond (e.g. 4.40 eV for H–H; 2.25 eV for C–C).

Electrons can no longer, however, be considered as small, hard spheres revolving in orbits which can be identified. The position of such an electron could in principle be determined experimentally only by its interaction with a beam of radiation by collision with a photon. This would involve a change in velocity which could not simultaneously be determined. It is therefore impossible to determine at one and the same time both the position and the velocity and this underlies the uncertainty principle. In its formal expression this states

$$(\Delta x)(\Delta(mv_x)) > h/2\pi = 1.05 \times 10^{-34} \text{ J s}$$

where 'Δ' means 'uncertainty in'. If we attempt to confine an electron in a small space Δx (defining the uncertainty in its position) the uncertainty in its momentum, and therefore the minimum possible average momentum, becomes very large. The kinetic energy and the total energy also clearly become very large. Calculations into which we cannot go here show that, for a hydrogen molecule, the uncertainty in the electron's position is about the distance of the electrons from the hydrogen nucleus found experimentally, and we have no option but to think of an electron as a smeared-out charge distribution. It is then possible to describe only the probability of finding an electron in a certain energy state within a certain volume of space. This volume is called an *orbital*. An orbital, or charge cloud, may be defined as a solid surface enclosing a volume of space containing (or having a certain probability, usually 95 per cent, of containing) 2, 1 or 0 electrons. The vast majority contain 2 and the 0 case allows reference to empty orbitals. Within any contour defining an orbital, there will be other contours enclosing smaller volumes within which the probability of finding an electron will be less; but within them the electron density will be higher because the volume is less. The probability of finding an electron decreases rapidly with increasing distance from zones of high electron density.

Before proceeding further we need to introduce one further principle. There are aspects of spectroscopy which can be understood only if the classical electron is spinning around a diameter as axis and this spin can occur in either direction to which quantum number $+\frac{1}{2}$ and $-\frac{1}{2}$ are given. Now we know already that not all the electrons associated with an atom are at the same mean distance from the nucleus and therefore in the same energy state. This is because of the principle we now need to invoke — the Pauli Exclusion Principle — which says that only *two* electrons can occur in the same energy state and that these must have opposite spins. In helium for instance, the two electrons in the ground state must have opposite spins — called a singlet state because the total spin has only one value namely 0; in the next element in the periodic table, lithium, the third electron must be in a higher state. We may now proceed to consider the various types of bonding possible.

2.2 Covalent bonds

As two hydrogen atoms approach each other the two electrons may have parallel or antiparallel spins. We consider only the case of antiparallel spins. Neither electron can be considered the exclusive property of the one nucleus or the other; both are associated with both nuclei. Solution of the Schrödinger wave equation gives a wave function ψ such that $\psi^2 dx$ gives the probability of finding an electron between x and $x + dx$. The appropriate distribution of ψ is given in Fig. 2.1. As the two hydrogens approach each other, the electrons have a larger space to occupy — and they can occupy it because they have opposite spins — and therefore their momentum and kinetic energy is lower than for widely separated atoms.

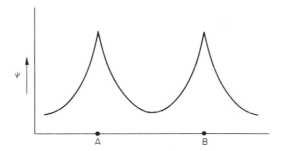

Fig. 2.1 The distribution of the wave function Ψ referring to two H atoms at A and B

This constitutes an attraction. If the nuclei move too close, the volume shrinks again and the kinetic energy rises. For the sake of simplicity we ignore potential energy changes, but this does not affect the general validity of an energy minimum at some particular distance. This interaction leading to an 'energy minimum' distance is the covalent bond. The corresponding orbital may be as represented in Fig. 2.2 and is valid for all atoms as a general way of understanding this bond. The energy required to separate two atoms at this minimum separation is called the *bond energy*. Some bond energies are given below, representing each single bond by a single line and expressing the energy in electron volts.

$$H-H, 4.40; C-C, 2.25; C-H, 3.80; C-N, 2.13;$$

$$C=C, 4.35; C=O, 6.30; C\equiv C, 5.35.$$

19

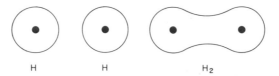

Fig. 2.2 Molecular orbitals for H and H₂

The minimal distance for bonds of the same order is remarkably constant over a range of compounds; for C—C, for instance, the lengths found (by X-ray diffraction analysis (Chapter 5) and otherwise) lie within the limits 1.52—1.55 Å for compounds as disparate as diamond and straight chain fatty acids. There is some variation, as in C—Cl for which values are found of 1.782 (CH_3Cl), 1.772 (CH_2Cl_2), 1.767 ($CHCl_3$) and 1.766 (CCl_4), but these are subtle differences and it is in part this close constancy which makes model building possible. Bond lengths are, moreover, *additive*; that is, for two atoms with covalent radii r_1 and r_2 the covalent bond length between them is $r_1 + r_2$. Some relevant covalent bond radii are as follows (Å):

$$-C, 0.77; \quad =C, 0.67; \quad \equiv C, 0.60; \quad -O, 0.66; \quad =O, 0.57.$$

When, however, several configuration of nuclei and electrons are almost equally possible (e.g. when the electrons shared between atom 1 and atoms 2 and 3 may at one time be more nearly associated with atom 2 and at another, with almost equal probability, be closer to atom 3), the total space available to the electrons is increased and the mean energy correspondingly reduced. The experimental evidence for this lies in the C—C data. Benzene may be considered as I or II (Fig. 2.3) in classical theory.

Fig. 2.3 Two possible situations for the double bond in benzene

In quantum theory the structure may not be taken as either but rather as a combination of both. One way of saying this is that the structure *resonates* between I and II. Correspondingly, the C—C bond in benzene places the atoms 1.39 Å apart as against 1.54 Å for the non-resonating C—C bond and 1.34 Å for the non-resonating C=C bond.

The angle between bonds when an atom has more than one (e.g. carbon with 4) is more sensitive to environment. It is, however, likely to be constant within 3° or 4° unless a very bulky group is present to open out the angle or ring-closure operates to contract it. The angle between the two bonds of an oxygen, for example, is 104.5° in H_2O, 105° in $CH_3.O.CH_3$ and 109.8° in solid HgO. None of these is far removed from the tetrahedral angle 109.4° which can be used for most purposes in model building.

The forces needed to distort covalent bonds are high, diminishing in the order (1) bond stretching (2) bond bending (3) bond twisting. Typical values for diatomic molecules for (1) are: single bonds, 5×10^5 dyn cm^{-1}; for double bonds 10×10^5 dyn cm^{-1}; triple bonds 15×10^5 dyn cm^{-1}.

2.3 Ionic bonds

In scarcely any molecule do the distribution of positive charge and negative charge coincide. A positive charge in any part of one molecule will tend to induce a shift towards it of the electron cloud of a neighbouring molecule, and this induces an attraction which is usually weak. In the extreme case — which can also be regarded as the extreme case of the covalent bond — an electron of one atom becomes more closely associated with the nucleus of the other than with its own. The force between the resultant *ions* can then be regarded as *electrostatic*.

2.4 Weak bonds

When the electrons are neither associated equally with both of two neighbouring atoms nor exclusively with one of them, attractive forces still exist though weaker than either covalent or ionic bonds. As already mentioned, these may be called collectively van der Waal's forces. They arise whenever the distribution of electron charges is uneven, so that a molecule possesses, permanently or temporarily, a net negative charge in one location and a net positive charge at the other. Although the resulting energies and the associative forces are small they are additive so that with large molecules with large numbers of such bonds the total bond energy can be high. These bonds are therefore at least as important as are covalent bonds in maintaining the integrity of living things. We begin by considering the simplest case to understand, the case in which each of two neighbouring molecules is characterized by a permanent separation of net charges.

2.4.1 *Permanent dipole-dipole interaction*

When two charges, q, of opposite sign are separated by a fixed distance l the result is called a dipole (Fig. 2.4) characterized by what is called the *dipole moment* $ql = p$. The molecule H_2O is such a dipole as shown in Fig. 2.5a in which the orbitals containing

Fig. 2.4 **Dipole; for explanation, see text.**

a proton (the protonated or bonding orbitals) are shown unshaded and the non-bonding orbitals (containing the 'lone pair' of electrons in the old electronic formula (Fig. 2.5b)) are shaded. The electron density is higher toward the heavier O atom leaving the proton partly unshielded. The dipole moment is 6.0×10^{30} C m.

(a)   (b)

Fig. 2.5 **Water as a dipole; (a) orbitals (b) classical electron notation,**
● = **electron**

When such a dipole is placed in an electric field (Fig. 2.6) it will tend to rotate into the minimal energy position in which the length l lies in the direction of the field, the turning couple and the resulting energy being proportional to p. For small molecules and for fields of the strength to be expected, the interaction is close to the average thermal energy so that in an assembly of dipoles there is at any moment a wide distribution about the minimal energy situation.

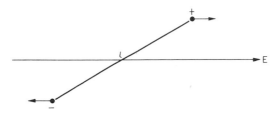

Fig. 2.6 For explanation, see text.

When two such dipoles come together, therefore, the effect of the electric field due to each of them on the other will be such that each tends to rotate to a minimal energy condition and the total energy will be less than for two dipoles far apart. The dipoles therefore attract each other. Since the field produced by each is proportional to the dipole moment, the energy of interaction varies as the product of the dipole moments, $p_1 p_2$. It may be seen intuitively that, if each dipole is completely free to rotate, the minimal energy situation occurs when the dipoles lie antiparallel and at a maximum when they are parallel. Since the energy of interaction is low, each dipole may, at any instant, lie in any orientation with respect to the other so that on an average there should be as much repulsion as attraction and therefore no net interaction. It turns out, however, that the attractive case has *smaller energy* than the repulsive case by an amount $2p_1 p_2 / 4\pi\epsilon r^3$ (where ϵ is the permittivity, i.e. ratio of the capacity of a condenser with the material as a dielectric to the capacity of a condenser of the same dimensions with a vacuum as a dielectric). The anti-parallel arrangement is therefore favoured. On these grounds, the interaction energy between two dipoles may be shown to be

$$V = -\frac{1}{3} \frac{p_1^2}{(4\pi\epsilon)^2} \cdot \frac{p_2^2}{kT} \cdot \frac{1}{r^6}$$

where r is the separation between their centres.

The interaction falls off rapidly with distance through two factors. Firstly, this is ensured by the inverse proportionability to r^6.

23

Secondly the interaction varies inversely as the square of the permittivity. If the dipoles are in water and relatively far apart, $\epsilon = 80$. If they are so close together that water cannot penetrate between them $\epsilon = 1$. These ensure that the forces are of short range, of the order of a few Ås. At very small distances the electron clouds of the dipoles begin to mingle and the dipoles then repel each other. There is again therefore a distance apart of a few Ås at which the energy of interaction is a minimum.

Dipoles of which one end is firmly fixed in a large immovable molecule, will not of course be able to rotate through thermal agitation. They will nevertheless develop mutual attraction if the positive end of one dipole can approach the negative end of the other. The attraction is still inversely proportional to the reciprocal if a high power of the separation, and the separation at minimum energy is of the same order. Exactly the same considerations apply to the remaining dipole interactions to be described.

2.4.2 Permanent dipole-induced dipole

When a molecule with a permanent-dipole moment such as H_2O approaches an ordinary non-polar molecule such as CO_2, the electric field of the dipole induces a displacement of the electron charge distribution of the non-dipole and therefore *induces* it to become a dipole. As a consequence, H_2O and CO_2 attract each other much as do permanent dipoles and this is why CO_2 is water-soluble. The value of the induced dipole moment is proportional to the strength of the inducing electric field and therefore to the moment of the permanent dipole. The energy of interaction takes the form

$$V = \frac{-\alpha p^2}{(4\pi\epsilon)^2 r^6}$$

where p is the permanent dipole moment and α is the *polarizability*, and represents therefore again a short-range force, the separation at minimum energy being dictated as for the permanent dipole interactions mentioned above. This type of interaction can also, of course, form part of the interaction between permanent dipoles.

2.4.3 Transient dipole-induced dipole

The dipole interactions so far described fail by far to explain even the simplest physical phenomena. Non-polar molecules such as H_2, which are symmetrical and without dipole moments, nevertheless attract each other. Such attractions are achieved by what are called *transient* dipole moments.

This type of van der Waal's attraction may be thought of as initiating from a *fluctuation* of the electron charge distribution of an atom about an average which is, approximately, the ionization potential. At any instant, one of two atoms may then have a large temporary dipole moment, with an associated electric field which induces a dipole in a neighbouring atom. The two effective dipoles then come under the influence of attractive forces much as discussed above. Since the fluctuation frequency is roughly the same as visible and ultraviolet frequencies, of the order of 10^{15} times s^{-1}, the molecules of the surrounding medium do not have time to rotate during the period of interaction. Just as in this way atom 1 can influence atom 2, so can atom 2 influence atom 1. For reasons into which we cannot go, the lack of rotation in the medium means that in the calculations of the energy of interaction the permittivity ϵ must be replaced by the square of the refractive index, n^2. In water as the medium, for example, ϵ^2/n^4 is about 2000 and this gives transient dipole interactions in dipole-dipole bonds an advantage by this factor. This interaction is usually, indeed, the predominant one in all dipole-dipole interactions. Because the fluctuation frequency is about that of visible light, the constants in the energy equation can be determined from the *dispersion* of light, i.e. from the change in refractive index with frequency, and the resultant forces are called *dispersion* forces. They are also sometimes called London forces to honour the man who first proposed them.

2.4.4 *Hydrogen bond interaction*

We are now in a position to consider the bond of major importance in biology, the hydrogen bond. The basic explanation of this bond is simple. Consider the covalently bonded group —OH in which the proton in the orbital bonding O and H is rather near the surface (Fig. 2.5a). This has, as we have already seen, a large dipole moment with the net plus charge in the neighbourhood of the proton. Other negative groups (or the negative ends of other dipoles) in the same or other molecules can get close to the proton before the electronic clouds intermingle and repel. Close interaction of charges vary as $1/(separation)^2$ and give interaction energies as high as $10\,kcal\,mole^{-1}$ or $0.5\,eV\,particle^{-1}$ (compared with $1/40\,eV\,particle^{-1}$ for thermal agitation) or $40\,kJ\,mole^{-1}$ (cp. $c\ 500\,kJ\,mole^{-1}$ for covalent bonds). Such an interaction is stronger than any van der Waal's interaction.

Hydrogen bonds can occur between molecules which contain protonated orbitals and non-bonding orbitals and are therefore almost exclusively confined to molecules which contain C, N, O or F. Another way of describing them is to say that they are derived from the powerful attraction of the dense region of electron density in the non-bonding orbital for the proton in the protonated orbital. The strength of the bond is a maximum when (1) the proton is nearest the 'surface' of the protonated orbital, (2) the non-bonding orbital is small with a high concentration of electron density. These conditions are best fulfilled when both orbitals are under the influence of a nucleus of high atomic number since this pulls in the region of high electron density and, in the protonated orbital, pulls with it the proton which is nevertheless left more remote from the region of highest electron density. Thus the $-NH----N$ bond has a mean bond energy of $13\,kcal\,mole^{-1}$ whereas the energy of the $-OH---O$ bond is $29\,kcal\,mole^{-1}$. The corresponding mean bond distances are $3.4\,Å$ and $2.75\,Å$.

The formation of a hydrogen bond modifies many of the physical properties and a few of the chemical properties of a substance in which it is formed. The most commonly observed physical modifications of importance here are altered solubility, dielectric properties, electrical conductivities and spectral frequency shifts. These latter refer to infrared frequencies. Covalently bonded molecules can vibrate about the bond in various ways, at frequencies in the infrared range. When a beam of infrared radiation is passed through an assembly of such molecules the wavelengths corresponding to these frequencies are absorbed and the vibrations can therefore be detected and quantified in the infrared absorption spectrum. The altered electron arrangement and positions of atoms in the neighbourhood of the donor group changes when a hydrogen bond is formed. This leads to a stretching vibration in the $O \leftrightarrow H$ direction of $-OH$ for example which is of lower frequency when the hydrogen bond $-OH---O$ is formed, and a bending vibration $(O - H)$ of higher frequency.

It is by virtue of this bond that all the compounds discussed later in this book are held together.

2.5 Molecular models

Although the chemical formulae of molecules can be written in two dimensions on paper, such formulae can never express the space relation since the molecules are three-dimensional. The formula H_2SO_4, for example, while satisfactory in defining a molecule and expressing some of its simple chemistry, conveys no impression of its size, no explanation of its reactivity and no way of telling how close the molecule can approach another molecule without interaction. In assessing the acceptability of a formula, or, for that matter, of the detailed structure of a molecule, it is always helpful — indeed sometimes unavoidable — to build a model. In this way it is possible, for instance, to determine whether a given structure is sterically possible, i.e. whether atoms in the molecule do or do not approach each other more closely than their 'size' will allow. Fortunately, the approximate constancy of covalent bonds, bond angles, van der Waal's forces and H bonds allow models to be made with a fair degree of accuracy.

The type of model to be used depends upon the purpose for which the model is required. There are broadly two types which may be called *skeleton models* and *space filling models*. These will be dealth with in turn.

2.5.1 *Skeleton models*

These are required whenever it is necessary to display the details of the covalent bonding and to visualize the relative positions of the constituent atoms; or, on a more sophisticated level, when it is desired to set out the atoms with correct co-ordinates relative to each other in order to refine a structure by further calculation. They take two forms — the ball and spoke model and the wire model.

(a) *The ball and spoke model*
The 'atoms' take the form of solid spheres, the diameter of which need bear no relation to the diameter of the atom (i.e. twice the

radius of the covalent bond) except that it must be smaller. Such spheres are available in a variety of sizes and are mostly of plastic. They are generally coloured following a widely accepted colour code — black for C, red for O, blue for N, white for H and so forth. Each ball is drilled with a number of holes, equal to the valency of the corresponding atom, lying at appropriate angles to each other (e.g. with C, four holes at the tetrahedral angle), and to such a depth that when the appropriate spoke is pushed home into two balls the centres are held at the correct distance on a scale, say, $1\,cm = 1\,Å$. Such a model of ethane, C_2H_6, is shown in Fig. 2.7a and other models, or drawings of models, will be found elsewhere in this book. As a variant, the spokes may be helical springs instead of rigid rods in order to allow some deformation. Hydrogen bonds

a

b

c

d

Fig. 2.7 a Ball and spoke model of C_2H_6
b Wire model of O (left) and C (right)
c Wire model of C_2H_6
d Space-filling model of C_2H_6

are made by slipping over the spokes on the oxygen models involved a plastic tube of the appropriate length (c 3 Å).

(b) *The wire model*

The models so designed are visually pleasing, but the only function of the balls is to hold the spokes together; they can therefore be discarded if the spokes are welded together. This has the advantage that the units of a model can be made in a laboratory workshop relatively cheaply; for although the individual balls and spokes described above are cheap, the corresponding model of a large molecule can be very expensive.

In this model the spokes representing carbon are welded together in a jig giving the correct bond angles and are cut at lengths equal to the covalent radius; oxygen atoms consist of spokes of the correct length bent centrally through the bond angle (the tetrahedral angle is often sufficiently accurate) and hydrogen is represented by single straight spokes again of the appropriate lengths. Examples are shown in Fig. 2.7b. These may then be joined together either by brass connectors or spring clips, exemplified in Fig. 2.7c by ethane. Hydrogen bonds are again made by fitting the appropriate plastic tube sheaths.

2.5.2 Space-filling models

Though the above models are enormously useful for display purposes, for setting out the spatial arrangement of even a complex model in such a way that the situation of every atom can be seen and for the purposes of structure refinement, they have certain disadvantages. Some essential features of a model are *not* displayed; for instance, no visual guidance is given to the separation allowable between two atoms not connected by a covalent bond without infringing van der Waal's separations, i.e. without steric hindrance. This is met by a model of which the units consist in principle of spheres with the appropriate van der Waal's radii rather than the covalent radii. In the models commonly used the scale is 0.8 in = 1Å so that divalent oxygen is in principle 2.24 in diameter, hydrogen is 1.6 in diameter and so forth. In order to allow atoms still to be bonded at the correct covalent distances, the spheres are cut away normal to each bond leaving flat faces such that when any two atoms are joined by special links the correct separation occurs.

Carbon is then represented by a tetrahedron, di-univalent oxygen by a segment, double-bonded oxygen and hydrogen by rather more than a hemisphere, and so on (Fig 2.8).

The atoms are connected by a link shaped at each end to fit press-stud sockets mounted in the plane faces of the models, so adjusted that the link may be inclined up to 12° from the normal (Fig. 2.8). A rubber collar fitted over the link ensures that the opposing faces of linked atoms return to the parallel position when stress is removed, and freedom is restricted by plastic rings surrounding the collars. A special hydrogen model is provided for the formation of hydrogen bonds (Fig. 2.8); the plane face contains a groove into which a ring of vulcanized fibre may be fitted so that when the model is looped by elastic bands to an oxygen model the H---- O distance is correct at 1.90 Å.

Fig. 2.8 Space filling models of
(a) —C=
(b) —$\overset{|}{\underset{|}{C}}$—
(c) O$\overset{/}{\underset{\backslash}{}}$
(d) O=
(e) H bond
(f) H—
(g) spacer and clip

A model of ethane built in this way is included in Fig. 2.7d for comparison with the other models. Other space-filling models will be found elsewhere in the book. As already mentioned, they have the clear advantages that with flexible molecules some otherwise possible conformations can be ruled out through steric hindrance, the positions of possible hydrogen bonds can immediately be detected, and for any molecule the overall shape is presented. They have the disadvantage that the inner parts of complex molecules are hidden and that accurate measurement is difficult.

CHAPTER 3

The chemical components

We now turn our attention to the molecular species comprising the cell walls of plants and take first the chemistry of these substances. With rare exceptions, plant cell walls are built of an assembly of polysaccharides which in special situations become associated with a variety of non-saccharide substances. Primary walls, and the equivalent walls of growing algal cells to which the term *primary* does not specifically apply, are, except for some protein and a little lipid, exclusively polysaccharide in nature. It is therefore necessary to consider the chemical constitution of these substances, particularly of the polysaccharides, and therefore of the constituent monosaccharides. Lipid and lipid-like substances are minor wall constituents in the sense that either they form only a small proportion of most cell walls with a structure and function which is little understood or, when they are more abundant, they occur in few cell types such as in the epidermal and cork cells. These will not be discussed here and instead the reader is referred to Frey-Wyssling (1959), Roelofsen (1959) and Pilet (1971).

3.1 Monosaccharides

Both hexoses and pentoses are found in plant cell wall polysaccharides, mostly in the six-atom ring now so familiar. Glucose, for example, may be drawn either as modification of the conventional formula (Fig. 3.1a) or as a hexagon (Fig. 3.1b and c). The hexagonal representation gives a closer approximation to the spatial configuration as determined by a variety of methods including the method of X-ray diffraction analysis to be described in Chapter 5.

32

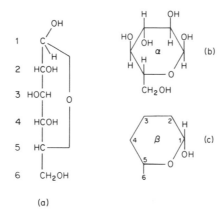

Fig. 3.1 Diagrammatic formulae for glucose. The groups attached to the carbon atoms 2—6 in β-glucose (c) are identical in orientation with those for α-glucose (b) and are omitted. In (b) and (c), −H or −OH groups lying above the carbon to which they are attached should be considered as lying above the plane of the ring; the rest lie below the plane.

It will be realized that, since the diameter of the oxygen atom is less than that of the carbon atom, the molecule should not in fact be drawn as a perfect hexagon; the perfection will be retained, however, throughout this book for ease of drawing. Just as in the conventional formula the carbon atoms are numbered 1 to 6, beginning with the potentially aldehydic carbon, so in the ring the carbon atoms are similarly numbered, carbon 6 standing outside the ring. In this form the sugar is called a pyranose sugar, so that glucose is formally glucopyranose, by analogy with pyran. Hexoses can occur in a 5-ring modification, and these are found in smaller quantities in cell walls. By analogy with furan they are called furanoses and the hexose a hexofuranose.

In the hexagonal representation there are two forms of glucose dependent upon the position of the −OH group attached to carbon 1. When this lies above the ring, as in Fig. 3.1b, i.e. on the same side as that on carbon 4, the molecule is called α-glucose, to distinguish it from the molecule of Fig. 3.1c in which this −OH lies below the ring and which is called β-glucose. Each molecule is dextrorotary (i.e. the plane of polarization of a beam of plane polarized light passing through a solution toward the observer is rotated in the clockwise direction) but the β-modification is much less active than is the α-modification. β-D-glucopyranose forms the basis upon which cellulose and other wall polysaccharides are built and we shall consider it further. α-D-glucose on the other hand, condenses to form starch, a polysaccharied with which we shall not be concerned.

The sugar rings depicted in Fig. 3.1 are not, of course, flat since the four bonds of each carbon atom lie at the tetrahedral angle to each other; this may be demonstrated by model-building and has been confirmed by X-ray diffraction analysis both of monosaccharides and of polysaccharides (Chapter 5). If they were flat, the distance across the molecule from the oxygen attached to carbon 1 to the oxygen attached to carbon 4 could be shown by simple geometry to be about 5.94 Å, whereas the experimentally observed figure is about 5.60 Å (p. 135). Because the rings must be puckered they can in principle take up one of two conformations, the *boat* form (Fig. 3.2a) or the *chair* form (Fig. 3.2b and c). This variation in conformation was first discussed in 1929 by Haworth, but it was not until almost 20 years later that it became possible to set up criteria dealing with the relative stability of the conformers. Hassel and Ottar (1947) demonstrated at that time that, of the various conformations that the six-membered pyranose ring can adopt, the chair form is the most stable. Somewhat later Reeves (see Reeves, 1958) was able to show that the pyranose ring does in practise adopt the chair form except when this is geometrically impossible; and the *boat* form does not occur in any of the polysaccharides considered here. As for the *chair* form, there are clearly two different conformations, called by Reeves respectively C1 (Fig. 3.2b) and

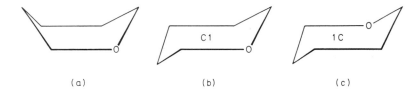

(a)   (b)   (c)

Fig. 3.2 **Hexose rings seen from the side.**

1C (Fig. 3.2c). These place the oxygen of carbon 1 at the same distance from that on carbon 4, namely at 5.60 Å but in other respects the two forms are different since the different dispositions of the carbon atoms within the ring leads to different dispositions of the hydroxyl groups and therefore different possibilities of H bonding to neighbouring molecules (though in each the —OH bonds on neighbouring carbon atoms still lie on opposite sides of the ring). The C1 conformation can be converted to 1C by flipping over carbons 1 and 4 and this means that the hydroxyls which lie almost in the plane of the ring (*equatorial* hydroxyls) in C1 (Fig. 3.3) lie normal to the plane (axial) in 1C. In this case the hydrogen bonding situation in glucose is very different between C1 and 1C conformations. Unfortunately from the standpoint of terminology, the mirror image of C1 is also called 1C and in this conformation hydroxyl groups which were equatorial in C1 still lie equatorially. According to Reeves, the chair form which a six-membered sugar ring assumes depends principally upon three factors; of these only one is major for the unsubstituted sugar with which we shall deal, namely, that *axial* groups (except —H) tend to promote instability.

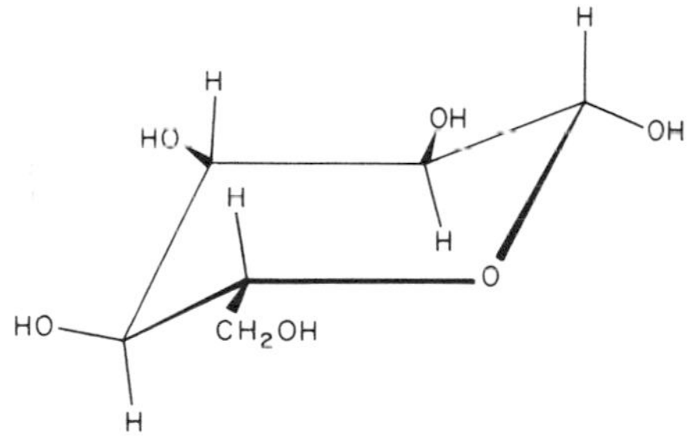

Fig. 3.3 The C1 conformation of β-D-glucose. Note that the —OHs lie almost in the plane of the ring (equatorial) and the —H almost normal to this plane (axial).

According to this, the most stable form of glucose is β-D-gluco-pyranose in the C1 conformation since all the hydroxyls are equatorial (Fig. 3.3). The conformation has been confirmed by infrared and nuclear magnetic resonance spectroscopy and by X-ray analysis. Fuller statements concerning conformation theory will be found in books by for example Hanack (1965) and Ramachandran (1968).

Other hexoses may be derived from glucose by switching —OH groups individually from one side of the ring to the other. Some of these are illustrated in Fig. 3.4; the conformation is regarded throughout as C1. The six —OH groups per molecule make all hexoses highly soluble in water by hydrogen bonding to water molecules. The structure of the appropriate pentose may similarly be envisaged by replacing —CH$_2$OH on carbon 5 of the relevant

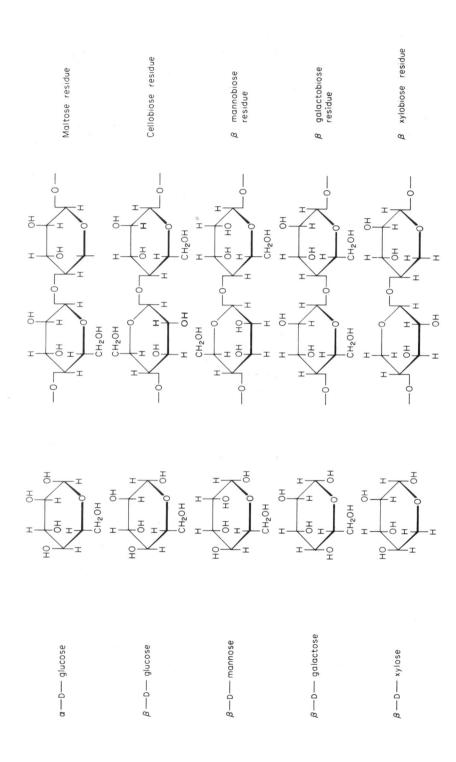

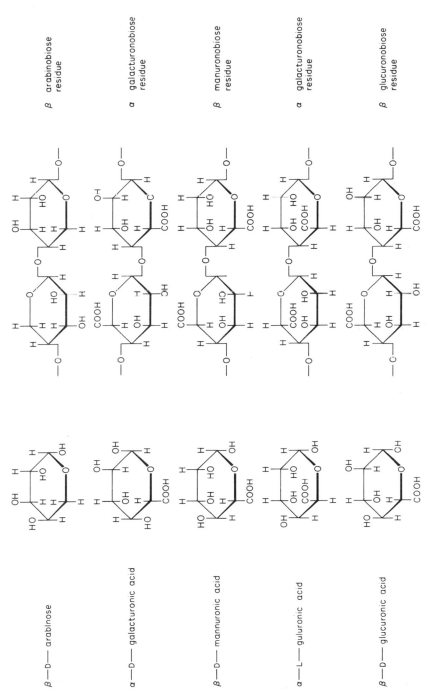

Fig. 3.4 A variety of sugars and disaccharide residues found in cell wall polysaccharides.

hexose by —H, and uronic acids by oxidation of —CH_2OH to —COOH; again the derivatives concerned in cell wall formation are illustrated in Fig. 3.4.

3.2 Disaccharides

Two hexose molecules may be persuaded to combine to a larger molecule in the presence of the appropriate enzyme. This occurs when the —OH group attached to a carbon on one molecule approaches sufficiently closely to the —OH on a carbon of the other that water is split off leaving a C—O—C linkage (Fig. 3.4). This is a process known as condensation; by removal of two —OH's, leaving only eight for the whole molecule, it reduces solubility. There is obviously an immense variety of possible disaccharides; fortunately we shall be concerned only with a few. When the two hydroxyls involved are in the carbonyl groups (carbon 1 in Fig. 3.1) the resulting $1 \rightarrow 1$ link results in a disaccharide which is non-reducing. One such disaccharide is, of course, sucrose which may be written α-D-glucopyranosyl-($1 \rightarrow 2$)β-D-fructofuranoside, noticing that in furanose sugars the carbonyl group is called carbon 2. In the di- and polysaccharides of plant cell walls, however, disaccharides are formed through the hydroxyl of carbon 1 of only one of the constituent sugars; the hydroxyl of the other is on carbon 3, carbon 4 or carbon 6, giving respectively a $1 \rightarrow 3$, $1 \rightarrow 4$ or $1 \rightarrow 6$ link. The nature of the link is determined also by the configuration of each sugar and is classed as α or β according to whether the separate sugars are in the α or the β form (Fig. 3.1). Usually a β link is stronger than an α link and this may be why reserve polysaccharides stored in the cytoplasm, and only those polysaccharides in cell walls which are labile, are α linked.

The nature of the linkage may be determined by methylation followed by hydrolysis and this is the method by which Haworth (1929) originally demonstrated the β-$1 \rightarrow 4$-link for the cellobiose isolated from cellulose. There are now, of course, other ways, even nuclear magnetic resonance having been used for this purpose (see references in Brown, Franke, Kleinig, Falke and Sitte, 1970). In this context, D-glucose forms two disaccharides, cellobiose (β-$1 \rightarrow 4$-linked) and maltose (α-$1 \rightarrow 4$-linked), the units of cellulose and starch respectively (Fig. 3.4). Note that the β-link requires that one of the glucose residues must, for stereochemical reasons, be rotated through 180° to the other in making the disaccharide (Fig. 3.5), whereas the α-link does not demand (and even dis-allows) this. With maltose, therefore, one sugar can be made to coincide with the other by a simple translation whereas in cellobiose a translation and a rotation is necessary. The axis of the cellobiose molecule is therefore a twofold screw axis, as confirmed by the X-ray diffraction method to be discussed on Chapter 5. There is a further complication with cellobiose in that, with this disposition of the two sugar residues, a hydrogen bond is formed between the ring oxygen of one residue and the —OH on carbon 2 of the other (Fig. 3.5). This is the Hermans configuration amply documented by X-ray methods both for cellobiose and cellulose and justified by conformation analysis (p. 47).

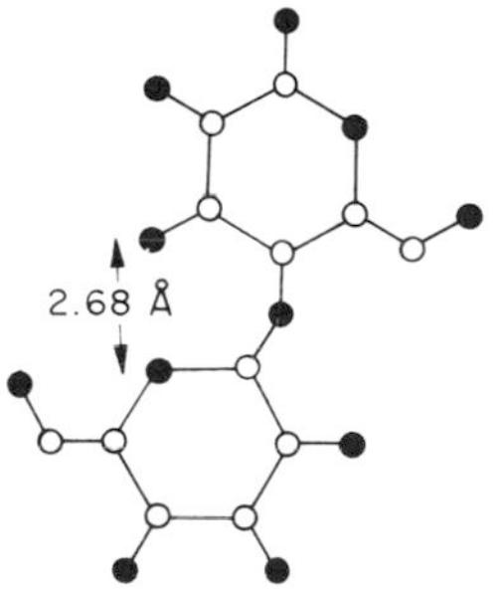

2.68 Å

Fig. 3.5 Cellobiose. (β-D-glucopyranosyl-(1 → 4)-β-D-glucopyranoside) ○ = carbon; ● = oxygen; hydrogen omitted (after Liang and Marchessault, 1959).

3.3 Polysaccharides

Further condensation of monosaccharides by 1 → 3 or 1 → 4 links leads through a series of oligosaccharides (through e.g. cellotriose, cellotetraose, etc.) with a reduction in reducing power per unit mass and reduced solubility (i.e. fewer —OHs per sugar residue) leading to high molecular weight compounds with negligible reducing power which either form gels or swell only slightly in water. These form, in wide variety, the major constituents of plant cell walls. In examining their structure chemically it is necessary first to extract them from the walls and methods whereby this may be achieved must first come under review.

3.4 Cell wall fractionation

The first problem is to isolate from a tissue, or from single cells in unicellular organisms, a fraction which may be regarded as 'pure' cell wall material, in the sense that there are present no contaminants from the cytoplasm which might lead to confusion. This is therefore not a severe problem as far as polysaccharides and other high molecular weight compounds go such as are known not to be present at or near the cell surface except in the wall. The increasing interests in cell wall proteins and lipids, however, means that with most isolations of cell wall material it is preferable to use a method which involves the much more difficult task of ensuring a minimal (or no) contamination with cytoplasmic debris. With large unicellular algae the protoplasm can often be removed by opening the cell and either brushing or washing, and perhaps the 'cleanest' walls have been obtained in this way. With multicellular tissue it has become the practice to freeze in liquid nitrogen followed by grinding, thawing and washing (Jermyn and Isherwood, 1956; Cronshaw, Myers and Preston, 1958) in, for example, a mixture of ethylene-glycol and glycerol (4:1). Centrifugation at about 10 000–12 000 g then separates out the wall fragments as a powder suitable for the fractionation of its components.

In the past, the situation from then on has been confused on account of the variety of methods in use for fractionation of the wall. These all depended upon the relative solubility of wall components and naturally the precise nature of the substances extracted from the heterogeneous mixture of polysaccharides composing the wall varied with the solvent sequence used. The following method developed by Jermyn and Isherwood (1959) has now become standard practice, with slight variations dependent upon the wall concerned.

The powdered wall material, dried either in an oven at 105°C or from acetone, is treated with boiling water for 12 h followed by centrifugation and washing. The supernatant contains mostly pectic compounds (p. 56) with some soluble protein. If it is required to isolate lipids, the pellet is then treated with an appropriate organic solvent. The residue is then sometimes treated with ammonium or potassium oxalate to complete the removal of pectic compounds (Rogers and Perkins, 1968) though this is usually unnecessary. Following this, the residue is subjected to mild chlorination, a method originated by Cross and Bevan. This has two effects. It removes any lignin (p. 57) which appears in a highly modified form in the supernatant, and tends to break any linkages between cellulose and the rest of the wall polysaccharides which are collectively called hemicelluloses (p. 53). The residue is, for historical reasons, called *holocellulose*. Treatment with 4N KOH at room temperature for a few hours then removes the alkali-soluble hemicelluloses leaving a residue called α-cellulose.

The various extracts and the final residue may then by hydrolysed in mineral acid (after reducing to small volume in the case of the extracts) and the constituent sugars separated by standard chromatographic methods such as those given by Cronshaw *et al.* (1958).

The composition of a variety of walls determined by these or similar methods is given in Tables 3.1 and 3.2. Hydrolysates of the pectic compounds yield galacturonic acid, galactose and arabinose

TABLE 3.1

The composition of some cellulosic cell walls in higher plants (% dry weight)

Material	α-cellulose	hemi-celluloses	Pectic compounds	Lignin	Protein	Authority
Avena coleoptiles	30	53	5	0	5	Setterfield & Bayley 1961
	25–42	38–51	1–8	0	9–12	Roelofsen 1959
Pine cambium	25	>26	24	0	–	Roelofsen 1959
Sycamore cambium	36.7	45.0	15.0	2.5	–	Thornber & Northcote 1961
Sycamore xylem	40.0	31.5	3.8	25.3	–	Thornber & Northcote 1961
Sycamore phloem	40.0	25.8	6.6	22.6	–	Thornber & Northcote 1961

and these are found in the hydrolysates of the water extract except in a few plants (e.g. *Chaetomorpha melagonium*) where they have not been found at all; it should also be noted that pectic compounds are present in the primary walls of higher plants in

41

TABLE 3.2

The composition of some cellulose cell walls in algae (% dry weight)

Material	Water-soluble*	4N KOH soluble[+]	Soluble on chlorination[+]	α-cellulose	Authority
Green Algae					
Cladophora rupestris	31.5	2	34.0	28.5	Frei & Preston 1961a
Chaetomorpha melagonium	41.5	8	9.5	41.0	Frei & Preston 1961a
Enteromorpha sp.	30	39	9	21	Cronshaw et al. 1958
Ulva lactuca	52	25	4	19	Cronshaw et al. 1958
Brown Algae					
Fucus serratus	44	29	14	13	Cronshaw et al. 1958
Pelvetia canaliculata	70	16	12.5	1.5	
Red Algae					
Griffithsia flosculosa	41.5	14	22.5	22	Cronshaw et al. 1958
Rhodymenia palmata	50	36.5	6.5	7	Cronshaw et al. 1958

*pectic compounds plus some hemicellulose

[+]hemicellulose

much lower amounts than was at one time thought. The hemicelluloses yield a range of sugars including glucose, galactose, arabinose, xylose and mannose, together with some methylated sugars. The composition varies from plant to plant and in some algae some of the sugars are sulphated (Percival and McDowell, 1967).

α-cellulose as defined in this way rarely, however, yields only glucose on hydrolysis. The exceptions are all green algae including *Valonia*, members of the Cladophorales (Frei and Preston, 1961a) and *Glaucocystis* for which sugars other than glucose have not been detected (p. 193 *et seq*). Celluloses from other sources yield non-glucose sugars (often xylose or mannose) ranging in amount from 1.5 per cent (xylose, Belford 1958) in cotton hairs, through 15 per cent (xylose, Belford 1958), in beech wood, even up to 50 per cent (xylose, Cronshaw *et al.* 1958) in *Rhodymenia*. This is a matter we shall return to again.

This crude separation of wall polysaccharide into solubility classes still leaves the problem of the isolation of individual poly-saccharides from the fractions. There are no standard methods whereby this can be achieved and each botanical material presents its own problem. Fractional precipitation by electrolytes or organic solvents, separation by column chromatography, isolation through treatment by specific enzymes are all used. Details will be found in the book by Percival and McDowell (1967).

3.5 The structure of polysaccharides

In determining the fine structure of a polysaccharide in the chemical sense it is necessary to make observations upon the following matters:

(1) The constituent monosaccharides. This is achieved by the chromatography of hydrolysates as already mentioned.

(2) The linkages between the monosaccharides. This resolves itself into the determination of the site of the —OH group on one molecule which is involved in the linkage with the —OH on carbon 1 of the next molecule. In homogeneous polysaccharides in particular, the linkage in the disaccharide obviously helps. In the polysaccharide itself, the linkage(s) may be deduced following methylation and hydrolysis (by then determining which —OH is not methylated; by periodate oxidation; by the use of specific enzymes (though suitable degradating enzymes have still to be found for many polysaccharides); by spectroscopic methods now available in some cases; and by a variety of other chemical processes (Percival and McDowell 1967).

(3) Whether the glycosidic linkage is α or β. This is associated with (2) and often comes out with it.

(4) The site of groups such as ester sulphate groups. This can

43

be determined by standard chemical methods (Percival and McDowell 1967).

(5) The average chain length, molecular weight and the overall shape of the molecule. The molecular weight lies in the range between, say, 5000 and many millions and can be determined by a number of methods. In some methods the effect of each molecule is proportional to its weight and therefore give the *weight average* molecular weight [light scattering (Stacey, 1956; Mackie and Sellen 1969); sedimentation equilibrium (Schachman, 1959; Bowen 1970); gel permeation chromatography (which also yields number averages) (Moore, 1964; Mackie and Sellen 1969)]. With other methods the number of molecules is, in effect, counted per unit weight, giving a *number average* [viscosity (Marx-Figini and Schultz 1966; Mackie and Sellen, 1969); osmometry (Donnan and Rose, 1950; Mackie and Percival, 1961; Mackie and Sellen, 1969); chemical methods, i.e. determination of reducing end groups]. Unless all the chains have the same length, and hence the same molecular weight (and this rarely, if ever, happens), the weight average is greater than the number average and the difference gives some idea of the spread of the molecular weights. It is often more instructive to determine the degree of polymerization, DP, by dividing the molecular weight by that of the anhydro sugar unit and molecular weights are therefore often reported in these terms. The overall shape of the molecule may be determined by light scattering or by ultracentrifugation.

We may now proceed to examine the situation individually of some of the more important polysaccharides of the wall.

3.5.1 *Skeletal polysaccharides*

(a) *Cellulose*

Cellulose is the most stable of the polysaccharides in those walls which contain it, as evidenced by its resistance to the solubilizing agents used in its extraction. This resistance is due, no doubt, in part to the very stable β-1-4-linkage, to the very great chain length, and to the marked tendency for the constituent chains to come together in bundles of crystalline order (p. 133). It was early shown by Haworth (1929) that on hydrolysis with mineral acids higher plant cellulose yields both glucose and cellobiose and, by methylation experiments, that the linkage is throughout between carbon 1 of one glucose residue and carbon 4 of the next. His experimental yield of 80 per cent glucose led him to the view that cellulose is a homopolymer consisting of glucose residues only. This has turned out to be correct, though the view that the remaining 20 per cent represented a loss or a conversion of glucose was probably not entirely correct. As we have seen celluloses, however they are extracted from the wall, usually contain molecular chains of sugar residues other than glucose and these cannot be removed without affecting considerably the structure and coherence of the glucan chains. Except, therefore, in the case of a few algal celluloses already mentioned, the substance called cellulose which is available for investigation contains chains of non-glucose residue to an extent ranging from say about 10 per cent to about 50 per cent of its total weight, depending on the source. Within any cellulose, however, the molecular chains of the glucan are in the main separated spatially from the other polymers into crystallites which can be studied directly by, for instance, X-ray diffraction analysis (p. 119) and the structures and other features derived thereby are referred to cellulose. The word 'cellulose' is therefore used in two senses; as a family of inseparable polysaccharides of which a major part is a long chain glucan, and as a polymer consisting *only* of long chains of β-D-glucose. This is a matter which will be referred to again later in another context. For the time being it should be noted that while the X-ray diagram and some other features of the natural product refer exclusively to the long chain β-1,4-linked glucan, the chemical properties of the same product may not. In the remainder of this chapter, as in most of the book, the word 'cellulose' will be used for the long chain polymer built exclusively of β-D-glucose residues, the cellulose of the crystallographer.

The chemical evidence of Haworth and later workers, coupled with the X-ray evidence of Sponsler and Dore and of Meyer and Mark (p. 134) supported the view that cellulose is a linear polymer of β-D-glucose residues linked together by 1,4-links so that neighbouring pairs of residues constitute a cellobiose residue (Fig. 3.6). It is now generally accepted that the bent-chain, or Hermans, configuration already described for cellobiose (Fig. 3.5) applies also to cellulose and the corresponding intra-chain hydrogen bonds are

45

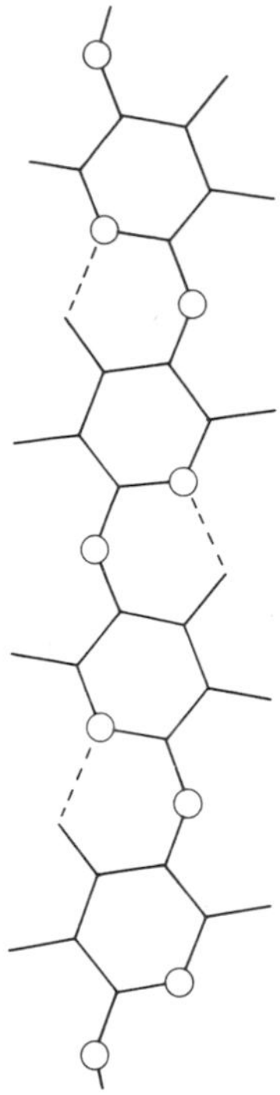

Fig. 3.6 **Section of a chain of cellulose. O = Oxygen; —OH represented by short lines, —H omitted.**

included in Fig. 3.6. The stability of this structure can be rationalized by application of the methods of conformational analysis introduced by Ramachandran and his colleagues (Rao *et al.*, 1967, Ramachandran, 1968). The reasoning is briefly as follows.

If the individual β-D-glucopyranose residues are each assumed to have the conformation shown in Fig. 3.3, then the co-ordinates of each atom with reference to any origin or axis may be calculated. If, in addition, it is taken that the bond angle at each glycosidic oxygen atom joining two residues together is fixed at some reasonable figure, then only two variables are necessary (and sufficient) to specify the shape of the chain — the angles of rotation ϕ and ψ shown in Fig. 3.7. Let the residue nearer to the reducing end of the chain be labelled R and that further away be called N. The set of co-ordinates for each residue is referred to its own axes with the origin at the centre of the glycosidic oxygen, and the axis $0x$ is defined by the respective bond to this atom. The axes $0y$ and $0z$

specify the state in which $\phi = \psi = 0$. As ϕ and ψ are allowed to change, the co-ordinates corresponding to any conformation (ϕ, ψ) may be derived by use of standard expressions for rotation of axes in two dimensions. For example, the axes for R may be rotated through ψ about O$-$C4' then, in the plane of C1$-$O$-$C4', through the supplement of the bond angles, and finally through ϕ about C1$-$O. The new co-ordinates for R are then expressed in terms of the axes for N. Together with N they now give the atomic co-ordinates of a disaccharide in conformation (ϕ, ψ). Inter-atomic distances are readily calculable and it is a simple matter to determine whether the conformation implies infringement of Van der Waal's bonds or whether any particular groups are in hydrogen-bonding positions. All possible conformations can in this way be examined by computer.

Fig. 3.7 Cellobiose, CH$_2$$-$OH marked by lines, $-$OH and $-$H omitted.

It has in this way been confirmed (Rea and Skerett, 1968) that the Hermans conformation is the only possibility that is free of steric clashes while obeying the demand (p. 135) that the chain should have twofold screw symmetry and a projected residue length of 5.15 Å. Further calculation has shown that the total Van der Waals energy then lies close to the minimum, displaced slightly from it to allow the intra-chain hydrogen bonds. This is another factor which relates to the high stability of cellulose.

It is by no means certain that the chain length of cellulose is as yet known with any accuracy; for the chain length can be determined only in solution and solvation of this resistant material inevitably brings the risk of chain breakdown. Moreover, the presence of polysaccharides other than cellulose, probably of shorter chain length, also reduces the average chain length determined. One is therefore inclined to place more reliance on results from, for instance, cotton rather than wood cellulose and upon methods which give extra weight to long chains than upon methods which count each chain as one unit irrespective of its length. The collected values presented in Preston (1952) show that from 1930 to 1947 reported degrees of polymerization increased from

TABLE 3.3

Molecular weights of celluloses from wood and natural fibres (from Goring and Timell, 1962)

Source	$(DP)_w$
Seed hairs	
Cotton (*Gossypium*) fibres from unopened bolls	15 300
The same, exposed to atmosphere	8100
Kapok (*Ceiba pentandra*)	9500
Milkweed floss (*Asclepias syriaca*)	8000
Textile bast fibres	
Flax (*Linum utisitatissimum*)	7100−8800
Hemp (*Cannabis sativa*)	9300
Jute (*Corchorus capsularis*)	8600
Ramie (*Boehmeria nivea*)	10 800
Wood	
Trembling aspen (*Populus tremuloides*)	10 300
Paper birch (*Betula papyrifera*)	9400
Red maple (*Acer rubrum*)	8300
Silver fir (*Abies amabilis*)	7500
Jack Pine (*Pinus banksiana*)	7900
Engelmann spruce (*Picea engelmanni*)	8000
Cinnamon fern (*Osmunda cinnamomea*)	8300
Bark	
Silver fir (*Abies amabilis*)	7200
Lodgepole pine (*Pinus contorta*)	10 300
Engelmann spruce (*Picea engelmanni*)	7100
Ginkgo (*Ginkgo biloba*)	8800
Paper birch (*Betula papyrifera*)	7500

175 [ramie fibres — end group analysis (Bergmann and Machemer)] to 10 000—15 000 [cotton hairs — viscosity in nitrogen (Galova and Ivanova)]. This reflects an improvement in methodology during these years. As far as the writer is aware, few figures higher than 15 000 have been recorded. Indeed, more recent demonstrations have yielded somewhat lower figures (Table 3.3). Marx-Figini and her colleagues (Marx-Figini 1963, 1964, 1966; Marx-Figini and Schultz, 1966) have concluded that the DPD of the cellulose in the primary wall of cotton hairs lies between 6000 and 7000 and in

the secondary wall between about 11 000 and 125 000. These figures will be considered again later when dealing with cellulose biosynthesis (p. 440).

The chemistry of cellulose has been treated *in extenso* (see e.g. Ott and Spurlin, 1954—55) and will not be considered here in detail; only those matters will be considered that are related to the identification of cellulose and to the behaviour of plant cell walls.

Cellulose is completely insoluble in water and hot dilute alkali, though it swells in these reagents. This is due to the particular circumstances that in nature the chains lie closely side by side, held together by innumerable hydrogen bonds which are therefore not available for attack by water. In KOH solutions ranging from 10 per cent to 22 per cent, depending upon the source of cellulose, the swelling is so enormous that structure is lost and the material disperses though it does not dissolve. On washing and drying, the cellulose then adopts a crystalline state different from that of the starting material (p. 141). It is then called cellulose II, to distinguish this form from the natural cellulose I. This is sometimes accepted as diagnostic for cellulose but such a diagnosis cannot be supported. It is attacked by hot dilute mineral acid, with chain degradation, but in cold acid only sulphuric acid in strengths of 70 per cent (w/w) achieves solution, again of course with very extensive hydrolysis. Similarly, cellulose is dispersed in some saturated salt solutions (e.g. $ZnCl_2$) and is soluble in cuprammonium (Schweizer's reagent). This yields the standard cytochemical colour reaction for cellulose — the development of a deep blue on treatment with, for instance, either saturated zinc chloride of 70 per cent sulphuric acid followed by aqueous iodine. As with many colour reactions this sometimes fails, both because substances other than cellulose may stain blue (see below) and because some cellulosic walls do not. In the latter category may be placed heavily lignified walls which do not stain until at least some of the lignin has been removed, and walls such as the *Valonia* wall (q.v.) which may not stain except after treatments which may be as mild as boiling in water.

(b) *Non-cellulosic polysaccharides*
It has long been recognized that, in most fungi, chitin occurs instead of cellulose. Until comparatively recently it was thought that cellulose was otherwise ubiquitous among plants. Some plants

have always, however, been recognized as presenting a problem and, of these, *Caulerpa, Acetabularia* and *Codium* are outstanding in the literature. Although unable to demonstrate the presence of cellulose in the trabeculae of *Caulerpa*, Nägeli (1844) thought of them as 'Zellulosebalken', a term still to be found in some modern textbooks (e.g. Strasburger, 1962). The cellulosic nature of the wall was first questioned by Correns (1894), who concluded that the skeletal material consisted of some substance at that time – and, in the event, during the next 60 years – unknown. Even in 1931, Preston (1931) while showing that neither the optical properties nor the X-ray diagram were consistent with the presence of cellulose, was unable to identify the polysaccharide and, even much later, Nicolai and Preston (1952) did no better with *Bryopsis*. Similarly, although cellulose was early considered to form the main cell-wall constituent in *Acetabularia* (Leitgeb, 1888) and *Codium* (Ernst, 1903 and 1904), this was shown to be an error by Mirande (1913). Nicolai and Preston (1952) were again unable to identify the skeletal polysaccharide of either plant, although they demonstrated by X-ray means the absence of cellulose, and Wertz (1957) had to be content with the view that, in *Acetabularia*, 'very probably the membrane does not consist of pure cellulose'.

In about 1959, however, three groups of investigators began independently to examine these plants by different modern methods which had by that time become available and the situation immediately became clarified. It became evident that cellulose does not occur in these plants [nor in others which are now known (Chap. 9).] and that, in terms of the skeletal polysaccharide which does occur, they can be divided into two groups, with a third overlapping the boundaries between the other two. The plants are all algae, members of the old order Siphonales among the green algae and of the Bangiales among the red algae; as far as is known, no higher plant falls in either category. These will now be taken in terms of the polysaccharide involved.

β-1, 3-linked xylan
The first significant report was made by Mackie and Percival (1959), showing that the walls of *Caulerpa filiformis* contains a xylan in the form of unbranched chains of xylose residues in the β-D-modification linked in the chain through carbons 1 and 3 (Fig. 3.8), together with a glucan, also β-1,3-linked. This is in

50

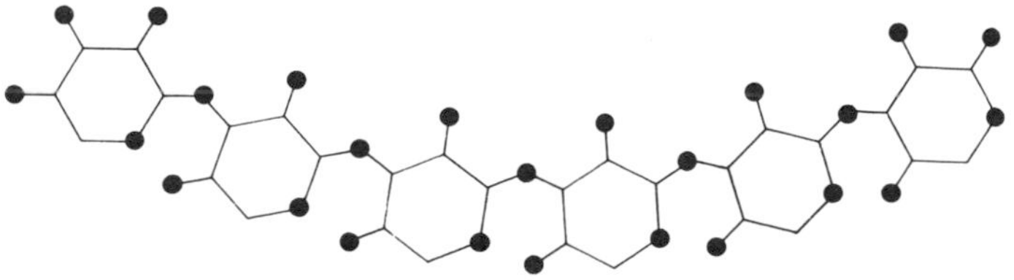

Fig. 3.8 Section of a chain of β-1,3-linked xylan, ● = O; H atoms omitted.

harmony with the earlier conclusion of Mirande (1913) and Escherich (1956) that the wall might contain callose, now recognized as a polymer of β-1,3-linked glucose residues (Aspinall and Kessler, 1957; Kessler, 1958). Subsequently Iriki and Miwa (1960) confirmed with another species the observations made by Mackie and Percival and extended them to cover a range of siphoneous green algae, many of which they found to have β-1,3-linked xylan as a major wall constituent. In no case could these workers demonstrate the presence of cellulose.

Meanwhile, Frei and Preston were examining a similar range of plants both by X-ray diffraction and by chromatographic methods. They demonstrated (Frei and Preston 1961 and 1964a) that in the Bryopsidaceae, the Caulerpaceae, the Udotaceae and the Dichotomosiphonaceae the crystalline skeletal polysaccharide in the wall is exclusively an unbranched β-D-1,3-linked xylan. This would have constituted the first definition of a polysaccharide linkage by X-ray diffraction had not the publications by the other two groups intervened. Frei and Preston (1964a) proceeded to determine the physical structure of the xylan, with later refinements from the same laboratory, and to examine wall architecture in these peculiar plants. The details will be considered later (p. 255).

The xylan can be 'isolated' from the wall by treatment in boiling water followed by chlorination. The resulting material — which may be called holoxylan in comparison with the equivalent holocellulose — contains glucose residues in addition to xylose residues to the extent of 1/4 (glucose/xylose) but treatment with caustic alkali does not remove the glucose. Holoxylan, unlike holocellulose, is completely soluble in alkali of strength ranging from 10 per cent to 18 per cent according to species. Frei and Preston (1964a) found themselves unable to support the claim of Parker, Preston and Fogg (1963) that the walls of *Dichotomosiphon* stain blue with iodine and sulphuric acid; indeed none of the organisms studied

51

yielded a blue stain except *Bryopsis* which contains a glucan (soluble in 6—8 per cent KOH and therefore not α-cellulose) responsible for the colouration. The holoxylan is soluble in cuprammonium, a somewhat surprising outcome in view of the reported insolubility of β-1,3-linked polysaccharide in this reagent. The solution can be fractionated to give a precipitate, mostly xylan (the bulk of the material) and a supernatant containing mostly glucose.

The degree of polymerization has been examined by Iriki, Susuki, Nisizawa and Miwa (1960) and by Mackie (1969) and Mackie and Sellen (1971). The last two workers used the methods of light scattering, viscosity, membrane osmometry and gel permeation chromatography on ethyl acetate solutions of the xylan after nitration under mild conditions and at low temperature. They found that there are xylan chains present in the wall with a DP in excess of 10 000, much higher than the figure of 50 given by Iriki *et al.* (1960). All their samples were contaminated by a glucan with a DP less than 1000 if it exists as a separate chain. There is no evidence from this or from any other work that part at least of the glucan does not exist as a copolymer in parts of an otherwise xylan chain.

Mannan

The first steps in the removal of the uncertainties concerning the second group of algae were taken by Iriki and Miwa (1960) in isolating from *Codium*, *Acetabularia* and *Halicoryne* a crude fibre which on hydrolysis yielded mainly mannose, with a little glucose, so that the major polysaccharide proved to be a mannan. There were strong indications that the mannan is β-1,4-linked in straight chain conformation identical with the mannan of higher plant cell walls (a hemicellulose) as exemplified by the mannan of ivory-nut endosperm. These results were confirmed by Frei and Preston (1961c) for *Codium*, *Acetabularia*, *Dasycladus* and *Batophora*, with the further information that the mannan in these plants is highly crystalline in the untreated walls, unlike the mannan of higher plants [except in ivory-nut endosperm (Meier 1956 and 1958)]. Subsequently, Kreger (1962) independently found the crystalline mannan in acid-washed *Codium* and Frei and Preston (1964a and 1968) gave a provisional structure for the mannan. About the same time, Love and Percival (1964) proved the linear, β-1,4-linked conformation. The structure of these walls, now so well known, will be considered later when the appropriate plants come under review (p. 239).

The walls of the algae which contain mannan contain little else. After treatment with hot water, the residue on hydrolysis gives only mannose with a trace of glucose (Frei and Preston 1968). Treatment with KOH does not effect any further fractionation. The mannan is partially soluble in alkali solution, the amount dissolving increasing with the concentration of the alkali solution up to 24 per cent. The switching of the −OH on carbon 2 of each glucose residue from one side of the ring (in cellulose) to the other (in mannan) has induced a considerable change in the reaction to alkali. Mannan does not give a blue stain with iodine and sulphuric acid, and this constitutes another difference from cellulose. On the other hand, treatment with KOH stronger than 12−14 per cent leads to a change in the X-ray diagram equivalent to that shown by cellulose so that, by analogy with cellulose, the new structure thus induced may be called Mannan II. Mannan walls take up metal ions from dilute solutions of the salts, recalling the behaviour of the hemicelluloses of higher plants.

3.5.2 *Hemicelluloses*

The nature of the hemicelluloses has been reviewed fairly recently for wood by Timell (1965) and for algae by Percival and McDowell (1967), and gums and mucilages have been dealt with by Rees (1969) and Aspinall (1969). These polysaccharides are therefore dealt with here only in outline. A common, though by no means universal, feature of these polysaccharides is the presence of acidic groups. In higher plants these take the form of D-glucuronic residues or their 4 O-methyl ester or of the corresponding D-galacturonic residues, as they also to some extent do in marine algae. In the latter, however, half ester sulphate groups are common (by replacement of $-CH_2OH$ on carbon 5 by $-CH_2OSO_2$) and these are characteristically absent in higher plants. While O-methyl sugar residues are common to both types of plants, O-acetyl groups and methyl esters of uronic acid are confined to land plants. A very common feature throughout is the presence in a polysaccharide of more than one type of sugar residue and, although there are these general distinctions between hemicellulose of different groups, there are many structural similarities (Aspinall, 1964). Polysaccharides of this general type probably occur also in fungi (Rogers and Perkins 1968). Hexosans including mannan (1-4 linked), some of which are $1 \rightarrow 3$ and $1 \rightarrow 6$ linked, and which must be included in this class of substance, are found in the walls of yeasts.

Xylan can be isolated from hardwoods in yields of 80—90 per cent by simple extraction with alkali and form the prominent hemicellulose. The xylose residues are in the β-form and are $1 \rightarrow 4$ linked in a chain which is linear except for occasional, short branches. When isolated under milder conditions, seven out of ten xylose units are found to be acetylated, mostly at carbon 3 but occasionally at carbon 2. Some of the residues carry a 4-O-methyl-α-D-glucuronic acid linked directly to carbon 2. The hemicellulose is therefore a O-acetyl-4-O-methyl glucurono-xylan (Fig. 3.9a). The unbranched sections are sufficiently long to allow crystallization and this has allowed Marchessault and Liang (1962) to demonstrate that the chain possesses a threefold screw axis. The DP is about 200 (Koshijima and Timell, 1964). This general type of polysaccharide is common among angiosperms and occurs also in conifers although with these an α-L-arabinofuranose residue may be attached to the carbon-3 of xylose; it is also found in ferns.

Combined xylose also occurs in polysaccharides in marine algae but usually comprises only a minor part of the molecule; fucoidan from *Fucus* sap, for instance, contains about 15 per cent of xylose residues, the rest being mainly fucose and galactose residues. The water-soluble xylan from *Rhodymenia palmata* (Percival and Chandra, 1950) appears to be exceptional in being a homopolysaccharide, though the linkage is 80 per cent 1-4 and 20 per cent 1-3 apparently distributed at random; this xylan is reported to have a DP of the order of 50—100, and it may be a reserve food rather than a wall material.

Mannan polysaccharides are common in higher plants and are the predominant hemicellulose of softwoods. They take the form of a family of closely related polysaccharides which may be termed O-acetyl-galacto-glucomannan with the ratio galactose : glucose : mannose usually about $1 : 1 : 3$ (Fig. 3.9b). Mannans also occur as hemicelluloses in those seaweeds in which the skeletal polysaccharide is mannan; here, however, the polysaccharide is more nearly homogeneous, only about 5 per cent of it consisting of glucose residues. Glucomannans are also the predominant hemicellulose in

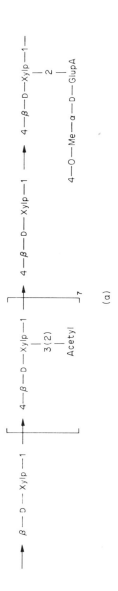

Fig. 3.9 Sections of hemicellulose chains
Xyl = xylose; Glu = glucose; Man = mannose; Gal = galactose; Me = methyl; Ara = arabinose; p = pyranosyl; f = furanosyl;
(a) = hardwood glucuronoxylan; (b) = O-acetyl-galacto-glucomannan; (c) = arabinogalactan.

ferns and in *Psilotum* and *Equisetum*. Substituted mannans have been isolated from a number of seeds, presumably as wall constituents. These are all galactomannans; they consist of a β-1,4-linked mannan backbone with α-D-galactopyranosyl residues as 1,6-linked single unit side chains (Smith and Montgomery, 1959). Enzymic hydrolysis indicates that the side chains occur in blocks.(Courtois and Dizet 1970) so that there are long mannan segments with no side chains.

Galactans are widely distributed among plants, but in higher plants appear particularly to be a feature of the tension wood of hardwoods and the compression wood of softwoods (Meier, 1962). In compression wood the link is uniformly $1 \rightarrow 4$ but tension wood galactan contains also $1 \rightarrow 6$ links. The galactans of most species of the red algae are different again in that the linkage is alternately $1 \rightarrow 3$ and $1 \rightarrow 4$. Galactans are also found as *arabinogalactans* in a wide variety of organisms. These polysaccharides are usually water-soluble and it is not absolutely certain that they are wall components. They occur notably in some larches though also in small amounts in other softwoods and in hardwoods. This is a highly branched polysaccharide with a backbone of $1 \rightarrow 3$ linked β-D-galactopyranose residue with a variety of side chains of L-arabinofuranose and D-galactopyranose and with occasional small amounts of glucuronic acid (Fig. 3.9c). The same polysaccharide has been found in a number of marine algae (e.g. *Cladophora, Chaetomorpha, Caulerpa, Codium*) where it is, however, sulphated.

3.5.3 *Polyuronides*

Polyuronides are widely distributed among plants. In land plants they are characteristically based upon pectic acid, in the main a linear chain of α-1,4-linked galacturonic acid residues (Fig. 3.4). Somewhat similar polysaccharides occur in brown marine algae and some bacteria in the form of alginic acid. These contain poly-D-mannuronic acid and poly-L-guluronic acid (Fig. 3.4) together with some poly-D-glucuronic acid. A notable difference is that in the pectic compounds many of the uronic groups are methylated while no methylation has been detected in alginates.

No residues other than the uronic acid residues have so far been proved to be present in alginic acids and the linkage is uniformly $1 \rightarrow 4$. With pectic compounds on the other hand, there is no doubt but that a variety of sugars (rhamnose, galactose, xylose, arabinose and fucose) occur either in the main chain or as branches (Aspinall 1964). In addition, the pectic compounds appear always to be associated with neutral arabans and galactans (Fig. 3.4) which may form part of a larger polymer broken down on extraction. Similar remarks apply to the pectic compounds of seaweeds (e.g. *Ulva, Enteromorpha, Ascophyllum*) although here again half ester sulphate groups occur.

The polyuronides, particularly the alginic acids, have been the subject of much structural investigation by X-ray methods. This aspect will be dealt with later when appropriate plant groups come under review.

3.5.4 *Lignin*

Among the non-saccharide components of cell walls lignin stands out as the unique aromatic polymer. It occurs to the extent of 15—35 per cent in most supporting tissues of higher plants and is therefore a major component of xylem and of the sclerenchyma of phloem. Lignin extractants (with the possible exception of dioxane) all react with the lignin, so that lignin isolated by extraction has been considerably modified. This effective insolubility, together with the absence of any regularity in structure and of any weak bonds, has presented chemists with a major problem in structure research, and still stands in the way of any precise structure formulation. Because the structure is so modified on solution, such scraps of information as can be found by physical techniques are particularly important.

The refractive index is high (1.61) indicating a closely-knit structure in the main covalently bonded. The absorption band in the ultraviolet at 2800 Å, and the infrared bands at 6300 nm and 6800 nm (Tschammler, Kratzl, Leutner, Steininger and Kisser, 1953) confirmed the earlier findings of Lange that lignin is aromatic in nature and removed once and for all the older belief that lignin might be a secondary product resulting from the extraction processes necessary for purification.

Lignin can be regarded as a polymer of phenylpropane with *p*-coumarylic acid and sinapylic acid. It is commonly recognized by the red colouration which develops on treatment with phloroglucinol followed by HCl. This colour does not develop if the plant material is first chlorinated because the coniferyl aldehyde group is then oxidized, and this group is solely responsible for the colour (Adler, Björkvist and Häggroth, 1948). Better criteria have been given by Kratzl (1965). The chemistry of lignin has been the subject of investigation by a number of workers, among whom Freudenberg is predominant, far too detailed to be discussed here. The resulting body of knowledge, coupled with the biosynthetic work of, for instance, Neish and Kratzl, as well as Freudenberg himself, has led Kratzl (1965) to present a 'formula' (Fig. 3.10). This

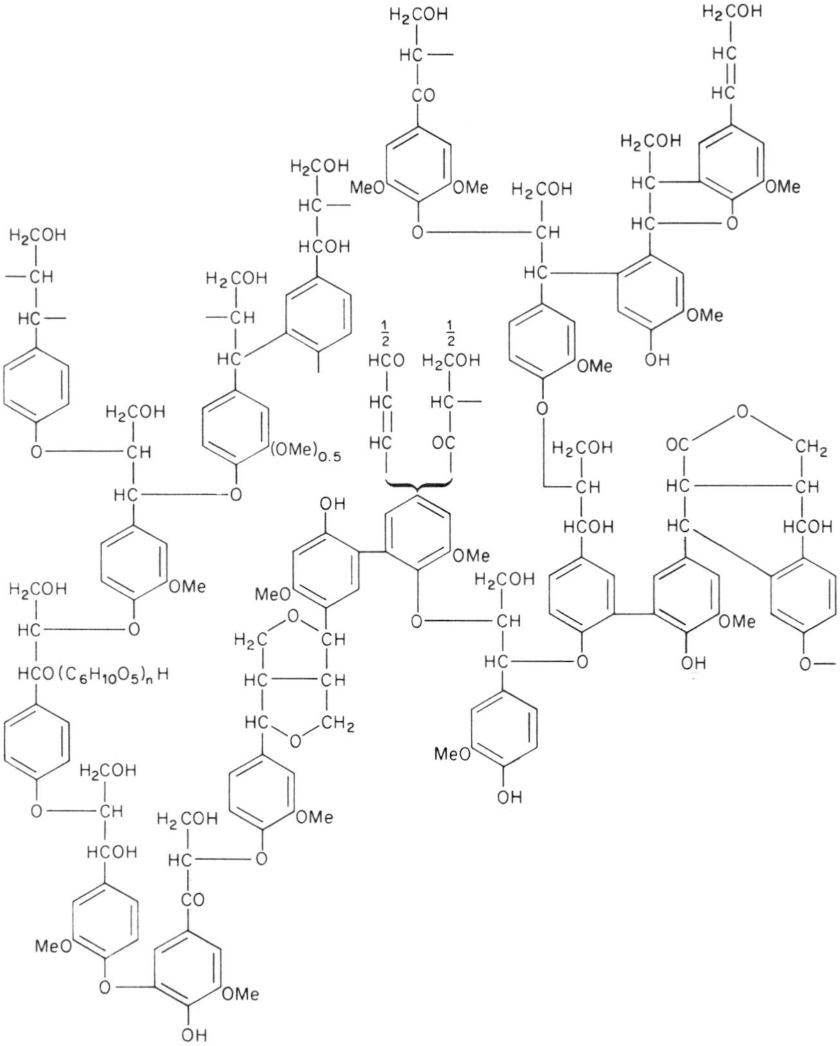

Fig. 3.10 The structure of lignin (after Kratzl, 1965).

is best regarded as one way in which units known to be present could be linked together. The precise arrangement may vary from plant to plant (and certainly between angiosperm and gymnosperm) and may not be constant even within one wall. As a three-dimensional polymer, lignin is almost certainly heterodisperse; mean molecular weights determined by osmotic pressure and by various diffusion and sedimentation methods have been given in the range of 800—12 000. Clearly the molecular weight in the intact wall may be even greater than the higher limit so far given.

3.5.5 *Proteins*

Hemicelluloses extracted from plant cell walls are often contaminated with protein material, which is very difficult to remove, sometimes to the extent of 8 per cent or more; examples are given freely in Percival and McDowell (1967). Tupper-Carey and Priestly (1924) were the first to demonstrate the presence of protein in a wall (the primary wall of meristematic cells) and, though this and later similar findings were invariably challenged as due to cytoplasmic contaminants, there can no longer be any doubt but that protein is present as an integral and important part of cell wall architecture, particularly in the walls of growing cells. This has arisen both through improved methods of cell fractionation and through techniques for the specification of cell wall proteins (see Lamport, 1965). It is convenient to divide these proteins between those known to be enzymes and those which appear to fulfil another function even though this distinction may not in the end prove valid. We will consider the latter first, as structural proteins.

(a) *Structural proteins*
Ever since the early work of Tupper-Carey and Priestly (1924) numerous workers had found proteins associated with isolated cell walls and, by 1951, the first chemical investigation had been carried out by Tripp, Moore and Rollins (1951) who found in primary walls the amino acids serine, glycine and aspartic acid. The problem of finding suitable criteria of cell-wall protein remained, however, until in the same year Lamport and Northcote (1960) and Dougall and Shimbayashi (1960) discovered hydroxyproline (hyp) (Fig. 3.11) as a major constituent of the hydrolysates of cell walls from tissue cultures. Each group of workers considered

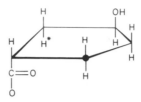

Fig. 3.11 L-hydroxyproline, ● = N. If the starred H is replaced by —OH, the molecule becomes *cis*-trans dihydroxyproline.

seriously the possibility that the wall-bound nitrogen might still represent a cytoplasmic contaminant by isolating the walls under a wide range of conditions (pH 2 to 10 in distilled water or solutions up to 40 per cent sucrose). The wall hyp remained throughout at the same level within the limits of experimental error. Since that time many other workers, using a wide variety of plants including algae, have satisfied themselves of the existence of a wall-bound nitrogenous compound containing hyp (e.g. Olson, Bonner and Morré, 1962; Crook and Johnstone, 1962; Brown, 1963; Edelman and Hall, 1964; Olson, 1964; Linskens, 1964; Bartnicki-Garcia, 1966; King and Bayley, 1965; Punnett and Derrenbacker, 1966; Cleland and Karlsnes, 1967; Thompson and Preston, 1967; Gotelli and Cleland, 1968; Stewart, Dawes, Dickens and Nicholls, 1969; Ridge and Osborne, 1970). Although there is still one dissentient, the cumulative evidence is clearly to the effect that substances containing hyp are widespread among plant cell walls and that they play a structural role (Thompson and Preston, 1968; see p. 415). They are intensely wall-bound and may be extracted only by hydrolytic processes. The hydrolysate contains many amino acids other than hyp (Tables 3.4 and 3.5) and the inference is that the substance is a protein or proteins. The evidence for the protein nature has been summarized by Lamport (1965, 1969) who, in view of the presumed function of the protein, has named it *extensin*. In some algal walls the protein contains hydroxylysine (Thompson and Preston, 1967), another amino acid which, like hydroxyproline, is typical of the animal structural protein collagen.

The protein appears to achieve its firm binding in the wall by attachment to specific polysaccharides. Enzymatic digests of tomato cell walls, using either cellulase or pronase, have been shown to contain a number of glycopeptides (Lamport 1967a, 1969) in agreement with earlier suggestions (Lamport 1962). These

TABLE 3.4

*Amino acid residues/10^5 g protein in cell walls of some higher plant cell cultures**

	Nicotiana tabacum	Lyco-persicon esculentum	Acer pseudo-platanus	Solanum tuberosum	Phaseolus vulgaris	Pisum (epicotyl)	sativum (root)	Rose	Ginkgo biloba	Centaurea cyanus
Hydroxyproline	181	142	182	85	58	64	102	52	50	43
Aspartic acid	49	50	59	61	104	71	81	70	87	69
Threonine	54	51	35	51	38	51	54	64	50	69
Serine	92	93	94	82	90	93	86	74	51	78
Glutamic acid	38	61	65	85	77	86	84	102	84	116
Proline	61	60	57	60	80	69	61	85	42	54
Glycine	48	85	51	88	66	112	95	105	146	50
Alanine	40	39	56	67	45	68	63	65	66	90
Valine	65	61	49	62	83	61	44	60	64	70
Methionine	4	4	2	15	9	trace	—	9	trace	20
Isoleucine	20	39	26	42	31	44	45	36	39	41
Leucine	33	41	44	73	57	74	73	89	61	74
Tyrosine	29	18	15	28	24	11	15	12	23	trace
Phenylalanine	14	20	24	37	24	26	34	19	37	28
Lysine	91	81	77	66	86	66	69	89	52	72
Histidine	23	18	21	17	19	10	9	9	26	29
Arginine	21	11	18	47	25	16	11	16	14	24
Cysteic acid	nd	nd	10	—	—	—	3	—	—	—

* from Lamport (1965).

nd, not determined

TABLE 3.5

Amino-acid compositions of various algal wall fractions Residues/1000

	Nitella		Chaetomorpha		Cladophora	Codium	
	A	B	Phenolacetic acid-water	Sporulated		Wall	Cytoplasmic fraction
Cysteric acid	85	64	65	1	64	23	22
4-Hydroxyproline	–	34	37	64	46	108	26
Aspartic acid	140	119	109	145	130	72	107
Threonine	35	41	45	49	39	60	68
Serine	53	49	57	55	34	68	75
Glutamic acid	48	109	108	106	117	88	114
Proline	106	75	69	80	52	53	54
Glycine	123	119	107	117	146	87	99
Alanine	95	73	70	74	65	97	102
Valine	39	57	34	41	55	91	70
Cystine (half value)	7	–	–	36	–	–	–
Methionine	Trace	2	6	10	–	7	3
Isoleucine	30	26	30	22	26	31	40
Leucine	38	75	59	65	46	61	73
Tyrosine	15	27	28	27	35	20	22
Phenylalanine	23	21	20	23	20	25	40
Lysine	108	70	75	63	71	61	52
Histidine	12	7	7	4	13	10	6
Arginine	36	30	25	25	nd	20	25
Peak 1	–	14	21	12	23	16	2

nd, not determined

all contain arabinose, galactose and hyp together with a number of other amino acids (valine, serine, threonine, lysine and tyrosine). By alkaline hydrolysis of these isolates, Lamport (1967b, 1969) was able to recover hyp-O-arabinoside, confirming a glycosidic link with the −OH of the amino acid. About 50 per cent of the wall hydroxyproline was accounted for by these types of compound and this is, of course, a minimal figure. W. Mackie (private communication) has found evidence for the presence of similar glycopeptides in the walls of the green alga *Codium*.

Most recently, Lamport (1971) has taken an enormous step forward in at last recording the amino acid sequence of three peptides resulting from a tryptic digestion of cell walls in which the hyp-O-arabinoside link had been cleaved by acid treatment. The sequences are: Ser hyp hyp hyp hyp thr hyp hyp val tyr lys; ser hyp hyp hyp hyp lys; ser hyp hyp hyp hyp lys. He has now also (private communication) recovered a fourth; ser hyp hyp hyp hyp val (?) lys lys. The possible significance of four hyp's in sequence will be discussed later.

As pointed out by Lamport (1969), the totality of the evidence available makes it possible to construct two molecular models. One, A, would consist of a polysaccharide backbone with peptide side chains (as in bacteria) and the other, B, (Fig. 3.12) of a polypeptide backbone with oligosaccharide side chains. Only model B appears to be fully consistent with all the data available. We will therefore adopt model B, and refer to the backbone as a protein.

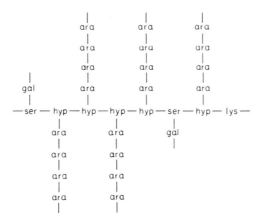

Fig. 3.12 **Association between hyp polypeptides and wall oligosaccharides (after Lamport and Miller, 1971).**

Most of the hydroxyproline is considered to be O-substituted by a tetra arabinose (Lamport and Miller 1971) and it has most recently been shown that with one of the peptides — NH_2 ser hyp hyp hyp hyp ser hyp lys COOH — the serine is also O-substituted to galactose (Fig. 3.12) [Lamport, Roerig and Katona (1972)].

This close association of hyp with oligosaccharides which are clearly wall components, possibly segments of polysaccharides, constitutes perhaps the clearest evidence yet found that the hyp protein is wall bound.

The protein content of cell walls is variable. Primary walls contain a higher percentage than do secondary walls and among them the extractable protein content varies from about 1 per cent to about 6 per cent. In the alga *Nitella* hyp is absent from the walls of growing internodal cells (Thompson and Preston 1967; Gotelli and Cleland 1968), as it perhaps also may be from other cells, though other amino acids are present in wall hydrolysates (Table 3.5). The data presented by Lamport (1969) is shown by him to be consistent with the view that tomato extensin consists of 50 per cent protein and 50 per cent polysaccharide; if therefore, the assumption is made that the pronase *specifically* releases extensin and about half of the hyp contained in the wall, then intact tomato cell walls must *contain* about 22 per cent protein. This is higher than the cellulose content.

The reality of the wall-bound protein appears to be accepted by all workers in this and related fields except by Steward and his school. They were the first (Steward, Thompson, Millar, Thomas and Hendricks, 1951) to draw attention to the existence of hyp in plants and concluded (Pollard and Steward, 1959) that the hyp-rich material was present only in the cytoplasm. More recently, their findings (Steward and Chang 1963; Israel, Salpeter and Steward 1968) that ^{14}C-proline or ^{3}H-proline fed to tissue cultures leads to accumulation of labelled hyp in cytoplasmic proteins and not in the wall, have convinced them that the so-called extensin of other workers is, in spite of all the precautions taken, a contaminant. More recently, Steward, Mott, Israel and Ludford (1970) have examined the giant vesicles of the alga *Valonia ventricosa* from this point of view. The advantage of using this particular alga is that the vesicle is so large that undoubtedly pure cell wall material is readily available. They have concluded that acid extracts of the wall of the mature vesicle can contain no more than 2 per cent protein at most (corresponding perhaps to a wall content of 4 per cent at most). This may be compared with the content of at least 9 per cent in sporulated *Chaetomorpha* walls (a related alga) which are equally certainly clean walls and free of epiphytes (Thompson and Preston, 1967). Steward *et al.* (1970) have further shown, however, that when *Valonia* is induced to form aplanospores and the sporelings are cultured continuously in a medium containing ^{14}C-proline, no radioactivity can be detected by microautoradiography in the considerable wall which develops during culture. Taken as a whole, this body of work appears at first sight to stand in stark contrast with the much larger bulk of evidence in favour of a wall protein. There is, however, more than one possible explanation of the results.

Except for the observations made on the mature walls of *Valonia* (which allows 4 per cent protein and this would be enough) the contrary findings stem from feeding experiments. These negative results might be due to (1) the absence of a protein from the wall, as claimed by Steward; (2) the absence of hydroxyproline, or any amino acid derivable from fed proline, from a protein in the wall (as for *Nitella*) (Thompson and Preston, 1967) or (3) an *indirect* biochemical route from fed ^{14}C-proline to ^{14}C-hydroxyproline bound in a wall protein whereby a good deal of the ^{14}C is side-tracked. Chrispeels and Sadava (1969) have repeated the work of Israel *et al.* (1968) and of Steward, Israel and Salpeter (1967) on carrot explants, but feeding with universally labelled ^{3}H-L-proline instead of 3-4^{3}H-L-proline (they point out that the latter loses half of its radioactivity on conversion to hyp). Moreover, again differing from the early workers, they plasmolyse the cells before attempting autoradiography. This separates the wall from the cytoplasm so that these can be 'counted' separately instead of via the complicated statistics necessary in the unplasmolysed cells used by Steward and his co-workers. The label is then found uniformly distributed about the wall; the deposition of the radioactive material appears to occur mainly on the inner wall face. Steward has not so far attempted to demonstrate the absence of *protein* from the walls he has examined, and his results perhaps have a closer bearing on the metabolism of proline than they do on the problem of the wall-bound protein.

(b) *Enzymes*

The presence of proteins in plant cell walls clearly points toward a concept of the wall as a part of the metabolic machinery of the cell rather than as a dead cell envelope. Indeed, purely structural investigations of the wall led long ago to this concept (Bonner, 1935; Heyn, 1940; Preston, 1955; Frey-Wyssling, 1959) and the phenomena of cell wall growth can hardly be understood without it (p. 406). The presence of enzymes gives a further pointer.

The observations presented by Newcomb (1951) on auxin-treated pith cells of tobacco form the first systematic investigation of wall-bound enzymes. He showed that most of the ascorbic acid oxidase activity of the cells rested in the cell wall fraction, with a specific activity (per mg nitrogen) several times as great as that for the whole cell homogenate, under conditions in which it could not be removed by repeated washing even in salt solutions. This has been confirmed time and time again with the same and with other cell types even up to the most recent times (Lamport, 1965; Pilet, 1971). Since 1951, three other groups of enzymes, all hydrolases have been reported as wall-bound even though none of them appear to have the ubiquity of ascorbic acid oxidase. These all occur in rapidly growing walls and constitute (a) invertase (Kivilaan, Beaman and Bandurski, 1961; Strauss and Campbell, 1963; Edelman and Hall, 1964), (b) pectin nethylesterase (Glasziou, 1959) and (c) phosphatases (Lamport and Northcote, 1960).

Among other enzymes more recently claimed as wall-bound enzymes may be mentioned glucanases, peroxidases, ATP-ase, DNA-ase, and RNA-ase. The standing of these claims is reviewed by Lamport (1970).

The claim by Kivilaan et al. (1961) that the wall contains enzymes concerned with its own synthesis – e.g. UDPG-pyrophosphorylase – has not stood the test of time; these are almost certainly associated with the plasmalemma and were therefore found in the wall as a contaminant (Hall and Ordin 1967). As far as the writer is aware, no claims have yet been made for the presence of a protease.

It is not as yet certain how the enzymes involved are bound in the wall. The fact that the enzymes cannot be washed out in salt solutions (whereas pepsin and chymotrypsin synthetically attached to the wall can be removed in this way) suggests that the binding is not ionic and is therefore rather covalent. The demonstrations by Stark and Dawson (1963) that ascorbic acid oxidase extracted from summer squash contains 10 hexosamine residues mol^{-1} points in the same direction. In this context it is interesting to note that Lamport (1969) has obtained, from a peroxidase, hydroxyproline-O-arabinosides similar to those he has found in the structural wall protein. Not only does this speak for a covalent link but opens the possibility that subunits of extensin may double up as enzymes.

CHAPTER 4

Structure determination – optical microscopy

4.1 The principles of microscopy

These various chemical constituents form a complex assembly in the wall which can be examined by a number of physical methods. Since there is always present at least one polysaccharide which is crystalline, methods designed for the elucidation of crystal structure are particularly suitable and the next two chapters will be devoted to two of the more powerful methods falling in this category. It should be noted that although the general structure of many walls has now been worked out by these methods an understanding of the techniques themselves remains important. The structure of walls is very variable, so that in investigating any particular piece of wall in hand the structure has often to be re-worked just as though nothing were yet known about it. We begin with techniques of microscopy.

Any study of a cell wall as of any other biological entity demands that the material should be examined at all possible levels of size. Observation by eye alone is of limited value, chiefly on account of the low resolving power of the eye, i.e. the relatively large separation between points in the object which can still be seen as two points as the points are moved closer together. An object can be focussed on the retina provided that the intervening distance is greater than a critical value known as the closest distance of distinct vision, about 250 mm with a normal eye. At this distance the resolving power is 0.2 mm and detail smaller than this cannot be seen. This is why most cell walls presented edgewise to the eye are invisible. Visualization is achieved by interposing between the object and the eye an optical system in such a way that the object is so magnified that detail smaller than 0.2 mm becomes at least this size, and the system must be such that its own resolving power is better than 0.2 mm. The system is, of course, a microscope. There is nowadays, a battery of microscopes available to the cytologist; these all have principles in common and these will be described first. The complications involved with the various microscope types will then be taken up in so far as necessary for the rest of the book.

It should be clear that the fundamentally important attribute of any microscope is its resolving power r. Clearly it is formally desirable only that a microscope should magnify an object to the extent that a distance r in the object should become 0.2 mm in the image; for then all the detail resolvable by the microscope is also resolvable by the eye. The useful magnification is therefore $0.2/r$ where r is in mm and, though greater magnification than this may be used for comfort of viewing, nothing further is gained. In light microscopy, simple lenses suffer from aberrations which materially reduce the resolving power below that theoretically attainable which will be derived below. For this reason microscope lenses are complex, consisting of trains of lenses, some of which are closely in contact and some of which are spaced. The reasons for such complexity will not be discussed further here; interested readers are referred to texts on microscopy. Lenses will be dealt with as though they were simple biconvex lenses.

Before considering the resolving power of a microscope we need to be reminded of some simple considerations concerning the nature of light and other electromagnetic radiations.

4.2 The wave nature of electromagnetic radiation

A beam of light passing through a homogeneous medium is associated with an electric vector and a magnetic vector, the magnitudes of which fluctuate at any point between limits, the end of the vector moving in simple harmonic motion. We will consider only the electric vector. If the vibration is conceptually halted, the locus of the vector in space is as shown in Fig. 4.1a, namely a sine curve. This sine wave can be specified by two parameters, (a) the *wavelength*, λ, between identical points on the curve, (b) the frequency, ν, i.e. the number of crests, say, which pass a given point per second

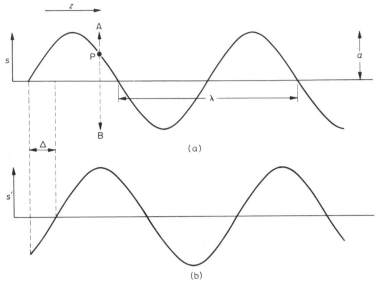

Fig. 4.1 Two sine waves, amplitude a, wavelength λ out of step by a path difference Δ.

when the vibration is again set in motion. Clearly the product $\nu\lambda$ equals the velocity of light, c. If the period is τ (i.e. the time for P (Fig. 4.1a) to pass from A to B and back to A) the wave vibration may be written as

$$s = a \sin 2\pi \left(\frac{t}{\tau} - \frac{z}{\lambda} \right) \qquad 4.1$$

A second vibration, passing from left to right parallel to the first and identical with it except that it is out of step by a distance Δ (Fig. 4.1b) may equally be written

$$s' = a \sin 2\pi \left[\left(\frac{t}{\tau} - \frac{z}{\lambda} \right) - \frac{\Delta}{\lambda} \right] \qquad 4.2$$

When $\Delta = \lambda$, this reduces to Equation 4.1 and the two vibrations reinforce each other. They are again in step. When $\Delta = \lambda/2$, on the other hand, the two vibrations are as far out of step as they can be, Equation 4.2 becomes

$$s' = -a \sin 2\pi \left(\frac{t}{\tau} - \frac{z}{\lambda} \right) = -s,$$

and at all points and all times the vibrations cancel. The intensity has become zero. This is destructive interference.

70

The path difference, Δ, may be considered in another way. Consider a point P, Fig. 4.2, moving uniformly along a circular path of radius a with angular velocity ω (= $2\pi/\tau$). As P moves, the foot N of the perpendicular from P to the vertical diameter AB of the circle moves in such a way that

$$ON = s = a \sin \theta = a \sin \omega t \qquad 4.3$$

For a second point P', an angular distance δ behind P, the corresponding motion along AB is

$$s' = a \sin(\omega t - \delta) \qquad 4.4$$

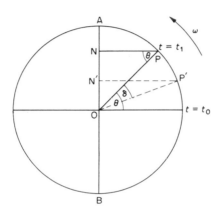

Fig. 4.2 **For explanation, see text.**

Comparison of Equation 4.3 with 4.2 (taking $z = 0$) shows that

$$\frac{2\pi\Delta}{\lambda} = \delta \qquad 4.5$$

The path difference Δ can therefore be expressed as a *phase difference* δ. When $\delta = 2\pi$ the two vibrations are in step; when $\delta = \pi$ they are out of step and destructive interference is complete. In considering microscopy in terms of the wave nature of light it is convenient to use Equation 4.4 in place of equations of the type of 4.2, with no loss of rigour.

4.3 Resolving power

The efficiency of an optical system may be judged by the accuracy

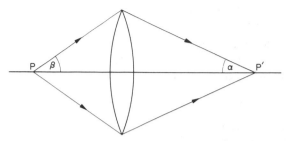

Fig. 4.3 A point object, P, producing a point image P' through a lens system.

with which a mathematical point P on the object, Fig. 4.3, (taken on the optic axis for simplicity) is recorded as a mathematical point P' on the image. Consider the space on the image side of the lens, the image space, and assume, as is usual in theories of microscopy, that the object P is self-luminous. Suppose further that the converging beam of light in image space is limited by a rectangular stop LMNO as in Fig. 4.4. Through F, the image position, draw a *very short* line FF' (= h) parallel to the plane LMNO and to MN or OL. Then since the wave trains centering upon F are parts of spherical sheets whose centre if F, FF' is part of a diameter of the spheres. We wish to enquire under what circumstances light of appreciable intensity will impinge upon F', i.e. how 'blurred' the image F will be.

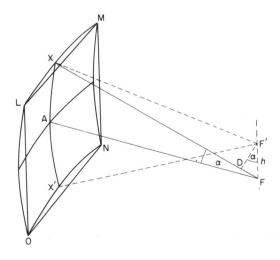

Fig. 4.4 For explanation, see text.

Since FF' is a diameter, LM and ON are parallels of latitude and all points on each of them are equidistant from F'. Light vibrations passing from all points on LM will arrive at F' in phase so that the phase from the whole strip may be taken as the phase from its mid-point X, and similarly for ON. The difference in phase between the light reaching F' from LM and ON is therefore equivalent to the path difference X'F' − XF'.

To evaluate this, note that if b is small, XF' = XD and

$$XF - XF' = XF - XD = b \sin \alpha$$

Similarly, $\qquad X'F' - X'F = b \sin \alpha$

Since XF = X'F, this means that

$$X'F' - XF' = 2b \sin \alpha$$

$$\text{or } AF' - XF' = b \sin \alpha$$

Clearly when AF' − XF' = $\lambda/2$ for any narrow horizontal strip above A, there is another below it, contributing opposite phase. Therefore, for all the light passing through LMNO there will be zero intensity at F' only when

$$b = \lambda/(2 \sin \alpha) \qquad\qquad 4.6$$

There will, on the other hand, be a maximum of intensity when

$$b = \lambda/\sin \alpha,$$

and additional minima at $3\lambda/(2 \sin \alpha)$, $5\lambda/(2 \sin \alpha)$, etc. and maxima at $2\lambda/\sin \alpha$, $3\lambda/\sin \alpha$, etc. Had a circular aperture been used, the calculation would have been more difficult, but the result would have been only to replace 1/2 in Equation 4.6 by 0.61; and if the whole operation had been carried out in a medium of refractive index n instead of *in vacuo*, as assumed, all distances would need to be multiplied by n [since this is the ratio of (velocity of light *in vacuo*)/(velocity of light in medium)]. The first minimum of intensity would then be at

$$b = \frac{0.61\lambda}{n \sin \alpha} \qquad\qquad 4.7$$

and this is the formulation we shall adopt.

The nature of light itself, therefore, however perfect are the lenses used, sets a limit of the accuracy with which an optical system can image a mathematical point. A point in the object becomes in the image a disc surrounded by a series of concentric annular rings of decreasing intensity, the Airey disc (Fig. 4.5). A second point in the object close to the first will be reproduced as a second Airey disc in the image and there is clearly a limit beyond which these discs may approach each other and remain visually separate. This limit is taken as the point at which the centre of the central maximum of one disc lies squarely over the first minimum of the other, when the overall intensity distribution (Fig. 4.5) is virtually that of a single broader image.

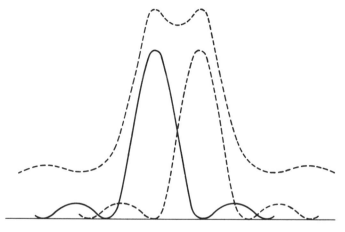

Fig. 4.5 Two overlapping Airey discs in image space, representing two close points in object space.

At this point the distance between the Airey discs is $(0.61)\lambda/n \sin \alpha$. The corresponding distance in object space (Fig. 4.3) is clearly $(0.61)\lambda/n \sin \beta$ and this is the resolving power of the optical system, r.

$$r = (0.61)\lambda/n \sin \beta.$$

This equation applies to all radiation and therefore also to the electron microscope. For visible light, remembering that β can at most be $\pi/2$ (and is of course in practice much less than this) so that the maximum value of $\sin \beta$ is 1, and taking $\lambda = 0.5\ \mu m$ (yellow light) and $n = 1.5$, then $r = 0.2\ \mu m$ and this is the limit below which no optical instrument can go. It follows that, since the resolving power of the eye is 0.2 mm, the maximum useful magnification of a system using yellow light is 0.2 mm/0.2 μm = 1000X.

The value of $n \sin \beta$ for the objective lens of a microscope (the lens which determines the actual resolving power) is called the numerical aperture, n.a., and is commonly marked on the lens (assuming use of the correct immersion oil in immersion objectives). In the critical use of a microscope it is always necessary to ensure that the n.a. of the objective is sufficiently high to resolve the detail required.

Other matters dealing with image formation are deferred until interference microscopy is discussed (p. 108).

4.4 The polarizing microscope

Just as the light microscope is still a very powerful tool in biology, so the polarizing microscope is still mandatory in all laboratories dealing with matter in the crystalline state, including biological tissues, organs or organelles which are crystalline. Under this micro-scope crystalline material can be recognized and the crystal axes determined even if the crystals are submicroscopic in size; and the nature of the crystalline material can often be defined. It is an ordinary light microscope built with certain precautions including devices to produce and detect *plane polarized light*.

When a ray of light passes through a crystal with symmetry less than cubic (e.g. orthorhombic, tetragonal or monoclinic) it is usually divided into two rays which travel through the crystal with different velocities and therefore different refractive indices (Fig. 4.6). Moreover, although the incident light vibrates in all directions at right angles to the direction of propagation, the refracted beams

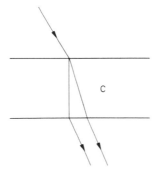

Fig. 4.6 **Double refraction by a crystalline plate, C.**

each vibrate in one direction only; and the vibrations occur in two planes at right angles to each other. The refracted beams are said to be plane polarized. If the incident beam falls normally on the face of the crystal, or otherwise if the crystal is thin enough, the two refracted beams overlap. Although they are statistically coherent since they are derived from the same beam, they do not interfere since the planes of vibration are at right angles. In all the material with which we shall deal, one vibration direction in the crystal always lies in the plane containing a crystal axis and the ray; the other, lying at right angles to this, also lies along a crystal axis in orthorhombic and tetragonal crystals. This forms the basis of the technique of polarization microscopy.

If one of the two beams is isolated (by deflecting the other away in a Nicol prism or by passing through a polaroid), the unique vibration direction of the other beam may be used as a probe to detect the axes in a second crystal.

Consider a hypothetical crystal consisting of a parallel array of diatomic molecules in which the atoms are joined by primary valences and the molecules are associated by secondary valences. Then the molecules lie further from each other than do the atoms in the molecule. Consider one of the atoms (Fig. 4.7). When a plane polarized beam passes through the crystal normal to the plane containing the primary valences and with the electric vector vibrating parallel to the direction of the bond (Fig. 4.7a), displacements within the atom separates negative from positive charges instantaneously as shown. The atoms are electrically polarized (using this term now in a different sense) and since the nearest neighbours are of opposite sign, the polarization is increased by induction. By the same token, when the electric vector lies at right angles to the

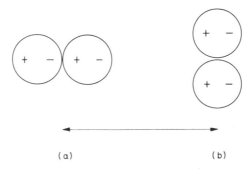

(a)   (b)

Fig. 4.7 **For explanation, see text.**

direction of the bond the polarization is reduced (Fig. 4.7b). The electric field in the visible region of the spectrum is changing its direction between 4 and 7.5 x 10^{14} times per second and only the electrons can respond to such rapid oscillations. We are therefore dealing here solely with polarization due to displacements of the electron atmosphere of the atoms.

Accepting the Lorenz-Lorenz equation connecting the refractive index n and the polarizability

$$\frac{n^2 - 1}{n^2 + 2} = k, \qquad k \text{ being constant for any one medium,}$$

it may be seen that a higher polarization is associated with a higher refractive index, and the crystal is optically *anisotropic*. The medium has the higher refractive index when the light is vibrating parallel to bond direction.

Transferring this concept to a real organic 'crystal' consisting of an assembly of parallel molecular chains, only the links within the chain being primary valences, then the allowed vibration directions within the crystal lie parallel and perpendicular to chain length; and the vibration direction lying parallel to chain length has the higher refractive index. Both the chain direction and the refractive indices can then be determined. We consider first the determination of chain direction.

4.4.1 *The extinction directions*

In a polarizing microscope, the light passing from the substage is plane polarized by passing through one of the devices already mentioned. In all modern microscopes this is a polaroid, a flat sheet of a chain polymer in which the chains lie parallel to each other and so constituted that one of the two vibrations passing through it, when light is incident on its lower surface, is absorbed. This plate is the *polarizer*. A second identical plate — the *analyzer* — is inserted in the body tube and one or both of the plates may be rotated about the optic axis of the microscope and the angular position read to 0.1°. Suppose — as is common — that the analyzer is the plate whose angular azimuth is measurable. As this plate is turned, keeping the polarizer fixed, the intensity of the light passing through the combination will vary. It will in principle fall to zero when the directions of allowed vibration in the two plates lie at right angles. The polaroids are then said to be crossed. The light intensity is usually not quite zero because in neither polaroid is the unwanted vibration completely absorbed and for other reasons which will be taken up later (p. 94). It is, however, very low even when the incident intensity is high. Suppose that a thin transparent ribbon containing parallel aggregates of parallel molecular chains — say a single lamella in the wall of a plant cell — is then placed on the stage between the polarizer and the analyzer. In general the lamella will appear bright (while the background still, of course, is dark) for the following reasons.

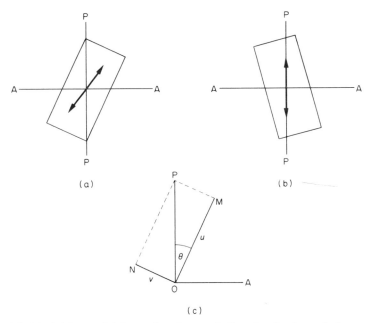

Fig. 4.8 (a), (b) PP and AA are the planes of vibration in the polarizer and analyzer respectively. The rectangle represents a cell wall, the arrow marking the direction of the constituent cellulose chains. (c) the optical situation represented in (a).

Looking down the microscope, let PP (Fig. 4.8) be the direction of vibration in the polarizer and AA that in the analyzer. In the general case the chain direction will lie at an angle θ to PP which is neither $0°$ nor $90°$ (Fig. 4.8a). The vibration PP is therefore resolved in the wall into two vibrations u and v (Fig. 4.8c) parallel and perpendicular to chain direction. Immediately upon entering the wall these vibrations are in phase. Since, however, they are propagated through the wall with different velocities and therefore different refractive indices, they are in general no longer in phase upon leaving the wall. They cannot then reconstitute the original vibration. We shall need later on to consider the nature of the vibration which they do combine to reproduce, but all that we need to note now is that it is not PP. Now PP is the *only* vibration which has no component along AA. Hence this new vibration must yield a component along AA and the specimen must be bright. In the special case, however, when the chain direction lies parallel to PP, the original vibration passes through unchanged and is extinguished by AA (Fig. 4.8b). The specimen is then as dark as the field — it is said to be in an *extinction position*. Since the specimen will also be dark when the chain direction lies parallel to AA (since PP again passes through unchanged) it will be seen that there are four extinction positions per revolution as the stage is turned. In two of these the chains lie parallel to PP, the refractive index of the specimen is high, and the position is called the *major extinction position* (or the *slow direction* since the velocity of transmission is low). The other two are accordingly the *minor extinction position* (or *fast direction*), a term which it is not in practice often necessary to use. The directions of the crystal axes can therefore be determined but, since visually all the extinction positions are identical, it is impossible to distinguish the major position from the minor and hence to determine chain direction. By the use of an accessory device, however, this can be done quite simply in a matter of seconds, though an understanding of how it is done takes a little longer to explain.

4.4.2 *The major extinction position (m.e.p.)*

Consider the specimen, thickness d, viewed edgeways (Fig. 4.9) with monochromatic light, wavelength λ, incident upon the face at L. Two vibrations are propagated from L to M, with velocities c_γ and c_α and therefore wavelengths λ_γ and λ_α (since $\nu\lambda = c$ and ν is constant). Let λ_γ represent the vibration parallel to chain length and therefore correspond to the m.e.p. The numbers of wavelengths of each kind in distance d are d/λ_γ and d/λ_α. There is therefore in d a difference in the number of wavelengths amounting to $d(1/\lambda_\gamma - 1/\lambda_\alpha)$ and this is why the two vibrations

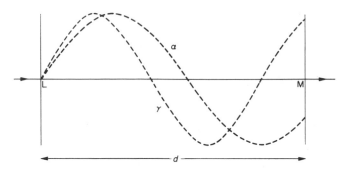

Fig. 4.9 For explanation, see text.

leave the specimen out of phase. Multiplication of this distance by λ_0 (the wavelength of the same light in air) gives the difference between the two paths, in the same units as d, as

$$p = d(\lambda_0/\lambda_\gamma - \lambda_0/\lambda_\alpha)$$
$$= (n_\gamma - n_\alpha)d$$

since $n = \lambda_0/\lambda_{(medium)}$.

This is called the *path difference*. It corresponds to a *phase difference* (p. 71) equal to $2\pi p/\lambda$.

We can now assess in a more formal way the intensity of the light passing from the crystal in Fig. 4.8c. Let us represent the vibration PP as

$$s = a \sin \omega t$$

where a is the amplitude OP.

Then on entering the crystal this is resolved into two vibrations of amplitude u and v, represented therefore by

$$u = a \cos \theta \sin \omega t$$

$$\text{and} \quad v = a \sin \theta \sin \omega t$$

On leaving the crystal each vibration has lagged behind the equivalent vibration *in vacuo* by, say, ϕ_u and ϕ_v where $(\phi_u - \phi_v) = \delta$, the phase difference. No rigour is lost by inserting δ in one of the vibrations instead of ϕ_u and ϕ_v in each vibration separately. Correspondingly, on leaving the wall, the vibrations have become, remembering that u is the m.e.p.,

$$u = a \cos \theta \sin(\omega t - \delta) \qquad \text{4.8a}$$

$$v = a \sin \theta \sin \omega t \qquad \text{4.8b}$$

These will combine to give a new vibration, the nature of which will be examined later. For the moment we maintain the equivalent position of keeping them separate and considering the sum of the resultant of each along AA. This is

$$s' = a \cos \theta \sin \theta \sin(\omega t - \delta) - a \sin \theta \cos \theta \sin \omega t$$

and must correspond to a single vibration of the form

$$s' = A \sin(\omega t + \Phi) \qquad 4.9$$

By equating terms in $\sin \omega t$ in Equations 4.8 and 4.9, repeating for terms in $\cos \omega t$ and squaring and adding the two expressions, we find

$$A^2 = \frac{a^2}{2} \sin^2 2\theta (1 - \cos \delta) \qquad 4.10$$

We note that A^2 is proportional to the light intensity. Clearly $A^2 = 0$ when $\theta = 0$, $\pi/2$, π, $3\pi/2$ or 2π and the specimen is extinguished four times per revolution. The intensity is also zero, irrespective of the value of θ, if $\cos \delta = 1$, i.e. $\delta = 0$, 2π, 4π, etc. and therefore $p = \lambda$, 2λ etc. This can also be derived from Fig. 4.9 since if the path difference is a whole number of wavelengths the two vibrations are in step on leaving the crystal as they were on entering it; they therefore re-form the original vibration PP.

Suppose, therefore, that the birefringent object under examination is a narrow wedge placed with one of the sloping sides on the stage of the microscope. Suppose that this is examined in blue light, $\lambda = 0.43 \, \mu$m. Then at the apex of the wedge $\delta = 0$ and the intensity is zero. As the point of observation passes along the wedge the illumination increases and then decreases to zero at $\delta = 2\pi(p = 0.43 \, \mu$m) and then increases much as in Fig. 4.10. The same behaviour occurs in yellow light ($\lambda = 0.56 \, \mu$m) and red light ($\lambda = 0.7 \, \mu$m) except that the second zero occurs further along the wedge (Fig. 4.10). If, therefore, all three wavelengths are presented together, the wedge will present various shades of grey up to a distance from the apex at which the path difference is about $0.2 \, \mu$m (A, Fig. 4.10) because the intensity of all three radiations is increasing. Beyond this, the wedge will begin to appear coloured as the intensity of the blue component falls sharply and of the yellow component slowly, while the intensity of the red component is still increasing. At point B, path difference $0.43 \, \mu$m, it will take on a brown to orange tinge. Similarly at C, path difference $0.56 \, \mu$m,

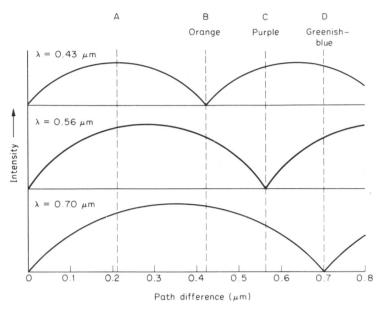

Fig. 4.10 **The relative intensity of light emitted from a crystal between crossed polaroids as a function of path difference, for each of three wavelengths.**

it will be purple and at D, path difference 0.7 μm, it will appear greenish-blue.

If the spectrum of wavelengths covered is now broadened to cover the whole visible region, i.e. if the wedge is observed in white light, clearly a very similar change of colour with path difference will be found, in a regular series. This is called Newton's series because the colour sequence is that found in the well-known Newton's rings phenomenon. The series is given in Table 4.1 both for crossed polaroids and parallel polaroids. Since the light corresponding to any wavelength is extinguished not only when $p = \lambda$ but also when $p = 2\lambda$, 3λ, etc., the colour sequence roughly repeats itself at intervals of $p = \lambda$. The sequences are therefore divided into orders. The point at which p is equal to the wavelength for green light ($p = 0.589\,\mu$m) is arbitrarily taken as the end of the first order colours; similarly $p = 1.178$, $p = 1.767$ marks the end of the second and third orders, and so forth.

These colour sequences have two major consequences. Firstly, a subjective estimate of the colour of a birefringent object gives a rough measure of the path difference. If, for instance, the colour is Red of the first order (Red I) (methods for determination of the

TABLE 4.1

The first two orders of Newton's colour scale (modified from Quincke)

Path difference (nm) ($\lambda = 589$ nm)	Order	Colour between crossed Nicols	Colour between parallel Nicols
0	I	Black	Bright white
40		Iron-grey	White
97		Lavender-grey	Yellowish-white
158		Greyish-white	Brownish-white
218		Clearer grey	Brownish-yellow
234		Greenish-white	Brown
259		Almost pure white	Light red
275		Pale straw-yellow	Dark reddish-brown
306		Light yellow	Indigo
332		Bright yellow	Blue
430		Brownish-yellow	Greyish-blue
505		Reddish-orange	Bluish-green
536		Red	Pale green
551		Deep red	Yellowish-green
565		Purple	Lighter green
575		Violet	Greenish-yellow
589		Indigo	Golden-yellow
664	II	Sky blue	Orange
728		Greenish-blue	Brownish-orange
747		Green	Light carmine
826		Lighter green	Purplish-red
843		Yellowish-green	Violet-purple
866		Greenish-yellow	Violet
910		Pure yellow	Indigo
948		Orange	Dark blue
998		Bright orange-red	Greenish-blue
1101		Dark violet-red	Green
1128		Light bluish-violet	Yellowish-green
1151		Indigo	Impure yellow

$1 \text{ nm} = 10^{-7} \text{ cm}$

order will be found on p. 92), then the path difference p is about $0.54\,\mu\text{m}$; if the specimen thickness d is $10\,\mu\text{m}$ then, since $(n_\gamma - n_\alpha)d = p$, $(n_\gamma - n_\alpha)$ is 0.054 with a possible error which should not be greater than ±0.003. Of more importance, the rapid change in colour for a relatively small change in p (e.g. a change from red at $0.536\,\mu\text{m}$ through purple and indigo to blue at 0.664) a test plate can be used to determine the m.e.p. of any specimen, as follows.

Test plates are available, (e.g. of cleaved mica) which give any desired first or early second order colour between crossed polaroids; Red I and Violet I are common, the latter being known as the *sensitive tint plate*. The m.e.p. is known (by an independent method described in p. 87) and marked on the plate mount. The plate is inserted in the microscope between the objective and the analyzer in a slot which in modern microscopes is rotatable about the optic axis (Fig. 4.11a). It is so positioned that the m.e.p. lies at 45° to the planes of vibration in the polaroids and runs from bottom left to top right in the field of view (Fig. 4.11b). This is done by extinguishing the plate and then rotating it through 45° in the correct direction. The field will now appear, say, violet. A crystalline specimen placed on the stage and viewed through the

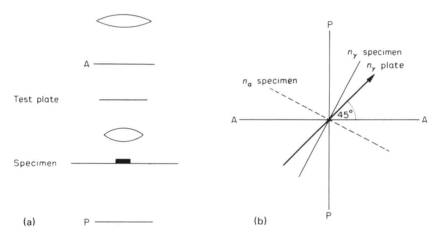

Fig. 4.11 (a) The point of insertion of a test plate in relation to the specimen, the polarizer, P, and the analyzer, A, in a polarizing microscope.
(b) The corresponding optical diagram.

microscope will generally appear some other colour. If the m.e.p. lies is the position shown by a full line in Fig. 4.11b then, since it lies at less than 45° from the m.e.p. of the plate, the path differences are additive and the colour of specimen plus plate will be higher in Newton's series than that of the plate alone. Biological specimens mostly are of low path difference (in the range of the low first order colours in Table 4.1). The *addition* colour will be therefore fairly close to violet, in the blue-green range. The specimen therefore appears green against a violet background. When, on the other hand, the stage is rotated so that the m.e.p. lies along the dotted line of Fig. 4.11b, the specimen shows a subtraction colour, say yellow. In the intermediate position when the m.e.p. is parallel to PP, the specimen has no effect on the light and therefore appears precisely the colour of the field. The position of the m.e.p. can now be determined.

Consider, for example, a tracheid or fibre of a higher plant over part of which one wall has been cut away so that the remaining single wall ABCD, Fig. 4.12, is available for examination. Suppose the m.e.p. lies along LM making an angle θ with the side wall BD. The position at which LM is parallel to PP, Fig. 4.11b, can be detected since the specimen will be precisely the same colour as

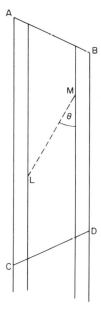

Fig. 4.12 **Diagrammatic representation of a tracheid with one wall removed (above CD); LM represents the m.e.p. of the (lower) single wall.**

the field. If the stage is now rotated clockwise through a few degrees the specimen colour will become an *addition* colour, say green. This verifies that the m.e.p. was parallel to PP, not AA. If the specimen shows a subtraction colour, say yellow, then the stage must be rotated through 90°. The method is therefore to begin by setting the stage so that the specimen is extinguished, i.e. the same colour as the field, but goes into addition colour on clockwise rotation. This extinction position is achieved with much higher accuracy than it can be when the field is dark because the eye is much more sensitive to colour differences than it is to intensity differences at low light intensity. The angular position of the stage, θ_1, is noted. The analyzer is then removed from the light path and the stage rotated until the side BD lies parallel to PP (marked by a cross-wire in the eyepiece) and the position of the stage again noted, θ_0. Clearly $(\theta_1 - \theta_0) = \theta$. Each of θ_0 and θ_1 must be determined several times until agreement between successive readings is better than ±0.5°. Even then the eye may be mismatching slightly the colour of the specimen and that of the plate and it is always advisable to turn the stage through 90° to obtain a match when LM, Fig. 4.12, lies parallel to AA, Fig. 4.11. The angular distance between the two matching positions should be 90° ± 0.5°. If this is not the case the whole series of observations must be repeated. With practice, θ may be determined in as little as 30 seconds.

4.4.3 Determination of refractive indices and birefringence

We are now in a position to consider the determination of $(n_\gamma - n_\alpha)$. This can be done in a straightforward way by determining n_γ and n_α separately by a method which is simple to explain but tedious to carry out. Alternatively, and usually of more use, the birefringence $(n_\gamma - n_\alpha)$ may be determined within a few seconds by a method which requires lengthy explanation. We consider the direct method first and exemplify it by considering the refractive indices of the wall of a fibre.

In principle the method used for objects of microscopical size is a very simple one. It involves nothing more than finding a liquid medium in which there is no bending of the rays of light at the edge of the fibre, so that the fibre becomes almost invisible. It never does in practice become completely invisible for a number of reasons into which we need not go at the moment. Since the refractive index varies with the wavelength of the light used, it is customary to use monochromatic radiation given by a sodium vapour lamp, i.e. to use the sodium D line. The method can be tedious, but if a set of liquids is prepared beforehand differing in refractive index by, say, 0.01, and a sufficient quantity of material in an adequate condition is at hand, then results come more quickly than might be imagined.

As a guide to the choice of the correct liquid a phenomenon first described by Becke and called the Becke line is employed. If the boundary between two media is observed under a microscope then a line of light is visible along the boundary if the refractive indices of the media are different. On raising the microscope objective, this line of light moves into the medium of higher refractive index. It is, therefore, a matter of a moment to decide whether the immersion liquid bathing a fibre has an index lower or higher than that of the fibre and thus to obtain a choice for the next liquid to be tried. The method, then, is briefly this. The fibre is mounted in a suitable medium of known refractive index on the rotating stage of a polarizing microscope and is set with the m.e.p. parallel to the direction of vibration of the light issuing from the polarizer. The analyzer is then moved out of the light path. Observation of the Becke line will show whether the refractive index of the medium is too high or too low. If it is too high, then a liquid of a lower index can be tried until the refractive index of the fibre is 'bracketed'; when by progressively narrowing the upper and lower limits, an increasingly close approximation to the refractive index of the fibre can be made. With care and under appropriate conditions, the refractive index can thus be determined to the third place of decimals. To obtain the necessary gradation in refractive indices it is necessary to use a mixture of liquids and a number of liquids are available. A very useful combination is α-monobromnaphthalene and liquid paraffin which fulfils the requirement that neither component shall swell the fibre (see below) and each shall have about equal volatility so that the refractive index of any mixture shall not change appreciably during the observation.

This determines the value of n_γ. n_α can be similarly determined with the appropriate orientation of the specimen.

Although, however, this technique presents no great difficulties the precise interpretation is not straightforward, on account of the physical and chemical heterogeneity of cell walls. The solid material of the wall consists of a family of polysaccharides and often other compounds such as lignin, varying from the completely amorphous to the truly crystalline; and there are 'void' spaces normally filled with water or an aqueous solution, some of which vary in size with water content. A prior decision has therefore

87

always to be made, in the light of the problem under investigation, as to the component whose refractive index is required, and how best to achieve the end in view. Consideration of these matters will be postponed, however, until we have before us the necessary details of wall architecture. Some, though not all, of these uncertainties, are avoided in determination of refractive index difference, the birefringence. The method here is to determine the path difference of a specimen and the specimen thickness in two separate operations. The path difference is normally low, in the range of 4 nm to 0.5 μm, so that the compensators used to determine path differences in inorganic crystals and petrological materials are too insensitive. There are, however, two compensators available, the de Sénarmont compensator and the Brace-Koehler or Elliptical compensator, effective in two ranges, which can give precise determinations of these low birefringences. To understand how these operate it is necessary to enquire into the form of the light vibration after it has passed out from the specimen into the body tube of the microscope.

(a) *de Sénarmont compensator*
Taking first the de Sénarmont compensator, the two linear vibrations just before leaving the specimen are as given in Equations 4.8a and 4.8b. With this, as with all compensators, the specimen is set so that $\theta = 45°$. The two vibrations are then

$$u = a\sqrt{2} \sin(\omega t - \delta) \qquad \text{4.11a}$$

$$v = a\sqrt{2} \sin \omega t \qquad \text{4.11b}$$

along the directions *ou* and *ov* (Fig. 4.13). These correspond to a single vibration figure which is the locus of points such as L (Fig. 4.13) in a curve referred to axes *ou* and *ov*. It is convenient to refer this curve to axes OA and OP. Equations 4.11 are therefore rewritten by substituting

$$x = (v - u)\sqrt{2}$$

$$y = (v + u)\sqrt{2}$$

giving

$$x = a(\sin \omega t - \sin(\omega t - \delta))/2$$

$$y = a(\sin \omega t + \sin(\omega t - \delta))/2$$

or,

$$x = a \cos(\omega t - \delta/2) \sin \delta/2 \qquad \text{4.12a}$$

$$y = a \sin(\omega t - \delta/2) \cos \delta/2 \qquad \text{4.12b}$$

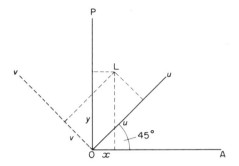

Fig. 4.13 **For explanation, see text**

Squaring and adding, we have

$$x^2/(a^2 \sin^2 \delta/2) + y^2/(a^2 \cos^2 \delta/2) = 1, \qquad 4.13$$

an ellipse with axes parallel to OP and OA and with axial ratio tan $\delta/2$ (Fig. 4.14). This is the form of vibration required. The original plane vibration on passing through the crystal is always transformed into an elliptical vibration invariably of this form. Determination of δ is therefore resolved into the determination of the slope of a diagonal of the rectangle circumscribing the ellipse (Fig. 4.14).

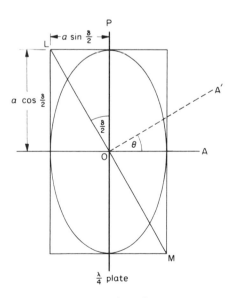

Fig. 4.14 **For explanation, see text.**

In devising a means to do this we note that Equations 4.12 may be rewritten

$$x = (a \sin \delta/2) \sin(\omega t - \delta/2 - \pi/2) \qquad \text{4.14a}$$

$$y = (a \cos \delta/2) \sin(\omega t - \delta/2) \qquad \text{4.14b}$$

These are the two vibrations parallel to axes OP and OA into which the elliptic vibration can be resolved; whatever the value of δ they themselves always differ in phase by $\pi/2$. If now we impose a phase lag on vibration y (∥ OP) of $\pi/2$ by inserting in the light path a crystal plate with phase difference $\pi/2$ (and therefore path difference of $\lambda/4$ so that this is called a *quarter wave plate*) with its m.e.p. parallel to OP, then this in effect removes the $\pi/2$ phase difference in Equations 4.14. They then become

$$x = (a \sin \delta/2) \sin(\omega t - \delta/2)$$

$$y = (a \cos \delta/2) \sin(\omega t - \delta/2)$$

i.e. $\qquad x/y = \tan \delta/2$

and the vibration is the straight line LM (Fig. 4.14).
This may be extinguished by rotating the analyzer through an angle θ to bring its plane of vibration perpendicular to LM, where

$$\theta = \delta/2$$

and the phase difference is determined. The path difference p is then $\delta\lambda/2\pi$.

It is perhaps more instructive — as well as being useful when we come to consider interference microscopy — to examine this compensator in a different way. Equation 4.13 shows that for a quarter wave plate ($\delta = \pi/2$)

$$x^2 + y^2 = a^2$$

and the ellipse becomes a circle. Any plane vibration incident upon a quarter wave plate with the direction of vibration lying at $45°$ to the crystal axes becomes transformed on passing through the crystal into a circular vibration. This may also be seen pictorially as follows. Let PP Fig. 4.15 represent a plane vibration incident normally on a crystal at $45°$ to the crystal axes lf, ci. The two vibrations in the crystal then correspond to lf and ci. We need to define path lengths along lf and ci of equal time interval. This may be done (cp. Fig. 4.2) by constructing a circle centre o, radius ol ($= oc$) and marking out equal lengths of arc along the circumference of this circle ($a - l$ in Fig. 4.15). Perpendiculars dropped on lf and ci from points a, b, c... then mark out the required lengths. When the vibration PP enters the crystal, corresponding points in the two vibrations will lie at l and i. If the crystal has a path difference of $\lambda/4$ then, on leaving the crystal, when a point in vibration lf is at l, the equivalent point on ci will

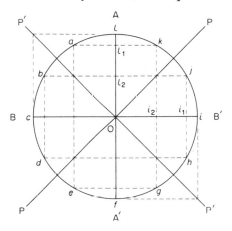

Fig. 4.15 For explanation, see text.

be at o and moving to the right; the net displacement is therefore at l. By the time the point has moved from l to l_1 the point at o has moved to i_2; the net displacement is at k. Similarly, by the time the point on lf has reached l_2, the point on li has reached i_1, and the net displacement is at j; and so forth. The resultant vibration is therefore the circle $lkj....a$, rotating clockwise. Similarly, a plane vibration P'P' on passing through the crystal becomes an identical circular vibration but rotating anticlockwise. This leads to a simple geometric explanation of the de Sénarmont compensator.

PP' (Fig. 4.16) represents the original vibration from the polarizer and AA', BB' the vibration directions in the specimen (at 45° to PP'), LL' and QQ' represent the vibration directions in the superposed quarter wave plate. The vibrations AA' and BB' become transformed into two circular vibrations of diameter AA' rotating in opposite directions. If these two vibrations are in phase, then when the point in one circle is at A, the point on the other must be at B' and two circular vibrations compound to give a plane vibration LL'. This is extinguished by the analyzer. If the vibration

91

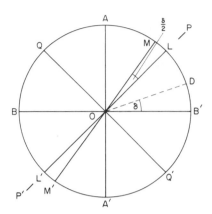

Fig. 4.16 **For explanation, see text.**

BB′ lags behind AA′ by a phase difference, δ, then when the point corresponding to AA′ lies at A that corresponding to BB′ lies at D, where $\angle$ DOB′ is δ. The two circles therefore collapse into a linear vibration MM′ making an angle of $\delta/2$ with PP′. It is therefore necessary to turn the analyzer through an angle of $\delta/2$ to extinguish the specimen.

The use of the compensator should now be clear. A $\lambda/4$ plate is inserted in the body tube of the microscope and extinguished between crossed polaroids (accuracy ±0.50°). The specimen is inserted, extinguished, and rotated through 45° to maximum brightness. The analyzer is then rotated until the specimen is extinguished and its position read; the angular displacement is $\delta/2$. Alternatively and better, the specimen may be rotated 90° and the analyzer again rotated to specimen extinction, when the angular rotation from one extinction to the other is δ. It should be noted that a rotation of 180° from the crossed position corresponds to a path difference of λ and that the measurement yields only the fraction of a wavelength; i.e. if the path difference is 3.4λ the measurement gives only 0.4λ; the whole number of wavelengths which must be added may be determined by a wedge compensator. Most biological objects which can be examined microscopically will, however, show a path difference less than λ. The compensator gives satisfactory results in white light but for highest accuracy the wavelength must be used for which the $\lambda/4$ plate is calibrated. It is not especially accurate for path differences of $\lambda/20$ or lower and for path differences below this the following compensator comes into use.

(b) *The elliptic or Brace-Koehler compensator*

The principle of this compensator follows from the discussion above leading to Equation 4.13. We begin with Equations 4.8a and b referring to Fig. 4.8c. Eliminating t from these and, for a reason which will be seen below, writing ϕ_0 instead of δ, we have

$$\frac{u^2}{a^2\cos^2\theta} + \frac{y^2}{a^2\sin^2\theta} - \frac{2uv}{a^2\sin\theta\cos\theta}\cos\phi_0 = \sin^2\phi_0$$

This is an elliptic vibration referred to axes u and v, inscribed in a rectangle of sides $2a\cos\theta$ and $2a\sin\theta$ (Fig. 4.17) and describes the form of the vibration passing out from a crystal in the general case. As before, when ϕ_0 is small this is an ellipse with axes closely parallel to PP and AA. To determine the form of this ellipse transform to co-ordinates axes PP and AA by using

$$x = u\sin\theta - v\cos\theta \qquad\qquad 4.15a$$

$$y = u\cos\theta + v\sin\theta \qquad\qquad 4.15b$$

Substituting from Equation 4.15a into Equation 4.8a and Equation 4.15b into Equation 4.8b, and taking the case for small values of ϕ_0 so that $\cos\phi_0 = 1$ and $\sin\phi_0 = \phi_0$, we have

$$x = (a/2)\sin 2\theta\cos\omega t.\phi_0$$

$$y = a\sin\omega t - a\sin^2\theta\cos\omega t.\phi_0$$

The amplitude of x is therefore $(a/2)\sin 2\theta.\phi_0$ and of y is a. The ratio of these, $\frac{1}{2}\sin 2\theta.\phi_0$, gives the ratio of the axes of the ellipse. When, therefore, a test plate, phase difference ϕ_0, is placed in the

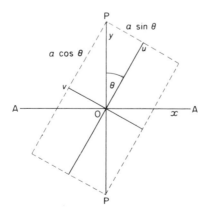

Fig. 4.17 **For explanation, see text.**

microscope with the m.e.p. lying at an angle θ to PP, the vibration passing from the plate is an ellipse, with axes closely parallel to PP and AA and with axial ratio $\frac{1}{2}\sin 2\theta.\phi_0$. We have already seen (Equation 4.13) that for a specimen placed with its m.e.p. at 45° to PP the resultant vibration is an ellipse with axes parallel to PP and AA and axial ratio $\tan \delta/2$ $(= \delta/2$ for δ small). If this is the same ellipse as the ellipse from the test plate, but rotating in the opposite direction, then the specimen is extinguished and

$$\delta = -\phi \sin 2\theta$$

In using this compensator, therefore, the specimen is set at 45°. The compensator, consisting of a plate with a path difference of say $\lambda/30$ and rotatable in its plane about the optic axis, is inserted and rotated until the field is extinguished. The angle θ is the angular movement of the compensator plate then required to extinguish the specimen.

In measuring such small path differences it is necessary to reduce reflection depolarization by blooming the objective lenses, to use the best polaroids available and to employ high light intensity.

4.4.4 *The interpretation of the m.e.p. and the path difference*

The determinations of the m.e.p. and the path difference in a biological specimen are therefore straightforward; their interpretation is, however, usually not so. Let us consider first the m.e.p. Only under the circumstance that, in the wall specimen under observation, the lengths of the submicroscopic crystallites — now known to be the cores of the microfibrils, a term we shall now use — lie in the plane of the wall, the wall is homogeneous as regards microfibril direction and the optical character of the microfibrils is known, does the m.e.p. in any sense determine the run of the microfibrils. If these conditions obtain then the m.e.p. corresponds to the mean molecular chain direction and therefore to the mean microfibril direction in cellulosic walls, or to the perpendicular to this direction in walls containing microfibrils within which the constituent chains run in slow helices as in xylan walls (Frei and Preston, 1964) (p. 255). Even then it should be remembered that the polarization microscope is highly sensitive to orientation; a small proportion of oriented microfibrils in a mass which is randomly scattered will give rise to a detectable m.e.p. Correspondingly, a slight departure from absolute randomness, detectable only through severe precautions in the electron microscope, can readily be detected under the polarizing microscope [as in *Bryopsis* (Frei and Preston, 1964)]. The degree of order can then roughly be defined only by determination of the refractive indices or the birefringence. The necessary discussion on this matter must, however, be postponed until the general structure of the wall has been presented.

A real specimen may depart from the above ideal either in that the microfibrils do not lie in the plane normal to the optic axis of the microscope or in that the microfibril orientation is not homogeneous throughout the specimen thickness. We examine the related complexities in turn.

(a) *Microfibrils tilted to plane of surface*

The effect of microfibril tilt can best be understood through a specific example. Consider a rectangular piece of a single wall lamella cut from the secondary wall of a fibre or tracheid (Fig. 4.18). Within this the microfibrils lie parallel to the surface of the lamella but usually tilted to cell length. Suppose the microfibrils lie perfectly parallel to each other. Observation in direction A yields an m.e.p. which correctly defines the microfibril direction. Equally observations in directions B and C do not; the correct orientation may nevertheless still be deduced in the following way.

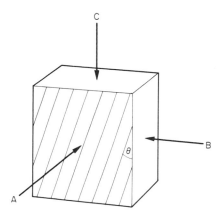

Fig. 4.18 **A segment of a cell wall, the striped area being one surface of the wall; the stripes are parallel to the m.e.p.**

It should first be recalled that for such a set of parallel molecular chains the refractive index for light vibration parallel to chain direction is greater than that for any other vibration direction. Equally, since the molecular chains of cellulose consist of a series of rings which, although somewhat puckered, are still rather flat and lie in the same plane, then for light propagated *along* the chains, the refractive index should be greater for light vibrating *parallel* to the plane of the rings than *perpendicular* to this plane. In this sense, cellulose should have three characteristic refractive indices, n_γ, n_β and n_α respectively. There are in the literature vague statements that this is true for cellulose but it is difficult to see how the relevant evidence can validly be obtained. It is usually taken that for cellulose $n_\beta = n_\alpha$ so that only two refractive indices need be considered. This places cellulose in the class of crystals which, for a reason given below, is called *uniaxial*.

As we have seen, n_γ and n_α can be determined directly only for light propagated through the wall lamella normal to microfibril length (or, for n_α only, parallel to this length). In light propagated in some other direction refractive indices intermediate between n_γ and n_α are yielded, the value depending on the direction of propagation. It is found that all these refractive indices can be correlated in the following way. Suppose an ellipsoid of rotation is constructed (Fig. 4.19) with half the major axis equal in length to n_γ and half the minor axis equal to n_α and the ellipsoid – the *index ellipsoid* – oriented in the wall with the major axis parallel to chain direction. Then for light propagated in any general direction *MO* (Fig. 4.19) a plane vibration will be resolved into two vibrations parallel to the axes of the ellipse formed by the intersection of the plane normal to the direction of propagation with the ellipsoid, with refractive indices equal in magnitude to the lengths of the major and minor axes, respectively, of the ellipse. Notice that the smaller refractive index is always n_α. The other refractive index, $n_\gamma{'}$, may be calculated. The method may be exemplified through the common case in which a transverse section of a cell is under examination, i.e. the direction of propagation is *A* (Fig. 4.19).

The geometry of the situation is set out in Fig. 4.20 representing a plane section through Fig. 4.19 parallel to the lamella surface; the refractive index required is n_γ^L (L signifying that the refractive index is the major refractive index in the transverse plane). Then

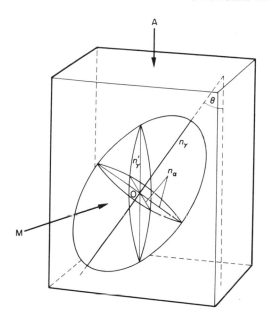

Fig. 4.19 The index ellipsoid.

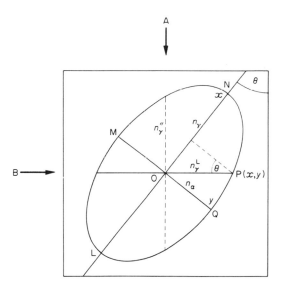

Fig. 4.20 The trace of the index ellipsoid on a plane parallel to wall surface, cell axis ↕.

97

calling the n_γ direction the x axis and the n_α direction the y axis the point P has coordinates x, y related by

$$\frac{x^2}{n_\gamma{}^2} + \frac{y^2}{n_\alpha{}^2} = 1,$$

the equation of the ellipse LMNP. This may be rewritten

$$\frac{(n_\gamma^L)^2 \sin^2 \theta}{n_\gamma{}^2} + \frac{(n_\gamma^L)^2 \cos^2 \theta}{n_\alpha{}^2} = 1 \qquad\qquad 4.16$$

whence $\qquad (n_\gamma^L)^2 = (n_\gamma{}^2 n_\alpha{}^2)/(n_\gamma{}^2 - (n_\gamma{}^2 - n_\alpha{}^2)\sin^2 \theta)$

What is usually measured (and all that can be measured for a wall lamella lying inside a wall) is $(n_\gamma^L - n_\alpha)$, from the path difference. n_α may usually be taken as 1.53. If n_γ is known, θ may be calculated; equally if θ is already known, n_γ may be calculated.

If neither n_γ nor θ is known then the refractive indices must be measured also in some other direction, e.g. in longitudinal section B (Fig. 4.20). The appropriate major refractive index is then n_γ'' for which

$$\frac{(n_\gamma'')^2 \cos^2 \theta}{n_\gamma{}^2} + \frac{(n_\gamma'')^2 \sin^2 \theta}{n_\alpha{}^2} = 1 \qquad\qquad 4.17$$

Equations 4.16 and 4.17 may be combined to give:

$$n_\gamma^2 = (n_\gamma'')^2 \, n_\alpha^2 (n_\gamma^L)^2 / [(n_\gamma'')^2 \, n_\alpha^2 + n_\alpha^2 (n_\gamma^L)^2 - (n_\gamma'')^2 \, (n_\gamma^L)^2] \, ;$$

$$\sin^2 \theta =$$
$$[(n_\gamma^L)^2 \, (n_\gamma'')^2 - n_\alpha^2 \, (n_\gamma'')^2] \, / [(2n_\gamma'')^2 \, (n_\gamma^L)^2 - n_\alpha^2 \, (n_\gamma'')^2 - (n_\gamma^L)^2 \, n_\alpha^2]$$

Since n_γ'' and n_γ^L are known and n_α is known (1.53) or measurable, n_γ and θ may be calculated. The structure of bamboo fibres was worked out in this way (Preston and Singh, 1951).

It should be noted that, for light propagated along the direction of the major axis of the index ellipsoid, the locus of intersection of the ellipsoid with the plane normal to this direction is a circle so that $n_\gamma = n_\alpha$. Viewed in this direction the microfibril array is isotropic; the direction is a unique direction, the optic axis, and this is why crystals of this kind are called uniaxial.

(b) *Superposed wall lamellae differing in m.e.p.*
The other complexity — lack of homogeneity in the wall — commonly takes the form that the wall consists of a number of thin

superposed lamellae different in chain orientation. Examples occur with *Valonia* (Preston and Astbury, 1937), the Cladophorales (Astbury and Preston, 1940; Frei and Preston, 1961) and in tracheids and fibres (Preston, 1947; Preston and Singh, 1951) and many other cell types. Indeed, any whole single cell always poses this problem. Microfibrils are normally tilted to the cell axis and therefore run helically round the cell; the microfibrils of front and back walls are therefore crossed (Fig. 4.21) and examination of the central regions A of the cell (but obviously not of the side wall B) presents the problem of two superposed crystal plates with crossed axes. In this case the problem is trivial since one wall can be (and must be) cut away to leave the other free for examination. It is nevertheless of profit to consider this problem first. The appropriate diagram is as in Fig. 4.22.

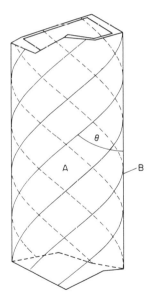

Fig. 4.21 **Diagrammatic representation of the run of the m.e.p. in a whole cell. B, side wall in optical section; A, two superposed walls with crossed m.e.p.'s.**

The analytical process already used for single crystal plates is first followed. The vibration along PP ($s = a \sin \omega t$) is resolved along ou_1 and ov_1 as before. Now, however, each is resolved in turn along ou_2 and ov_2 and the component of these vibrations along AA calculated. The procedure is then as before; the calculation

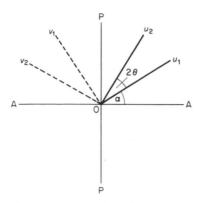

Fig. 4.22 For explanation, see text.

is more tedious but leads straightforwardly to

$$I/I_0 = -\sin 2\alpha \cos 2(2\theta + \alpha) \sin 4\theta \sin^2(\phi_1/2)$$
$$+ \cos 2\alpha \sin 2(2\theta + \alpha) \sin 4\theta \sin^2 \phi_2/2$$
$$+ \sin 2\alpha \sin 2(2\theta + \alpha) \cos^2 2\theta \sin^2 [(\phi_1 + \phi_2)/2]$$
$$- \sin 2\alpha \sin 2(2\theta + \alpha) \sin^2 2\theta \sin^2 [(\phi_2 - \phi_1)/2]$$
$$= (\sin 2\alpha \sin(\phi_1/2))^2 + (\sin 2(2\theta + \alpha) \sin(\phi_2/2))^2$$
$$+ 2 \sin 2\alpha \sin 2(2\theta + \alpha) \sin(\phi_1/2) \sin(\phi_2/2) \cos \phi'$$

where I and I_0 represent respectively the intensity of light passing through the analyzer and of the incident beam, ϕ_1 and ϕ_2 are the two phase differences and ϕ' represents the third side of a spherical triangle as in Fig. 4.23.

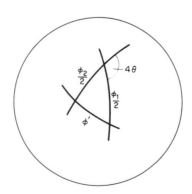

Fig. 4.23 For explanation, see text.

100

The second version of this equation is of the form $(a^2 + b^2 + 2ab \cos \phi')$. The specimen can be extinguished at some value of α only if (1) $a = b$ and $\cos \phi' = -1$ or (2) $a = -b$ and $\cos \phi' = 1$ or (3) $a = b = 0$. If ϕ_1, ϕ_2 and θ are given none of these can be true; indeed, if 2θ is not zero or $\pi/2$ or π, extinction may be achieved only if the path differences ϕ_1 and ϕ_2 are multiples of λ (i.e. $a = b = 0$). As the specimen is rotated on the stage, therefore, its intensity does not fall to zero at the 'extinction positions' but to a minimum less than zero. For a typical cell $\phi_1 = \phi_2$ and the minimum of intensity then occurs when $\alpha = 90° - \theta$ or $-\theta$, i.e. when cell length lies either parallel or perpendicular to PP. If $2\theta < 45°$ the m.e.p. lies parallel to cell length; if $2\theta > 45°$ and $< 90°$ it lies perpendicular to cell length.

The double wall of a whole cell does not therefore behave like a single, homogeneous wall (note that the side walls B, Fig. 4.21, do so behave and therefore extinguish).

The physical reason for the lack of extinction may best be seen through the use of a model — the Poincaré sphere — which has the advantage that a multi-lamella problem with a single wall can also be dealt with in an elegant way such as would be unacceptably tedious by the analytical method used above. The tedious algebra is built into the model and automatically dealt with by the construction. The properties of the sphere will be given without the proof which involves mathematical considerations too complex to handle here; the proof may be found in Poincaré (1889–92) or Pockels (1906) (see also Ramachandran and Ramaseshan 1961; Harsthorne and Stuart 1970). The sphere may be represented as a globe with poles, N, O (Fig. 4.24), an equator ABPXY, lines of latitude EGF and of longitude NGPO. The significance of these lines is that they represent at every point two directions in real space normal to the direction of light waves travelling toward an observer outside the sphere. All points on the equator represent *linear vibrations* whose directions vary on passing round the equator such that diametrically opposite points (e.g. P, L) represent vibrations at right angles. We adopt the convention that on passing anticlockwise round the equator (viewed from above) the vibration also rotates anticlockwise. Clearly to pass through two points such as P and X with a difference in vibration direction α we need to pass round the sphere through an angular distance 2α. All points on the sphere away from the equator except the poles correspond

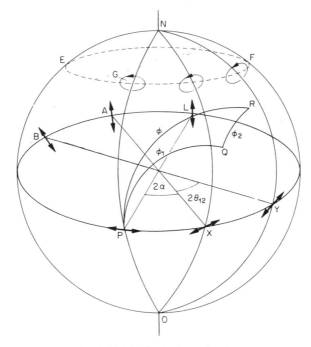

Fig. 4.24(a) The Poincaré sphere.

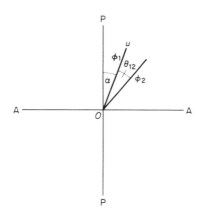

Fig. 4.24(b) For explanation, see text.

to elliptic vibrations. Lines of latitude such as EG represent ellipses rotating in the same direction, anticlockwise above the equator and clockwise below, with equal axial ratio. Lines of longitude represent ellipses of equal orientation but decreasing in ellipticity from the equator (at which they collapse into linear vibrations) to the poles which represent circular vibrations. Consider now the problem illustrated in Fig. 4.24b.

Let point P, Fig. 4.24a, on the equator, represent the incident plane vibration PP, Fig. 4.24b. In order to plot the effect of the passage of this vibration through the plate of phase difference ϕ_1, move around the equator from P to X through an angle 2α. The plane vibration at X lies at an angle α to that at P and represents *ou*, one of the vibrations in the first crystal plate; the diametrically opposite point A represents the other vibration. Now rotate the equator circle round the axis XA through an angle ϕ_1, bringing P along a small circle to Q. Q represents precisely the elliptical vibration passing out from the first plate. To determine the effect on this of the second crystal, move further along the equator from X to Y, through an angle $2\theta_{12}$, and rotate about the equatorial diameter YB through an angle ϕ_2, bringing Q to R. R represents the effect of the passage of the plane vibration PP through both plates. Clearly the passage from P to R may be achieved directly by rotation about a single diameter, but this is not an equatorial diameter. Since rotation only about an equatorial diameter corresponds to passage through a simple birefringence plate, the path difference ϕ equivalent to the arc PR does not represent such a plate only. What is does represent may be seen by noting that rotation about any diameter of the sphere is equivalent to rotation about an equatorial diameter plus a rotation about a polar diameter. Rotation about an equatorial diameter represents a component of the plate combination which is itself a simple birefringent plate. Rotation about a polar diameter clearly corresponds to a second component which is a circularly polarizing plate. It is this second component, exercising its effect whatever the orientation of the crossed crystal plates to the plane of vibration of the polarizer, which ensures that the crossed plates are never extinguished.

Rotation about a third, fourth... equatorial diameter corresponds to passage of the light vibrations through a third, fourth, etc. crystal plate so that any number of crossed plates can be dealt with by the Poincaré sphere. Note that if the resulting polygon is closed, the combination represents only a circularly polarizing plate.

The construction required therefore is very simple. For the plates indicated in Fig. 4.25b, one begins by drawing an arc, length ϕ_1 ($D_1 D_2$) (Fig. 4.25a) corresponding to the Plate 1. A second arc $D_2 D_3$ is constructed, length ϕ_2, making an external angle of $2\theta_{12}$

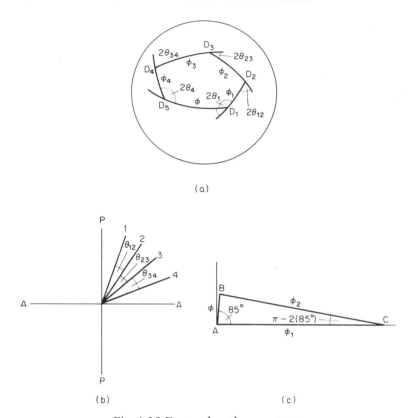

(a)

(b)   (c)

Fig. 4.25 For explanation, see text.

with D_1D_2. Similarly for D_3D_4 and D_4D_5. The arc required to close the polygon gives ϕ, the phase difference of the corresponding birefringent plate and θ_1 gives the angle between the m.e.p. of this plate and that of plate 1 (the internal angle this time, because formally the phase difference ϕ is that required to close the polygon and therefore with m.e.p. lying at right angles to that of the resultant of the plate combination).

Since the resulting polygon is drawn on a sphere,

$$2\theta_{12} + 2\theta_{23} + 2\theta_{34} + 2\theta_{41} = 2\pi - \beta$$

β is an index of the strength of the circularly polarizing component.

In most instances in biology the ϕs are all small. This means that in effect the sphere is of large diameter, the polygon is plane, and the circularly polarizing component disappears; the polygon can be drawn on flat paper. The order in which the individual plates

104

are taken is then immaterial (it is not so in the general case). An illustrative example may serve to show how the method may help in otherwise recalcitrant problems (a second example will be found on p. 295). The wall of a cell of *Chaetomorpha melagonium* (p. 203) 15 μm thick consists of a number of superposed cellulosic lamellae alternating in chain direction through an angle of 85° given by optically visible striations on the wall. These lamellae are very thin (say about 20 nm) and their phase differences and *a fortiori* that of the whole wall, is small. Since therefore the order in which the lamellae are taken is irrelevant, the wall may be taken as two layers each 7.5μ thick. The path difference of the whole wall is 0.035 μm and the m.e.p. bisects the acute angle between the striations. The question is, does this harmonize with the presence of cellulose as the only crystalline material? Set out line AB, (Fig. 4.25c), length 0.035 units making an angle of 85° (twice the angle between one striation direction and the m.e.p.) with the abcissa taken as one striation direction. AB represents in magnitude and direction the path difference of the whole wall. Draw a line from B making an angle of $2(\pi/2 - 85°) = 10°$ to the abcissa, intersecting the abcissa at C. BC and AC are the path differences of each layer separately. These are naturally identical and each is found, either graphically or by calculation, equal to 0.202 units. The path difference of each is therefore 0.202 μm and the birefringence is $(0.202/7.5) = 0.027$. Since the wall thickness contains only 45 per cent cellulose, the birefringence of the constituent cellulose should be $0.06 = (0.027/0.45)$ which is correct. The path difference of the whole wall therefore harmonizes with its cellulosic nature.

4.4.5 *Birefringence and its interpretation*

It will be clear that birefringence is derived by dividing the path difference by specimen thickness. The latter must be determined, of course, under conditions as closely like those used with the former. With a cylindrical cell it can often be assumed that the wall is uniform so that the side walls (measurable by an eyepiece micrometer) may be used. Similarly, with care sections may be stood on edge and the thickness measured in a similar way. The best method, though time-consuming, is to use an interference microscope for direct determination at the point at which the path difference has been determined (p. 117).

The significance to be attached to birefringence of a wall lamella depends upon the conditions under which it is determined, in terms of the heterogeneity of the lamella. To a first approximation crystalline components in the form of long thin rods — the microfibrils — may be taken as immersed in a matrix of non-cellulosic polysaccharides (p. 188) and other substances which are optically amorphous and possess voids of submicroscopic dimensions (p. 375). In nature the lamella is saturated with water and therefore swollen. Determination of the refractive indices n_γ and n_α separately demands the use of organic embedding liquids. There must therefore, be no excess water on the specimens and the liquids must be such as not to extract water. The method is in any case not easy. Birefringence $(n_\gamma - n_\alpha)$ may, however, be determined without ambiguity and may be interpreted (see below).

If the birefringence of the intact wall is required, drying by any method which induces shrinkage has at least two undesirable consequences. Firstly, the mere act of shrinking is liable to change the orientation of the microfibrils, and therefore the birefringence, by an amount which is unascertainable. Secondly, the change in volume and increase in refractive index of the non-crystalline component relative to the crystalline component (by the withdrawal of water from the non-crystalline component only) automatically changes the birefringence. This involves the principle that an array of isotropic parallel rods of submicroscopic diameter immersed in a liquid will appear optically anisotropic if the refractive index of the liquid differs from that of the rods. The theory of such a system was worked out by Wiener (1912) who showed that the birefringence of such a body is

$$n_\gamma^2 - n_\alpha^2 = \frac{\delta_1 \delta_2 (n_1^2 - n_2^2)^2}{(\delta_1 + 1)n_2^2 + \delta_2 n_1^2}$$

Where δ_1 and δ_2 are the relative volumes of the rods and the liquid respectively $(\delta_1 + \delta_2 = 1)$, n_1 is the refractive index of the rods and n_2 that of the liquid, n_γ is the refractive index for light vibrating parallel to the rods and propagated in a direction perpendicular to their lengths and n_α is the refractive index for light vibrating at right angles to their lengths. The system is clearly uniaxial and the birefringence is referred to as *form birefringence.* For present purposes we note only that the birefringence varies with δ_1 or δ_2. In a wet cell wall containing parallel microfibrils, n_1 may be taken as 1.5 and n_2 as 1.45 at least. Taking $\delta_1 = \delta_2 = 0.5$, then $(n_\gamma - n_\alpha) = 0.001$. On drying this may drop no lower than zero (if e.g. n_1 then equals n_2). The total change in birefringence is then 0.001 and this may be ignored if the microfibrils are cellulose with an intrinsic birefringence of about 0.06. With an unknown wall, however, it is always best to check the presence of form birefringence by determining the birefringence in a number of liquids of different refractive indices which penetrate the wall but do not swell or shrink it.

As far as the individual refractive indices n_γ and n_α of the wall are concerned, it is usually safer to determine these in dried specimens and for purposes of comparison it has become standard practice to make the determinations on material dried over P_2O_5 and to accept some small error due to re-orientation. Even then the refractive indices cannot of course be accepted as the refractive index of e.g. cellulose in a cellulosic wall. We need to add here the complication that the wall is not a two-phase system as assumed above but a multiphase system consisting at least of (a) cellulose (b) non-cellulosic polysaccharides, some chains of which are parallel [and parallel to the microfibrils (p. 165)], (c) randomly arranged polysaccharide chains and other non-crystalline components such as lignin, (d) an aqueous solution in voids. Denoting the relative volumes by f and the refractive indices by n and assuming that the wall behaves as a liquid (as it does in this context to a fair approximation) then

$$n_{\gamma_{\text{wall}}} = f_a n_{\gamma a} + f_a n_{\gamma b} + f_c n_c + f_d n_d$$

$$n_{\alpha_{\text{wall}}} = f_a n_{\alpha a} + f_a n_{\alpha b} + f_c n_c + f_d n_d$$

$n_{\gamma a}$ and $n_{\alpha a}$ (which are usually what are required) may be determined only if the other refractive indices and the relative volumes are known. Note that

$$n_{\gamma_{\text{wall}}} - n_{\alpha\,\text{wall}} = (n_{\gamma a} - n_{\alpha a}) f_a + (n_{\gamma b} - n_{\alpha b}) f_b$$

The second term on the right is difficult to assess and it is customary to assume $n_{\gamma a} = n_{\gamma b}$ and $n_{\alpha a} = n_{\alpha b}$, to add $f_a + f_b (= f)$ and write

$$n_{\gamma_{\text{wall}}} - n_{\alpha\,\text{wall}} = (n_{\gamma a} - n_{\alpha a}) f$$

f is taken as the volume of polysaccharide, relative to the total volume, after extractable polysaccharide and other material have been removed. $n_{\gamma a} - n_{\alpha a}$, which refers to cellulose (or other skeletal polysaccharide) may then be calculated.

4.5 Interference microscopy

Interference microscopy is dealt with here only in broad principle and only in so far as it forms a useful adjunct in cell wall studies as, for instance, in determining wall thickness for birefringence determination, and in checking the condition of material for electron microscopy. Further details will be found in a number of text books (e.g. Françon, 1961; Ross, 1967; Oster and Pallister, 1956; Barer and Mellor, 1955), and detailed methods of use are readily available in the manuals referring to individual commercial microscopes. Phase contrast microscopy will be considered as an imperfect interference microscope.

Most biological material in thin section is transparent, that is to say that light vibrations passing through it are not reduced in amplitude or therefore in intensity. The vibrations are nevertheless affected by passage through the material since the various parts of the section — or a cell — have different refractive indices which are all normally different from the refractive index of the embedding medium. This induces differences in phase which are related to refractive indices. Consider a thin sheet of cell wall, thickness t and with mean refractive n_s surrounded by a liquid of refractive index n_o. Then the optical thicknesses of wall and liquid, for light propagated in a direction normal to the surface are $n_s t$ and $n_o t$ respectively. The light transmitted through the wall then lags behind that transmitted through the liquid by a path difference $\Delta = [(n_s - n_o)t]$, equivalent to a phase difference of $2\pi\Delta/\lambda$ for monochromatic light of wavelength λ. The eye cannot detect differences in phase so that the contrast with such a *phase object* — so-called in distinction from an absorbing object called an *amplitude object* — is zero. The purpose of interference microscopy is to convert such small phase differences into amplitude differences, so that contrast is achieved and, more relevant to the discussions here, to use this contrast in determining Δ and hence both the refractive indices and the thickness. This is achieved in different ways in the phase contrast microscope and in interference microscopes so that these must be considered separately. We take phase contrast microscopes first.

For the cell wall plate mentioned above, let the vibration o, Fig. 4.26, represent the light transmitted by the surrounding liquid and s that transmitted by the wall with a lag of Δ. In principle o has been converted to s by passage through the specimen. This conversion could be achieved in another way. Consider a third vibration d, Fig. 4.26, of the same wavelength as o and s but with a smaller amplitude, a' and lagging behind o by a path length $\lambda/4$. Then even by inspection it can be seen that d and o can be added to give s. This can be shown formally in the following simple way.

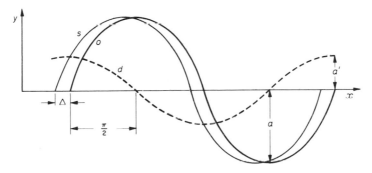

Fig. 4.26 **For explanation, see text.**

Vibrations s and o are of the form

$$y = a \sin 2\pi x/\lambda$$

Hence
$$a' = a \sin 2\pi\Delta/\lambda$$

$$= a2\pi\Delta/\lambda$$

since the phase difference is small.
Vibrations o and d may therefore be written:

$$y = a \sin 2\pi x/\lambda$$

$$y' = a2\pi\Delta/\lambda \sin(2\pi x/\lambda - \pi/2)$$

$$= a2\pi\Delta/\lambda \cos 2\pi x/\lambda$$

Adding y and y' give a vibration

$$y'' = a \sin 2\pi x/\lambda + a2\pi\Delta/\lambda \cos 2\pi x/\lambda \qquad 4.18$$

$$= A \sin(2\pi x/\lambda + \phi) \qquad 4.19$$

since this must be a sine wave, where ϕ is a new phase difference. Equating terms in $\sin 2\pi x/\lambda$ and $\cos 2\pi x/\lambda$ between Equations 4.18 and 4.19 and squaring and adding gives

$$A^2 = a^2(1 + (2\pi\Delta/\lambda)^2)$$

$$= a^2,$$

and dividing gives

$$\tan\phi = 2\pi\Delta/\lambda$$

i.e.

$$\phi = 2\pi\Delta/\lambda$$

since $2\pi\Delta/\lambda$ is small.
The resultant vibration is therefore

$$y'' = a\sin 2\pi/\lambda(x - \Delta)$$

which is vibration s.
Without for the moment considering the physical nature of d, suppose now a second path difference of $\pi/2$ is imposed on vibration d (by passing it through a plate of optical thickness $\lambda/4$ for example). Then d becomes

$$y' = a2\pi\Delta/\lambda \sin 2\pi x/\lambda$$

whereas o remains

$$y = a\sin 2\pi x/\lambda$$

These are now in opposition and may be directly subtracted giving

$$A^2 = a^2(1 - 2\pi\Delta/\lambda)^2$$

$$= a^2(1 - 4\pi\Delta/\lambda) \qquad 4.20$$

where A is the resultant amplitude. This converts a phase difference into an amplitude difference with the object darker than the field. Note that if the vibration d is *advanced* through $\lambda/4$ then $A^2 = a^2(1 + 4\pi\Delta/\lambda)$ and the specimen is brighter than the field. Both conditions can be realized.

It is therefore necessary to consider the physical nature of vibration d so that this relation may be used. d is clearly the sum of the rays diffracted by the specimen, and the proposition is that the vibrations transmitted by the object are the vector sum of the vibrations transmitted through the background and the vibrations diffracted by the object. Since the paths of these two vibrations are not the same, then clearly in principle the diffracted rays may be passed through a $\lambda/4$ plate to give the condition leading to

Equation 4.20, so that the specimen is darker (or brighter) than the field.

The principle is illustrated in Fig. 4.27 for a specimen consisting of a line grating and the practical situation in a phase contrast microscope is shown in Fig. 4.28. The direct light passing through

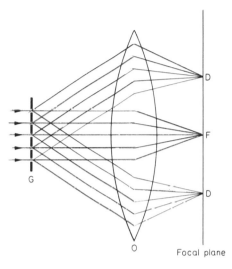

Fig. 4.27 **The diffraction image of a line grating G at the back focal plane of the objective, O.**

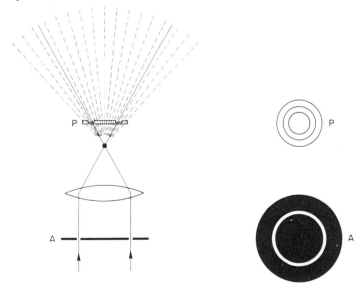

Fig. 4.28 **The principle of construction of a phase contrast microscope.**

the microscope is confined to a hollow cone and this light passes through an annulus in a plate P (Fig. 4.28), above the object, of depth equivalent to $\lambda/4$. The diffracted light mostly passes through the rest of plate P, inducing a phase lag of $\pi/2$. In view of the discrepancy in intensity between the incident and diffracted beams, the annulus is usually part-silvered in order to reduce the intensity of the direct beam. Most manufacturers provide phase plates with 70–90 per cent absorption and these have some advantages as will be seen below; however, they have also some drawbacks.

Equation 4.20 refers only to conditions under which Δ is small compared with λ, and under these conditions the intensity varies linearly with Δ. This is no longer true for higher values of Δ as can be seen from a (very simplified) vectorial treatment of the situation. Let the incident vibrations be represented by a line OM (Fig. 4.29) of length equal to the amplitude a and with a phase determined by the direction of OM. On passing through the specimen, OM is changed in phase by an amount ϕ and may be

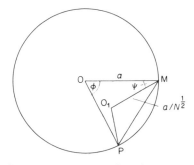

Fig. 4.29 **Simplified vector diagram for phase contrast microscopy.**

represented as OP, making an angle ϕ with OM. The line MP is the vectorial sum of OM and OP and is therefore equivalent in amplitude and phase to the diffracted beam. When ϕ is small, the angle OMP (the phase) is almost 90° as we have seen. MP and OM are together equivalent to OP. In the microscope OM is reduced in intensity and changed in phase by an amount ψ. This can be represented by rotating MO about M through an angle ψ and reducing its length to $a/N^{\frac{1}{2}}$ (i.e. reducing the intensity by a factor N). O then moves to O_1 and the resultant intensity of the specimen is O_1P. This intensity can obviously be calculated from triangle O_1MP and it is clear that in the general case the relation between

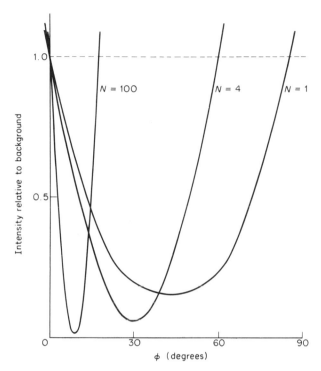

Fig. 4.30 **The relative intensity of light from objects of difference phase retardation in terms of the absorbancy of the wave plate annulus.**

phase and intensity is not linear. The corresponding relations for various values of N are presented in Fig. 4.30, because they present a dilemma of which users of microscopes with a fixed phase plate must be constantly aware.

A phase plate of high absorption gives high contrast for small phase changes but is completely insensitive to large phase changes. Unless this is realized, the appearance of a specimen may be misleading. On the other hand, a phase plate of lower absorption, while covering a wider range of phase, is not as sensitive at low phases as is the highly absorbing plate. The real dilemma is, however, that with any phase plate the intensity of the specimen is the same for two values of the phase and the observer cannot distinguish between these unless a second phase plate of different absorption is available.

Phase contrast microscopes have other disadvantages in addition, deriving from the circumstance that some of the diffracted light

113

passes through the annulus of the phase plate so that these microscopes are in effect only imperfect interference microscopes. This has two effects upon the image (Fig. 4.31). Firstly, the microscope is sensitive only to abrupt changes in phase and these are associated with a halo, bright for dark specimens and dark for bright specimens, which obscures edges and makes measurement of phase difficult. Secondly, specimens show what is known as the 'shading effect'; in this, an extended area of uniform refractive index which would be expected to be uniformly dark appears brighter in the centre than at the edges and can even appear brighter than the field. This effect is not normally noticeable with a cell containing numerous organelles but is marked when cell walls are observed in surface view, precluding any accurate measurements.

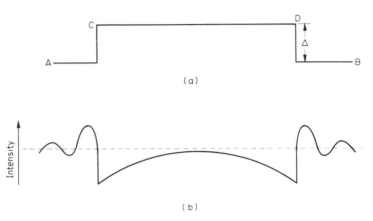

Fig. 4.31 (a) The path difference Δ between the wave front CD passing through an optical uniform object and the corresponding wave front in the medium. (b) The corresponding intensity distribution observed in a phase contrast microscope.

Nevertheless, phase contrast microscopes have certain clear advantages over the more perfect interference microscopes. They are simple to use and relatively cheap, perfectly adequate for checking the condition of material being processed for examination by other techniques such as electron microscopy. The image is more 'pleasing' than it is with interference microscopes. The microscope's insensitivity to moderate variation in phase imposes no demands with regard to variations in microscope slide thickness and objects of extended area can be observed.

Interference microscopes do not make use of the separation of the diffracted and direct beams. Instead, the incident beam is split into two so that two statistically coherent beams pass through the microscope and are brought together so that they interfere. Only one of these beams passes through the specimen so that the visible field consists of two overlying fields, one containing the background only and the other with the specimen -superposed. The beams can be separated only by a small distance so that only small objects, or the edges of large objects, are displayed under conditions in which the phase is measurable.

The principle of this microscope may be understood from Fig. 4.32, analogous to Fig. 4.29. OM and OP are again the incident illumination and the illumination transmitted by the specimen. The second beam introduced into the microscope and passing through the background only is OO_1, differing in phase from OM by a

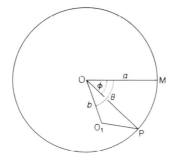

Fig. 4.32 **Simplified vector diagram for interference microscopy.**

variable phase angle θ and perhaps in amplitude. Again the vectorial addition of OO_1 and OP is O_1P and this gives the amplitude of the specimen beam. Again this amplitude depends on ϕ only, for any given value of θ. The situation expressed by Fig. 4.30 applies also here but is less serious since θ may be varied at will and ambiguity in the value of ϕ can be avoided.

ϕ may be determined in either of two ways. In the more sensitive approach the microscope is set, following the instruction in the manual, so that the phase difference of the empty field is constant over the whole field. The incident intensity of the specimen beam is then OM (Fig. 4.33), the transmitted intensity OP and the comparison beam intensity OO_1 (now made of the same amplitude as OM). As the phase θ is increased, O_1 moves round the circle MPV'.

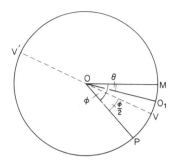

Fig. 4.33 For explanation, see text.

The background intensity is therefore always O_1M^2 and the specimen intensity O_1P^2. When $\theta = 0$, O_1 coincides with M and the specimen intensity is MP^2, i.e. the specimen is bright against a black background. As θ is increased up to $\phi/2$ the specimen darkens and the background lightens until at $\theta = \phi/2$, when O_1 is at V, a cross-over point occurs at which the specimen and background are of the same intensity and beyond which the specimen continues to darken. The position $\theta = \phi/2$ can be used as a sensitive measure of ϕ since θ is known. Note that there is a second match at V' but this is very insensitive and can readily be distinguished from the match at V. When $\phi = \theta$, O_1 is at P and the specimen is black against a bright background. This position can also be used to determine ϕ.

If, on the other hand, the field is set to a constant gradient of phase difference, the field is covered by a series of equidistant fringes. The phase difference of a specimen can again be determined as in Fig. 4.34 but this is a much less accurate procedure, and is not to be recommended.

When white light is used the field, and the fringes, are coloured. A matching of colours then means a matching of phases and this proves useful in examining cell walls.

The more readily available interference microscopes are of one of two types. In one, the splitting of the incident beam into two is achieved by the Dyson method in which internal reflection in a narrow wedge below the specimen achieves the splitting and the two beams are recombined by reflection in a second identical wedge placed above the specimen. In the other the illumination is divided by passage through a thick crystal, separating the beam into two beams polarized at right-angles to each other. These are

116

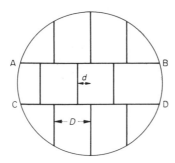

Fig. 4.34 A strip of material, ABCD, viewed in an interference microscope with a gradient of phase difference across the field. The fringes, spacing equivalent to one wavelength, are displaced in the object by a distance d, proportional to $\phi\,[\equiv (d/D)\lambda]$.

subsequently recombined in a second identical crystal after passing through a $\lambda/2$ crystal plate which turns the direction of each beam through a right angle. Since beams of light vibration at right-angles to each other cannot interfere, a $\lambda/4$ plate is inserted in the light path which converts the two plane polarized beams into two circularly polarized beams rotating in opposite directions (see p. 90), which then interfere.

The thickness of a cell wall may be determined in either of these microscopes, but care is needed with those using polarization-splitting since the light passing through the specimen plane is plane polarized; the object needs to be suitably oriented.

Since

$$\Delta = (n_s - n_o)t$$

and n_o is known and Δ measurable, t is known if n_s is known (this may be taken as the average of n_γ and n_α). If n_s is not known then two observations of Δ need to be made in media of different refractive index. Then

$$\Delta = (n_s - n_o)t$$

$$\Delta' = (n_s - n_o')t$$

where Δ' and n_o' are the second values. Hence

$$t = (\Delta' - \Delta)/(n_o - n_o')$$

whence n_s may be calculated.

Alternatively, if t is known, n_s may be calculated in one operation. This may be used to give a rough estimate of composition. If, for instance, a wall has a relative volume of cellulose Σ and a relative volume of lignin $(1 - \Sigma)$, then

$$n_s = \Sigma n_c + (1 - \Sigma) n_e$$

where n_e is the refractive index of lignin (1.46). Hence Σ may be determined. Or, if a wall consists entirely of substances of refractive index n_w but there is a relative void volume Σ, then

$$n_s = (1 - \Sigma) n_w + \Sigma$$

Hence, taking n_w as 1.55, Σ may be calculated and hence the density, $(1 - \Sigma)_{\rho_c}$, where ρ_c is the density of cellulose.

CHAPTER 5

Structure determination – X-ray diffraction

It is now clear that the individual long molecular chains which for the most part constitute the cell walls of plants already show some order. The sugar residues in a homogeneous chain occur in regular sequence along the chain so that in most β-1,4-linked glucans, for instance, the glycosidic oxygen bridges occur regularly at intervals along the chain of 5.15 Å. All that is needed to specify the chain is the nature of the residue and the spacings of 5.15 Å. If a number of chains can be induced to lie parallel to each other then a higher degree of order is achieved whether or not the chains are spaced regularly the same distance apart. This higher degree of order can be detected, and to some extent quantified, by optical methods which were described in the last chapter. We now confine attention to the situation in which the chains lie not only parallel to each other but spaced regularly side by side, that is the situation in which the polysaccharide achieves full three-dimensional order. In all cell walls there is at least one polysaccharide which is synthesized in three-dimensional order and which therefore occurs in intact walls in a form open to examination by any method which can interpret the order. In the large bulk of cell walls this is cellulose and this particular polysaccharide will be taken as an example of the use of the most powerful tool available for the study of three-dimensional order — X-ray diffraction analysis. As already mentioned, it is now known that in some plants cellulose does not occur and its place is taken over by another polysaccharide, and equally it is known that many polysaccharides which do not occur naturally in three-dimensional order can be induced to take up this order by various extraction and precipitation methods; these will, however, be discussed in the appropriate place in ensuing chapters.

Three-dimensional order is a characteristic which defines the *crystalline state*. With inorganic crystals the order is visually evidenced by the occurrence of crystal faces and it is through the beautiful forms which crystal faces induce that crystals are normally recognized. The absence of crystal faces with the crystals with which we shall deal should not be allowed to mislead; a parallel array of molecular chains of different length cannot produce faces but can nevertheless lie in the three-dimensional order basic to the crystalline state.

Naturally occuring crystalline polysaccharides are clearly of great importance in determining the physical properties of a plant cell wall and, as it turns out, in defining the mechanism which must have synthesized these polysaccharides and hence in interpreting some details of the structure of the plasmalemma. It is therefore imperative that their structure should be known in detail. Equally, it is imperative that those polysaccharides which are not crystalline in the natural condition but which can be crystallized should be studied with equal intensity; for it is only when a polysaccharide lies in three-dimensional order and can then be examined by diffraction analysis that some details of structure in individual chains can be determined such as are relevant for an understanding of the wall as a whole. Even then, of course, it still remains a question whether a chain which is part of a crystalline array adopts the same conformation as a chain which lies isolated.

5.1 The unit cell — two dimensions

A crystal consists of an array of identical molecules regularly arranged in space. Let us consider each molecule represented by a point so that for the moment we do not need to enquire about the nature of the molecule or the disposition of its atoms. We therefore, for example, replace the cellulose chain by a line of points spaced 5.15 Å apart. A two-dimensional crystal would then be represented by a flat sheet (Fig. 5.1) consisting of rows of points spaced a apart in columns distance c apart staggered through an angle β. The whole array is known if a, c and β are known and therefore these three parameters alone are required to specify the array. The points can equally be represented as the intersections of lines ABC, DEF, etc. with lines ADH, BEI, etc. and the required parameters replaced by d_1 ($= a \sin \beta$) and d_2 ($= c \sin \beta$), the perpendicular distances between ADH and BEI and between ABC and DEF, and β. In principle, the distances d_1 and d_2 can be determined by methods which will be explained later, analogous to but more complex than, the determination of the spacing between lines in a line grating by optical methods. In fact, however, the points can be joined by lines other than ABC etc. and ADH etc., some of which are shown in Fig. 4.1, and each of these corresponds to an interline spacing d_r. A determination of spacings would therefore yield a series $d_1, d_2 \ldots d_r \ldots d_n$ where n is large. The problem therefore resolves itself into the mathematical problem of deriving a c and β from thcsc spacings.

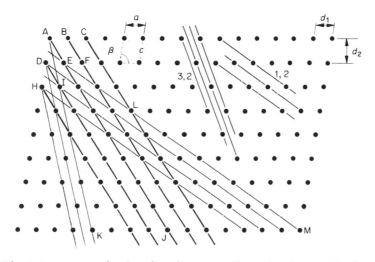

Fig. 5.1 An array of points forming a two-dimensional crystal lattice.

For this purpose it is necessary in some way to specify the relationship between any set of parallel lines drawn through the points and the lines ABC and ADH. This could be done, for instance, by choosing point D as origin and specifying the set of lines by the intercepts which the line nearest the origin makes with the lines ADH and DEF. Lines parallel to AEJ then make an intercept of $1a$ along DEF and $1b$ along ADH and could be called the (1 1) line. Similarly lines parallel to AIK would be (1 $\frac{1}{2}$). For ease in mathematical treatment, the lines are not specified in this way; instead the *reciprocals* of the intercepts are taken. Lines AEJ then retain the indices (1 1) but AIK becomes (1 2) and ELM (2 1). Note that lines parallel to ABC are (0 1) and parallel to ADH (1 0). In general any set of parallel lines may be indexed $(h\ k)$, with a spacing d_{hk} which can readily be shown to be given by:

$$(1/d_{hk})^2 = (h^2/a^2 + k^2/c^2 - 2hk \cos \beta/ac)/\sin^2 \beta. \qquad 5.1$$

In principle therefore the process is to obtain the list of spacings, assign indices to three of them and calculate a, c and β. If this gives values which are acceptable, the spacings of all $(h\ k)$ lines are calculated and checked against the list of spacings. If all is still well, the values can provisionally be accepted; it is then necessary to apply still other criteria, discussion of which will be postponed until three-dimensional crystals are considered. The area of ABED contains all the elements typifying the crystal and the whole crystal can be built up by laying similar areas alongside in the two directions. It is called the *unit cell*.

This is a convenient point at which to consider an implication of Equation 5.1 which also turns up in the three-dimensional case. In Equation 5.1 d, a and c appear as reciprocals, and if $1/d$ is replaced by d^*, $1/a$ by a^* and $1/c$ by c^* we have the simpler expression:

$$d^{*2} = (h^2 a^{*2} + k^2 c^{*2} - 2hk \cos \beta a^* c^*)/\sin^2 \beta.$$

d^* is then a *single* line drawn in reciprocal space corresponding to a *set* of planes spaced d apart in real space. This gives the representation of the lattice shown in Fig. 5.2. OPQR is the unit cell of the lattice identical with ABED in Fig. 5.1. With O as origin, the line OU is drawn equal in length to $1/d_{01}$ $(= d_{01}^*)$ and normal to OR. The point U represents in reciprocal space the set of lines in real space parallel to OR, the (O1) lines, and may be labelled O1. Continuation of this line through the same distance d_{01}^* leads to the point V corresponding to (O2) lines, halving the distance between PQ and OR, which do not exist in this particular lattice. The point V still has meaning since the values of d are determined by a diffraction method; it corresponds to the second order of diffraction. Similarly to O3 etc. Again, the line OL is drawn normal to the lines parallel to OP and $1/d_{10}$ long. The point L represents (1 0) lines, M, (2 0) lines and N, (3 0) lines. Completion of the net as in Fig. 5.2 leads to points (1 2), (2 1) etc. and a little thought will reveal that the distance from the origin of any point is the

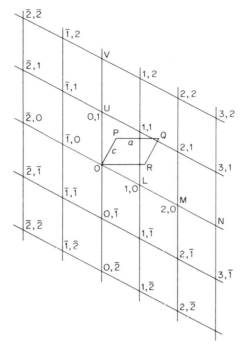

Fig. 5.2 The two-dimensional reciprocal lattice. The bar over a number indicates a negative index.

reciprocal of the perpendicular distance between the lines in real space with the same index. The reader should have no difficulty in showing that, using the reciprocals of the spacings d_r and allocating indices to three of these, it is easy to construct the reciprocal net graphically, to determine therefore a c and β, and to check all possible $(h\,k)$ spacings.

5.2 The unit cell — three dimensions

The three-dimensional case is of course more complex but the principles are the same, (Fig. 5.3). Now, in general, six parameters are needed to specify the unit cell since $a \neq b \neq c$ and $\alpha \neq \beta \neq \gamma$. Fig. 5.3 illustrates the simpler case in which $\alpha = \gamma = 90°$, representing the crystal class known as *monoclinic*, the class to which cellulose belongs. The other five possible crystal classes are given in Table 5.1. It will be clear that with an unknown crystal the

123

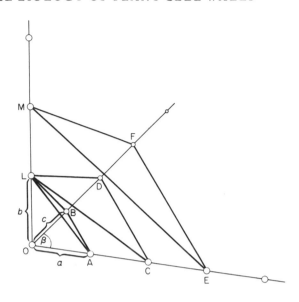

Fig. 5.3 The derivation of the Miller indices for a crystal lattice.

TABLE 5.1

The crystal classes

Class	Unit cell translations	Angles between crystal axes (degrees)
Triclinic	$a \neq b \neq c$	$\alpha \neq \beta \neq \gamma \neq 90$
Monoclinic	$a \neq b \neq c$	$\beta \neq 90; \alpha = \gamma = 90$
Orthorhombic	$a \neq b \neq c$	$\alpha = \beta = \gamma = 90$
Tetragonal	$a = b \neq c$	$\alpha = \beta = \gamma = 90$
Hexagonal		
Hexagonal division	$a = b \neq c$	$\alpha = \beta = 90; \gamma = 120$
Rhombohedral division	$a = b = c$	$\alpha = \beta = \gamma \neq 90$
Cubic	$a = b = c$	$\alpha = \beta = \gamma = 90$

assumption that it is monoclinic and the determination of *abc* and
β will lead automatically to the correct crystal class if the crystal
is of higher symmetry than monoclinic; if it is triclinic, no monoclinic cell will be found to fit.

Just as with the *lines* in two-dimensional crystals, so the *planes*
passing through the points in a three-dimensional lattice are specified by the reciprocals of the intercepts on the crystal axes of the
plane nearest to any point chosen as origin, O. Thus, places parallel

124

to ABL, Fig. 5.3, are the (111) planes, LCD are the (121) planes, MEF the (232) planes and so on. These are called the Miller indices of the planes, The sides of the unit cell correspond, of course, to (100) (OMF), (010) (OEF) and (001) (OME).

The reciprocal lattice is obtained as before by drawing lines normal to the (100), (010) and (001) faces equal in length to $d_{100}(a^*)$, $d_{010}(b^*)$ and $d_{001}(c^*)$ respectively (Fig. 5.4). Note that:

$$a^* = bc/V$$

$$b^* = ca \sin \beta/V$$

$$c^* = ab/V$$

where V is the volume of the unit cell.

The complete reciprocal lattice can then be drawn as in Fig. 5.4 in which each intersection of three lines represents the set of parallel planes in real space whose indices (hkl) are given at the point and whose perpendicular spacing is the reciprocal of the distance of the point from the origin. Notice that all points with the same value of k themselves lie in a plane parallel to the ac plane (and similarly

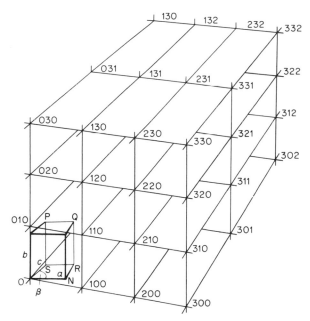

Fig. 5.4 The three-dimensional reciprocal lattice referring to a monoclinic unit cell.

for all points with the same h or the same l though we shall not use this). Again this reciprocal lattice is conceptually simpler than the real lattice, since each *point* in reciprocal space represents a *set of planes* in real space.

5.3 The determination of interplanar spacings

Before fully demonstrating this simplicity it is necessary now to consider how the interplanar spacings are determined. This is done by the diffraction of X-rays since the wavelength of X-radiation is comparable in size with the distance apart of atoms (a few Ås). X-rays are generated when high energy electrons impinge on a metal target so that electrons are ejected from the inner (usually K or L) shells in the metal atom. The only metal we shall use is copper. As outer electrons fall back into the vacated shell, a spectrum of X-rays is emitted which, if the electron energy is high enough (e.g., for copper accelerated over a potential drop of say 40 kV), consists of a broad band (the *background* or *white* radiation) upon which is superposed with copper two peaks, the Kα and the Kβ peak. The Kβ peak and the background are normally removed through absorption by a piece of nickel foil. The remaining peak, Kα ($\lambda = 1.54$ Å) is itself double but the wavelength interval is so small that with polymers this can be neglected. When such a beam of X-rays is directed at an atom the electrons become secondary emitters and we say that the electrons 'scatter' the X-rays. The scattering may be *coherent*, with no change of wavelength, or *incoherent* when the wavelength is changed.

The major part of the scattered energy goes into coherent scattering and this is used in structure determination. The radiation scattered by any one electron in an atom is generally not in phase with that scattered by any other so that there is an angular dependence of scattered intensity; the intensity falls off as 2θ, the angle between the incident and scattered beams, increases. The shape of the curve relating the scattering power to 2θ is called the *scattering* or *form* factor (f) of the atom; the scattering factor f_0 for $2\theta = 0$ is set equal to the numbers of electrons in the atom.

Suppose a parallel beam of X-rays falls upon a row of atoms considered as point scatterers set a Å apart (Fig. 5.5) at an angle α. The secondary waves generated by the points will be in phase along

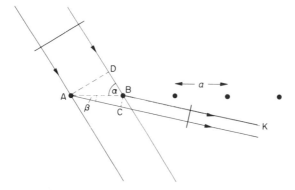

Fig. 5.5 Diffraction by X-rays by a row of atoms. The incident waves are in step at AD. If BK is a 'reflection' at angle β to AB the waves must also be in step at B and C; this gives the condition stated in the text.

any direction BK when

$$AC - BD = m\lambda$$

where m is an integer. That is,

$$a \cos \beta - a \cos \alpha = m\lambda$$

whence a can be calculated if β is measured.
With a three-dimensional lattice with axial translations a, b and c, three equations of this type must be satisfied, namely

$$a(\cos \beta_1 - \cos \alpha_1) = m\lambda$$
$$b(\cos \beta_2 - \cos \alpha_2) = p\lambda$$
$$c(\cos \beta_3 - \cos \alpha_3) = q\lambda$$

These are the Laué equations (Laué, 1912). They can be solved to give a relation between λ, d and θ (half the angle between the diffracted ray and the incident ray)

$$n\lambda = 2d \sin \theta, \qquad\qquad 5.1a$$

the Bragg law (Bragg, 1913), where n is an integer.

This equation can be derived in a much simpler way by assuming that planes of atoms reflect X-rays much as mirrors reflect light. Fig. 5.6 represents a plane normal to the diffracting planes (only two planes, PQ and RS are represented) containing both the incident and diffracted beam. The 'reflected' beams from O and B are in phase only when

127

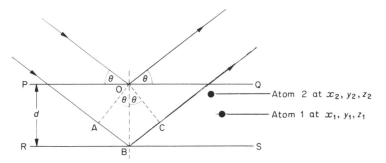

Fig. 5.6 **Diffraction of X-rays by molecular planes in a crystal. PQ and RS are two of a set of parallel planes all d units apart, with a beam of X-rays incident at angle θ.**

$$n\lambda = AB + BC$$

i.e. $$n\lambda = 2d \sin \theta$$

where n is an integer, and d is the interplanar spacing.

Determination of d is therefore resolved into the determination of θ. The principle of the method is illustrated in Fig. 5.7. The specimen O is placed in the X-ray beam at a distance D from a photographic plate, insulated from the beam by a lead cup at L. If a reflection occurs at P, distance r from the centre of the plate, then clearly

$$r/D = \tan 2\theta$$

and θ and hence d can be determined.

5.4 Diffraction by plant fibres

The conditions under which a reflection will actually occur for any spacing d_{hkl} may be readily seen by consideration of the reciprocal lattice, in a manner which leads readily to an understanding of the X-ray diffraction by plant fibres. Let the crystal and the origin of the reciprocal lattice be placed at O, Fig. 5.8, and construct a sphere centre C with radius $1/\lambda$. This is the *sphere of reflection*. Suppose the crystal is tilted so that a point in the reciprocal lattice corresponding to (hkl) planes touches the sphere at P. Then OP = $1/d$ and, from triangle QPO,

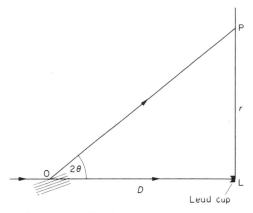

Fig. 5.7 Determination of θ ($\equiv d$). The specimen at O is represented by a series of parallel equidistant planes at an angle θ to the beam OL. LP is the photographic plate. The diffraction arc on the plate, corresponding to the planes at O, is at P.

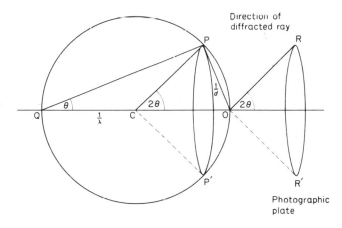

Fig. 5.8 The condition of reflection. For description, see text.

$$1/d = (2/\lambda) \sin \theta$$

or

$$\lambda = 2d \sin \theta$$

again. θ is therefore the Bragg angle and the direction of the diffracted beam is OR, parallel to CP. The condition for reflection is therefore that, with this construction, the reciprocal lattice point shall touch the surface of the sphere of reflection.

It is instructive to note in passing that, if the specimen is a

129

powder consisting of myriads of crystals at all orientations, some crystals will be so arranged that overall any *hkl* plane is in a position for reflection i.e. all possible P_{hkl}s will lie on the sphere. The reflection for any particular value of $1/d$ e.g. that represented by P in Fig. 5.8 can be deduced by rotating the single crystal in this figure around the direction of the X-ray beam. The point P then describes the small circle PP' and the reflected ray OR produces the hollow cone ORR' which will be recorded on a flat plate as a circle. The X-ray diagram of a powder therefore consists of a series of concentric rings (Fig. 5.9) from the diameter of which ($= 2r$) the spacings d can be determined. The construction in Fig. 5.8 suggests that P is a mathematical point and the ring is infinitesimal in thickness. In practice, and particularly with polymers, the individual crystallites are not infinite in extent and have other imperfections and the X-radiation is not truly monochromatic. These imperfections have the effect that the sphere of reflection is a shell of appreciable thickness and the reciprocal lattice points are small domains; the rings therefore have finite widths.

In the bulk of cellulose cell walls, the crystallites of celluloses are long and thin and tend to be aligned along a direction which

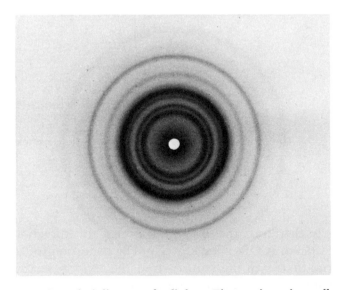

Fig. 5.9 A 'powder' diagram of cellulose. The specimen is a pellet of disoriented *Chaetomorpha* cellulose (p. 203).

bears some relation to the cell axis; they are never, however, precisely parallel to each other so that their orientations are spread over a finite angle and the orientation is referred to as a *preferred orientation*. Attention is here confined to those particular plant fibres in which the preferred orientation is approximately parallel to the fibre length. As a particular example the fibres of ramie (*Boehmeria nivea*) are chosen since these were the fibres from which the structure of cellulose was first deduced. Some attention will also be paid to the cellulose of *Valonia* and the Cladophorales in order to take in more modern developments but the situation with other organisms will be examined only in later chapters as appropriate.

The specimen to be examined is a bundle of parallel fibres (which we remember are hollow tubes) in which the preferred orientation is axial. The bundle is placed in the X-ray beam at right-angles to the beam and the diagram is recorded on a flat photographic film some 4 cm away and also normal to the beam. Both because each individual fibre can be represented by a longitudinal strip of its wall rotated about its length, and individual fibres are inserted in the bundle without reference to any particular orientation around the fibre axis, the resulting X-ray diagram resembles that of a single crystal rotated about one axis. It is called a *fibre diagram*. The genesis of the diagram can easily be understood from the corresponding reciprocal lattice.

It has already been noted that for a single monoclinic crystal the planes in the reciprocal lattice lying parallel to the *ac* plane in the real unit cell correspond to a unique value of k namely 1,2,3... in succession. Three of these planes are set out in Fig. 5.10 for $k = 0,1,2$, in relation to the reflection sphere of Fig. 5.8. Rotation of the crystal about the vertical axis corresponds to rotation of the reciprocal lattice around axis ROR'. Points for which $k = 0$ can cut the sphere of reflection only in the great circle EO and therefore, from Fig. 5.8, all reflected rays from planes $(h\,0\,l)$ lie in a plane which intersects the photographic plate in a straight horizontal line. Reflections from $(h\,1\,l)$ planes equally lie on the surface of a cone of semi-angle $(90° - 2\alpha)$ such that

$$\sin 2\alpha = \frac{\lambda}{b}$$

similarly for $k = 2$,

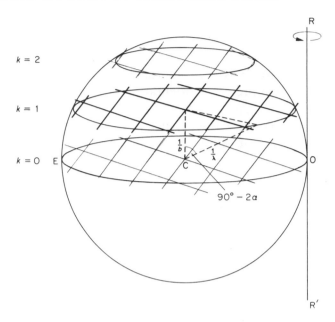

Fig. 5.10 The genesis of the fibre diagram. A parallel bundle of fibres is placed at O, parallel to RR′ and normal to the direction of the beam CO. The reciprocal lattice of a single longitudinal strip of the wall of one fibre is drawn together with the reflection circle. The fibre diagram may be generated by rotation of the reciprocal lattice about RR′.

$$\sin 2\alpha = \frac{2\lambda}{b}$$

and for $k = r$

$$\sin 2\alpha = \frac{r\lambda}{b} \qquad\qquad 5.2$$

These various cones intersect with the plane of the photographic plane as hyperbolae so that reflections corresponding to planes with the same value of k lie on a single hyperbola. In the diagram of the bundle of ramie fibres (Fig. 5.11) these hyperbolae can be clearly made out and each can be indexed for its value of k. In a perfect crystal, lattice points vertically above 0 [$(0k0)$ planes] should not reflect. In practice they do so, if not dis-allowed for other reasons, because the crystallites are not perfectly aligned. To allow for this, the axis ROR′ should be allowed to wobble during the rotation and this is sufficient to bring $(0k0)$ planes into reflection which occurs along the meridian in Fig. 5.11.

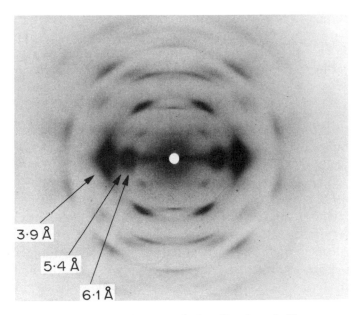

3·9 Å

5·4 Å

6·1 Å

Fig. 5.11 **X-ray diagram of a bundle of ramie fibres.**

5.5 The X-ray-derived structure of cellulose

The b dimension can now be determined directly. The apex of the rth hyperbola is separated from the primary beam through an angle 2α which can be measured and b follows from Equation 5.2. Alternatively, the first strong reflection on the meridian occurs near the fourth layer line and its spacing can be calculated to be 2.58 Å from Bragg's law. The parameter b is therefore (4 x 2.58) i.e. about 10.3 Å. Note that this reflection does not coincide with the apex of the hyperbola since it lies at an angle given by

$$\lambda = 2d \sin \theta$$

whereas the apex lies at an angle given by

$$\lambda = b/4 \sin 2\alpha$$

in which d and $b/4$ are the same.

Determination of ac and β is not so straightforward since the form in which cellulose is presented by nature precludes rotation about the a and c axes. The procedure is in principle that already mentioned in the two-dimensional case (p. 122). Three strong arcs on the equator ($k = 0$) are assigned arbitrary values of h and l, remembering that in the real lattice the most densely populated planes (giving the most intense reflections) have low values of the Miller indices. In monoclinic unit cells the spacings have values given by

$$(d_{h0l})^2 = \sin^2 \beta / [h^2/a^2 + l^2/c^2]$$

whence a c and β may be calculated. Taking the spacings at 3.9 Å, 5.4 Å and 6.1 Å (Fig. 5.11) as the (002), (10$\bar{1}$) and (101) planes respectively it can thus be shown that $a = 8.35$ Å, $c = 7.9$ Å, $\beta = 84°$. These are the values derived by Meyer and Mark (1928, 1929, 1930); they checked sufficiently well with the layer line spacings then recorded as to be provisionally acceptable. It will be clear that similar values could be reached graphically by setting out the basal plane of the reciprocal lattice, as in the two-dimensional case. A more sophisticated method is to work via the reciprocal lattice to Bernal charts but there is no point in discussing this here; the relevant treatment will be found given by Alexander (1970) in a simple and elegant way.

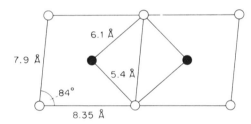

Fig. 5.12 The basal plane of the unit cells due to Meyer and Mark, and Sponsler and Dore.

At an earlier date, Sponsler and Dore (1926), working from first principles, had proposed a different unit cell; $a = 6.1$ Å, $b = 10.3$ Å, $c = 5.4$ Å, $\beta = 88°$. As shown in Fig. 5.12 this is geometrically (though it is not crystallographically) equivalent to the Meyer and Mark unit cell.

In positioning the cellulose chains in the unit cell, Meyer and Mark took note of certain features of the X-ray diagram and of cellulose itself. The density of cellulose is about 1.57 and this means that the unit cell must accommodate two cellobiose units. The length of the b axis, 10.3 Å, moreover, lies close to the distance between alternate glycosidic oxygens in the cellulose chain. It was therefore possible to align the cellulose chains parallel to the b axis with one chain at each corner of the unit cell (contributing 4/4 ($= 1$) cellobiose residues to each unit cell) with another chain running centrally (contributing the other cellobiose residue). Since that time, closer definitions of bond distances have made it certain that the distance between alternate glycosidic oxygen bridges in the fully extended molecule is greater than 10.3 Å. This means that the chain must be slightly kinked and this is attributed to the intrachain hydrogen bond between the —OH on C3 of one glucose residue and the ring oxygen of the next in the Herman's configuration mentioned in Chapter 3 (Hermans, 1943). It converts the cellulose chain into a true ribbon-like molecule. There is now much experimental evidence supporting the Hermans configuration and this is adopted quite generally (see Rees, 1969). A plausible structure which has wide agreement is shown in Fig. 5.13. The absence of an 030 reflection suggests that the b axis is an axis of twofold

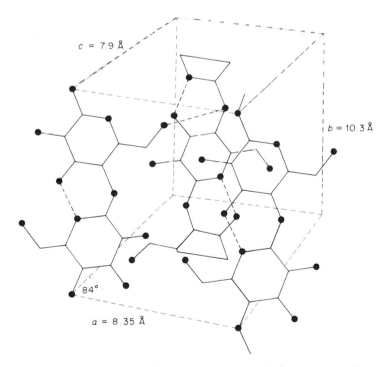

Fig. 5.13 The unit cell of cellulose I. ● = oxygen, hydrogen omitted; dotted lines represent hydrogen bonds.

135

screw symmetry and for this reason, as well as the ease with which a molecular model can then be made and to conform with the known structure of cellobiose, alternate glucose residues are rotated through about 180°, (Fig. 5.13). The central chain is displaced vertically through a distance $b/4$ with respect to the corner chains in order, among other things, to account for the strong 040 reflection and the whole structure is clearly in harmony with the absence of a 010 reflection (since this reflection is quenched by reflection from the interposed 020 and 040 planes). Note that, by the definition of the unit cell, the central chain *must* in some way be different from the corner chains and it is sufficient that this difference should consist of the simple translation along the b axis. Nevertheless, in the model of Meyer and Misch (1936) a second difference is introduced in which the central chain lies upside down with respect to the corner chains in what is called an antiparallel arrangement, included in Fig. 5.13. The crystallographic evidence for this is weak and the general argument which led Meyer to this situation is not strong either. In spite of many attempts at checking this arrangement by a wide variety of techniques no final conclusion has been reached and one is inclined to accept the conclusion of Jones (1968) that not enough information is available to make an unequivocal decision with regard to a parallel $v.$ an antiparallel arrangement (see, however, p. 140). The antiparallel structure is generally accepted partly on the grounds that so many other polymers are known to occur in this form.

The comparatively highly crystalline celluloses of the marine algae *Valonia*, *Chaetomorpha* and *Cladophora* have more recently been objects of intense study in the expectation that these celluloses would throw light on the structure of cellulose in general. Honjo and Watanabe (1954) obtained 400 reflections from *Valonia* cellulose using electron diffraction (as against *c.* 40 reflections from ramie cellulose) and found that indexing required a unit cell with a and c dimensions twice as great as in the Meyer and Misch model i.e. $a = 16.7$, $b = 10.3$, $c = 15.8$, $\beta = 84°$. This is an eight-chain unit cell and soon received support from Frei and Preston (1961) using *Chaetomorpha melagonium* and Fischer and Mann (1960) using *Valonia* cellulose itself. If an eight-chain unit cell is accepted — as it seems it must be — then a considerably increased complexity is introduced since all eight chains (instead of two chains as in the Meyer and Misch model) must in some way differ from each other.

There are severe limits to the information which can be extracted from an X-ray diagram simply by determination of spacings and the presence or absence of reflections, and the standard procedure, once the unit cell is defined, is to determine the intensities of the reflections and to check these against the intensities to be expected from the model. This has been attempted for cellulose by a number of workers without, unfortunately, spectacular advances. Most recently, Nieduszynski and Atkins (1970) have used this treatment, and others, to make specific proposals and to underline some difficulties and on this account and because this general process has yielded information with other crystalline polysaccharides which will be taken up later, the principles of the method are now outlined. A fuller treatment will be found in standard works such as those of Alexander (1969) and Nyburg (1961).

Consider first the scattering of X-rays by two atoms, 1 and 2 (Fig. 5.6), scattering factors f_1 and f_2 ($f_1 > f_2$), in terms of the scattering from an origin placed arbitrarily at B. Note that while the choice of origin is arbitrarily the resultant intensity of scattering is not and it is therefore intuitively clear that the intensity is the same for any chosen origin; this can be proved (Nyburg, 1961). It has already been shown that, when radiation is 'reflected' from the planes shown in Fig. 5.6, the phase difference between the radiation scattered from neighbouring planes PQ and RS is 2π. Now any point with fractional* co-ordinates (x, y, z) with reference to B is at a perpendicular distance $d(hx + ky + lz)$ from plane RBS and waves scattered from an atom at this point will have a phase $2\pi d(hx + ky + lz)$ with respect to those from B. The scattered waves from 1 and 2 may therefore be represented as

$$s_1 = f_1 \sin(\omega t - 2\pi(hx_1 + ky_1 + lz_1))$$

$$s_2 = f_2 \sin(\omega t - 2\pi(hx_2 + ky_2 + lz_2))$$

These must be compounded to a wave of the form

$$s = A \sin(\omega t - \Delta_{hkl})$$

whence, by a simple process demonstrated on p. 81 for a somewhat different circumstance, it can readily be shown that

---

* i.e. if the real co-ordinates are x', y' and z', then $x = x'/a, y = y'/b, z = z'/c$.

137

$$A^2 = [f_1 \cos 2\pi(hx_1 + hy_1 + lz_1) + f_2 \cos 2\pi(hx_2 + hy_2 + lz_2)]^2$$
$$+ [f_1 \sin 2\pi(hx_1 + ky_1 + lz_1) + f_2 \sin 2\pi(hx_2 + ky_2 + lz_2)]^2$$

and

$$\tan \Delta_{hkl} = \frac{f_1 \sin 2\pi(hx_1 + ky_1 + lz_1) + f_2 \sin 2\pi(hx_2 + ky_2 + lz_2)}{f_1 \cos 2\pi(hx_1 + ky_1 + lz_1) + f_2 \cos 2\pi(hx_2 + ky_2 + lz_2)}.$$

In generalizing these expressions to cover n atoms in the unit cell the above terms must be summed over all the atoms when the total intensity $I(hkl)$ (equivalent to A^2) is related to what is called a *structure factor* $F(hkl)$ in such a way that $I(hkl)$ can be derived from $|F(hkl)|^2$ where

$$|F(hkl)|^2 = \left[\sum_{r=1}^{r=n} f_r \cos 2\pi(hx_r + ky_r + lz_r)\right]^2$$

$$+ \left[\sum_{r=1}^{r=n} f_r \sin 2\pi(hx_r + ky_r + lz_r)\right]^2$$

The symbols $|\ |$ mean 'the magnitude of'.

For any proposed model of a structure, therefore, the structure factors can be calculated and compared with the experimentally determined intensities. If the two do not correspond then the model is progressively altered to give the best fit and the resulting refined structure may be accepted as possible. The process is not tedious with simple structures such as diamond but with structures even as complex as cellulose (and there are of course many much more complex structures which have been solved) recourse needs be made to a computer.

Nieduszynski and Atkins (1970) in a re-examination of *Chaeto-morpha* cellulose confirmed that the reflections in the *Valonia* diagram (Fig. 8.3) could be indexed only on an eight-chain unit cell ($a = 16.43$ Å, b(fibre axis) $= 10.33$ Å, $c = 15.70$ Å, $\beta = 96°\ 58'$) a 'super-lattice' with respect to the Meyer and Misch lattice (and quoting the complement of the angle ($84°$) used by Meyer and Misch). A model of the unit cell was devised on the basis that (1) each chain should have the same environment of nearest neighbour interactions (favoured by energy considerations) (2) the equatorial reflections can be indexed on the one-chain unit cell ($a = 5.30$ Å; $b = 10.33$ Å; $c = 6.02$ Å; $\beta = 93°$) closely similar to that of Sponsler and Dore so that each chain must have the same equatorial projection and relative rotation of the chains about the b axis is apparently excluded.

138

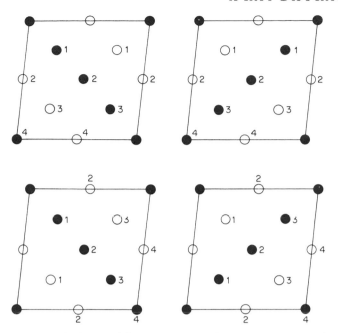

Fig. 5.14 The basal planes of four possible eight-chain unit cells of cellulose. Solid circles and open circles represent chains in antiparallel arrangement. The numbers represent the multiples of $b/4$ by which the chain is translated upwards with respect to the corner chains. There are another four such cells with negative displacements (after Nieduszynski and Atkins, 1970).

Assuming the presence of antiparallel chains, eight 8-chain unit cells could thus be constructed (Fig. 5.14). These share certain attractive features. The alternate sheets of the same polarity meet the requirements of Warwicker and Wright (1967) reached by an independent method. Again, hydrogen bonds within a sheet of one polarity would give rise to inter-chain distances different from those between chains of opposite size and hence explain $\beta \neq 90°$. Moreover, this model would give rise to only six hydroxyl stretching bands in the infrared spectrum as observed. Unfortunately, a structure factor calculation on the lines mentioned above does not give an acceptable fit with observed intensities for any of these models, and this is the outcome indeed of all such calculations for cellulose. The meridional, equatorial and even-order layer lines were well matched, as were all the reflections indexable only on the super-lattice. No reasonable fit could be found, however, for the first and third layer lines.

It is disappointing that this should be so, and a little difficult to understand. The immediate prospect of a closer approximation to an acceptable model does not appear bright. Relaxation either of the concept that each chain must lie in the same environment or of the demand that the chains may not be rotated relative to each other, gives far too wide a range of possible structures to be distinguishable with only eight equatorial X-ray reflections and only thirty-four in all. The present position, which might be expected to remain for a long time, is that the only acceptable unit cell for these algal celluloses is an eight-chain unit cell but that there is no certainty as to the relative arrangements of the chains; it is still not certain, even, whether the chains lie parallel or anti-parallel. It remains, indeed, possible that cellulose chains are units which can form a whole range of structures between which the differences are rather subtle. The X-ray spacings of algal cellulose vary appreciably from one species to another particularly in the algae (Frei and Preston, *unpublished*) (Table 8.3) just as similar variations are known among higher plants (Welland, 1954). This is perhaps why the diagrams are not richer and the search for a unique structure could be a search for a chimaera. Nevertheless, Gardner and Blackwell (1974) have had the courage to open this problem again in an important paper described briefly on p. 479.

5.6 Regenerated cellulose – cellulose II

The X-ray diagram of wet cellulose is precisely the same as that of dry cellulose which means that water cannot enter the lattice. When, however, native cellulose is treated with solutions of caustic potash, varying in strength among the celluloses but about 17% for higher plant celluloses, water does enter and the lattice is destroyed by swelling. When the alkali is washed out, the cellulose is left in a different crystallographic form, regenerated cellulose, or cellulose II with therefore, a different X-ray diagram. This is therefore the stable form at room temperature and native cellulose (cellulose I) is metastable. If cellulose II is then heated to 300°C in the absence of air, the lattice changes back to a form similar to that of cellulose I, but orthorhombic instead of sligthly mono-clinic and termed cellulose IV. It has been claimed that both cellulose II and cellulose IV occur in nature but consideration of these claims will be delayed until the appropriate plants come under review. The structure of cellulose II must be mentioned here, however, since current treatments for the extraction of cellulose from plant cell walls are often such as to convert cellulose I to cellulose II and because the structural differences between these forms are relevant to discussions on cellulose biosynthesis.

The X-ray spacings of cellulose II vary with the humidity of the specimen; water enters the lattice. The unit cell is always, however, strongly monoclinic, and the parameters of the dry cell are $a = 8.14\,\text{Å}$; $b = 10.30\,\text{Å}$; $c = 9.14\,\text{Å}$; $\beta = 62°$ (see references in Meyer, 1942). The proposed orientation of the chains in the unit cell is illustrated in Fig. 5.15. The same unit cell is achieved when cellulose is dissolved and reprecipitated.

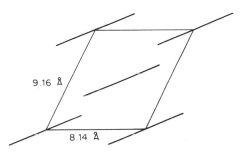

9.16 Å

8.14 Å

Fig. 5.15 Basal plane of the unit cell of cellulose II. Oblique lines represent the trace of the planes of the rings of the glucose residues.

5.7 The structure of chitin

The structures of other polysaccharides relevant to plant cell walls will be taken up later in the context in which they appear important. It is convenient to separate chitin for special treatment here partly because the plants in which it occurs will only receive peripheral mention but mostly because a comparison of this structure with that of cellulose can help in an understanding of each.

Chitin occurs in the walls of fungi as well as in invertebrates and insects and is often highly oriented. It is a poly-N-acetylglucosamine which, like cellulose, occurs in more than one crystal form. The more common form, α-chitin, has been extensively studied (Meyer and Pankow, 1935; Darmon and Rudall, 1950; Carlström, 1957; Dweltz, 1960) and is the form of most consequence here. Nevertheless another form, the β-modification, is well recognized and closely documented (Lotmar and Picken, 1950; Rudall, 1955; Dweltz, 1961).

The generally accepted structure for α-chitin is still that presented by Carlström in 1957. The unit cell is orthorhombic with $a = 18.85$ Å; $b =$ (fibre axis) 10.28 Å; $c = 4.76$ Å, (interchanging a and c from Carlström's treatment in order to conform with the usage for cellulose) a fibre repeat identical with that of cellulose. The repeating unit along the chain is therefore chitobiose (Fig. 5.16), for which a Hermans configuration is assumed (as for cellulose) more because this is the minimum energy form rather than because the X-ray data demand it. As with cellulose, $0k0$ reflections are absent for k-odd so that again the b axis is a twofold screw axis. Additionally, however, $00l$ reflections are also absent

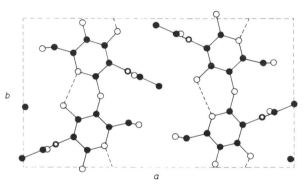

Fig. 5.16 The ab plane of the unit cell of α-chitin
● = carbon; ○ = oxygen; ◐ = nitrogen

when l is odd and this is highly significant. It means that the a axis is also a twofold screw axis. This has the important consequence that, since the chains are polar and therefore have direction, then within one unit cell the same number of chains must run in each direction. Since the unit cell has only two chains (for exactly the same reason as for cellulose) then one must lie in one direction and the other lie upside down. An antiparallel arrangement here is therefore unavoidable. In the Carlström scheme, parallel chains are bonded to each other by hydrogen bonds while antiparallel chains seem to be linked by hydrophobic bonds between neighbouring $-CH_3$ groups. Using Carlström's data for the atomic co-ordinates, Marchessault and Salko have computed theoretical intensities for all reflections except meridional reflections (for which the

142

experimental determinations are unreliable) and have found an exceptionally good match with experimentally determined values for all strong reflections and most weaker reflections. In a more recent refinement of structure, Parker and Haleem (*private communication*) have achieved an even better fit of observed and calculated intensities by rotating the chains through about 7° about the *b* axis and in other minor ways, while at the same time achieving a structure in which all the hydrogen bonds which could be made are made. They find reason to believe that, as for cellulose, the unit cell represents a statistical average of innumerable 'real' unit cells differing among each other slightly; and that at least some of the chains of one kind are hydrogen bonded to those of the other.

With regard to the significance of antiparallel chains, Rudall, (1962) has called attention to the significance of β-chitin ($a = 4.7\,\text{Å}$; $b = 10.3\,\text{Å}$; $c = 10.5\,\text{Å}$). This polymorph has only one chain per unit cell and therefore all chains must be parallel. Since α-chitin is derivable from β-chitin by swelling, Rudall presents the apparently undeniable argument that during the swelling the parallel chains fold back on themselves to give the α structure. During the process there is a 50 per cent contraction in length as would be expected, and, of course, an approximate doubling of the unit cell in the *c* direction for which the evidence is also direct. If it is accepted that this occurs by folding, then the chains must fold in the diagonal plane.

Precisely what this means for cellulose is not clear. If the structure of α-chitin is analogous to that of cellulose then the chains in cellulose must be antiparallel. If the analogy is, on the contrary, between β-chitin and cellulose (with a two chain unit cell rather than a one chain cell on account of relative chain translation) then the chains in cellulose must be parallel. It may be significant that introduction of a fold in a chain represents a higher energy state which can be stable only if it helps to achieve a high degree of crystallinity or a crystal of lower free energy. Transformation from parallel to antiparallel via chain folding should be irreversible as is, indeed the transformation of β- to α-chitin. The question of the possibility of chain folding in cellulose will be taken up later (p. 181); for the moment we may take note of the fact that the transformation from cellulose I to cellulose II is also irreversible.

5.8 Crystallite size

Attention has so far been confined to those features of the X-ray diagram which are informative with respect to the internal structure of crystals. There are other features, however, which give at least an indication of the size of the crystals. One of these is a feature already mentioned, namely that the reflections recorded on the diagram are not the narrow lines or small spots expected from large crystals (and found with inorganic crystals) but rather broad domains or arcs. The radial breadth of an arc gives a guide to the size of the crystal concerned by a study of what is called *X-ray line broadening*. The other is connected with the circumstance that reflections closer to the primary beam (θ small) correspond to larger spacings; if the crystals are of the order of 10s or 100s of Ås broad, then information may be found in reflections in the appropriate position close to the primary beam. This is called *low angle scattering* as against the *high angle scattering* dealt with already. Each of these methods have been applied to the cellulose of cell walls and we shall now consider them in turn.

5.8.1 *Line-broadening*

The radial breadth of the reflections in an X-ray diagram is due to two factors, lattice imperfections (including small crystal size, lattice strain faults and mistakes) and instrumental conditions governed by the experimental arrangement necessary for recording the diagram. It may be determined from the profile produced in a densitometer when a fine pencil of light is scanned across the arc and the intensity of the transmitted light recorded at each setting of the spot within the arc. Instrumental broadening must be subtracted from the observed broadening and ways of achieving this have been derived by Jones (1938) and Stokes (1948). The crystal reflections, moreover, are superposed on a background which decreases from a relatively high level of intensity near the primary beam to a minimum at $2\theta = 90°$. This again must be subtracted from the crystal reflection in some way, and the method of Hermans and Weidinger (1948 and 1949) may be used. We are then left with a profile corresponding to imperfections in the specimen itself. The relationship between crystallite size and broadening is then as established by Scherrer (1918).

$$\beta = \frac{k\lambda}{Nd \cos \theta}$$

where β is the angular width of the arc (viewed from the specimen) at half the maximum intensity of the profile, k is a constant, N is the number of reflecting planes, d is the spacing and θ is the Bragg angle. Since $Nd = t$, the thickness of the crystal in a direction normal to the reflecting planes,

$$t = \frac{k\lambda}{\beta \cos \theta}$$

k may be taken as unity (though a value of 0.89 is sometimes used).

This approach was used very early by Hengstenberg and Mark (1928) with ramie cellulose. They concluded that the crystallite, as it must be called on account of its small size, is about 50 Å wide and at least 600 Å long. A number of workers since then have reached similar figures for higher plant celluloses. Even visual comparison between the X-ray diagram of ramie cellulose (Fig. 5.11) and *Valonia* (Fig. 8.3) makes it reasonable to suppose that the crystallites of *Valonia* cellulose are much broader than 50 Å, because the equatorial arcs are so narrow. This situation has been examined quantitatively by Nieduszynski and Preston (1970) using the closely similar cellulose of the alga *Chaetomorpha melagonium*. A summary of their results is presented in Table 5.2 for the three equatorial reflections giving measures, therefore, of crystallite thickness in three directions normal to crystallite length. The dimensions for ramie and cotton are in harmony with those

TABLE 5.2

Transverse dimensions of crystallites in a variety of celluloses (Å)

Source	Equatorial X-ray reflection investigated		
	6.1	5.4	3.9
Chaetomorpha melagonium — alga	114	169	153
Cionia intestinalis — animal	34	84	76
Acetobacter xylinum — bacterium	70	84	73
Sea Island Cotton — higher plant seed hair	49	62	61
Ramie, Chinese green — higher plant sclerenchyma cells	—	—	59

145

recorded by earlier workers but the crystallites of *Chaetomorpha* cellulose are clearly much broader. The figure given for *Chaeto-morpha* has been found by Caulfield (1971) for *Valonia*. It is a very material figure which will be taken up again later after the electron microscopic appearance of the various celluloses has been described.

Nieduszynski and Preston (1970) took regard also, not only of the angular width β of the arcs, but the shape of the profile as a guide to the statistical spread of crystallite size in any one cellulose. They came to the conclusion that a distribution of size centring on about 170 Å for *Chaetomorpha* with a standard deviation of 50 Å would explain the profile. In principle this would mean that the chance of finding in this cellulose a crystallite 50 Å thick is minute though positive.

It is vital to notice that the figures presented in Table 5.2 are *minimal* values for no account has been taken of the superposed broadening effects of for instance, lattice distortions. Differentiation of Bragg's equation (p. 127) leads directly to the expression

$$d\theta \;=\; \tan\theta \; dt/t$$

replacing the spacing d by t for obvious reasons. An increase of the spacing by dt therefore causes diffracted intensity at $(\theta \pm d\theta)$ and therefore increases the width of the arc. Accurate assessments of crystallite size can be achieved only if broadening of this kind may be subtracted from the observed broadening. The difficulties involved are, however, formidable and, with cellulose, prohibitive. Fortunately, as will be seen later, (p. 127) the *minimal* values of Table 5.2 are still highly significant.

5.8.2 *Low-angle scattering*

Low-angle scattering from an inhomogeneous system such as a cell wall may be of two kinds, either *diffuse* — as a streak running out from the primary beam (Fig. 5.17) or discrete. Long spacings derived from discrete reflections *via* the Bragg law may be regarded as the mean distance between neighbouring crystallites but these are seldom observed in cellulose. With diffuse scattering, the angular width over which scattering is observed is related inversely to the size of the inhomogeneities in the electron density distribution within the specimen. Diagrams such as that of Fig. 5.17

Fig. 5.17 Small angle X-ray diagram of water swollen *Chaetomorpha* cellulose, microfibrils parallel to longer edge of page (photo by I.A. Nieduszynski).

therefore reveal immediately that inhomogeneities are more frequent in a direction normal to chain length than they are parallel to it; that if the scatter is a reflection of particle size then the particles are long and thin. As with line broadening, the sizes of the inhomogeneities are calculable from the profile of the scatter determined on a densitometer but the mathematics is too involved to be discussed here. Derivations will be found in Alexander (1970).

Unfortunately, diffuse low-angle scattering never leads to an unequivocal structural determination on the X-ray data alone. The two chief reasons for this are (1) that it is impossible to distinguish between discrete particles in space and a complementary system of micropores in a solid (2) it is often difficult to estimate how far the scattering curve is affected by interparticle interference. Since other evidence (Chapter 7) proves directly the presence of long thin particles in cellulose it is reasonable to assume that the scattering reflects the particle size, and on this basis Heyn (1953, 1958, 1966) and others have published scattering curves for higher plant celluloses from which they have calculated that the crystallites are long and with a mean transverse diameter of about 40 Å. More recently Nieduszynski (1969) and Nieduszynski and Preston (1970), using the same method, have concluded that for *Chaetomorpha* cellulose the transverse diameter of the crystallites is about 170 Å, in harmony with the figure given by high angle line-broadening. Both methods suggest that the celluloses of plants like *Chaetomorpha* (and *Cladophora* and *Valonia* and a few other algae to be discussed later) differ from that of other plants including higher plants in their much broader crystallites.

CHAPTER 6

Structure determination – electron microscopy

6.1 Historical development

The methods of investigation developed in the last two chapters are informative concerning details down to 0.1 μm (100 nm) and from about 10 nm downwards. They leave the range between 10 nm and 100 nm, however, open only to indirect attack. This gap is now filled by the method of electron microscopy developed as a working tool during the 1940's. It is dealt with here in a separate chapter from light microscopy because emphasis will need to be placed on specimen preparation techniques rather than on the theory of the microscope and its design which will not normally be the concern of biologists. The theory will nevertheless be dealt with in brief in terms of the general theory of the microscope already discussed in Chapter 4.

Formally, electron microscopy found its beginning in the development by de Broglie in 1924 of theories on the wave nature of electrons and the first steps in the derivation of the geometrical optics of electron beams in 1926 by Busch. There had, before then, been a somewhat haphazard use of solenoids for the deflection or concentration of electron beams, by Wiechert in 1899 and Ryan in 1903 for example, and Fleming had already obtained magnified images by this means in 1900. These early findings could not, however, be exploited until the necessary body of theory had developed.

de Broglie showed that an electron of velocity v and mass m is associated with a wave of velocity λ given by

$$\lambda = h/mv$$

148

where h is Planck's constant. A beam of electrons can therefore be considered as a beam of radiation of the equivalent wavelength, and Busch indeed considered a magnetic coil as an optical lens and derived the lens equations for it. He did not use his coils as imaging elements, but appears to have provided the stimulus for the study of the axially symmetrical magnetic fields produced by excited solenoids as lens elements for electron beams. This began in about 1929 and led to the first studies of magnetic lenses published by Knoll and Ruska in 1931 and experimental electron microscopy could be said to have begun. It is always difficult to date exactly the beginning of a new technique and it needs here to be noted that, judging from publications by Gabor, the whole idea had been very much in the air around Berlin for three or four years prior to 1931.

The first technical instruments were, however, built in 1931-32 by Knoll and Ruska (who have priority and used magnetic lenses based on the lens developed during 1926-7 by Gabor) and by Brüche and Johannsen (with electrostatic lenses). These were, however, emission microscopes giving only an image of the source. The first transmission microscope came in 1933, one by Knoll and Ruska (magnification 12 000x) and a simpler one by Marton (2000x) and the first biological observation was made in this instrument in 1934. The prospect of obtaining micrographs at a resolution higher than that of the light microscope stimulated from then on intense efforst to increase the resolving power. By 1938 Borries and Ruska had designed a laboratory instrument with a resolving power of 10 nm, initiating manufacture by Siemens, a practical microscope was being developed in Canada by Prebus and Hillier; and Martin, Whelpton and Barnum had commenced the development of a British microscope. During the war years Le Poole, who introduced limited area electron diffraction, was developing another version in Holland which became available shortly after the cessation of hostilities. By 1944 Ardenne had described a microscope providing resolution down to 1.2 nm and a year later Hilliers reduced this to 1 nm. Subsequent improvements have taken the resolution down to 2—3 Å though this is not by any means usually reached with biological specimens.

6.2 The Instrument

It is not proposed to deal here with any particular instrument but rather to discuss in broad outline those features which affect the user. The electrons are accelerated in a beam by emission from a white hot filament (bent in a hairpin bend, the source being at the bend which may be pointed) and acceleration toward a plate at a high (positive) electric potential relative to the filament, pierced with a hole through which the electrons pass. If the potential difference across the path is V then:

$$\tfrac{1}{2}mv^2 = eV$$

where e is the charge of the electron. The corresponding wavelength is then:

$$\lambda = \sqrt{\frac{150}{V}}\,\text{Å}$$

if V is in volts. At $V = 100\,\text{kV}$, $\lambda = 0.04\,\text{Å}$. The resolving power is then

$$r = \frac{(0.61)(0.04)}{\sin\alpha}\,\text{Å}$$

(p. 74) since *in vacuo* $n = 1$. The apertures of electron lenses are small, for design and other reasons, so that $\sin \alpha$ is never even remotely close to unity. This is no matter, however, since the wavelength is so small. If $\alpha = 1/100$ radians (when $\sin \alpha \doteq \alpha$) the theoretical resolving power is still 2.44 Å.

The realizable resolving power depends upon lens aberrations which cannot be as nearly removed as are the corresponding aberrations in the optical microscope. The main aberrations concerned are *chromatic aberrations, astigmatism,* and *spherical aberration.* These will be dealt with briefly in turn. They have the effect of transferring energy from the centre of the Airey disc (p. 74) to the rings without necessarily changing their diameter. The above criteria for resolving power then no longer hold.

The focal length of an electron lens depends both on the current (I) passing through the coil and the wavelength (or electron energy). If either of these varies except within very narrow limits the resolving power falls drastically. The effects of wavelength change are, by analogy with similar effects in visible light, called chromatic aberration. It can be shown that the stability requirements are roughly

$$\delta V/V < 0.5 d^2/\lambda C_e$$

$$\delta I/I < 0.25 d^2/\lambda C_e$$

where C_e is a constant less than unity, − the chromatic constant − expressing the relationship between the focal length and V. These equations show that it is necessary, if high resolutions (low values of d) are to be achieved, that both V and I must be controlled within fine limits. For a value $d = 5$ Å, for instance, neither can be allowed to vary more than a few parts per million. This has, however, been achieved and chromatic aberration need not be considered as a factor reducing resolving power, provided that the instrument is functioning correctly. Similarly, methods are available for removing astigmatism, caused by assymmetry either in the geometry or the magnetic quality of the pole pieces of the lenses, and correction of astigmatism is a routine matter for electron microscopists. There remains spherical aberration. Unlike glass lenses, electron lenses show only positive aberration i.e. marginal rays are refracted more than are axial rays; it depends partly upon lens design and partly upon lens aperture. For high resolution the aperture needs to be as large as possible as far as diffraction effects go but it needs to be as small as possible in terms of spherical aberration. The choice of aperture size is therefore a compromise. The optimum value is obviously reached when, from a point object, the circle of confusion in the image due to aberration is identical with that due to diffraction. This leads to the relations:

$$d = 0.43 C_s^{\frac{1}{4}} \lambda^{\frac{3}{4}}$$

$$\alpha = 1.4 (\lambda/C_s)^{\frac{1}{4}}$$

where α is again the angular value of the aperture and C_s is the spherical aberration constant, $C_s \alpha^3$ giving the radial error in object space of a ray leaving the object at an angle α to the object. The aperture is usually set in the back focal plane of the objective and these equations set its diameter at a few tens of μm. The equations give the optimum *theoretical* resolution.

Apart from the objective lens there are at least three other electron lenses in the optical path. The condenser lens (there are two in modern microscopes) focuses the beam on the specimen precisely as in light microscopy; a projector lens receives the image from the objective lens and enlarges it onto a screen. There is in addition a lens, the intermediate lens, between the objective and the projector and the current through all of them is of course accurately controlled. Although positioning of the object, and focussing, is achieved at the screen, working results are recorded photographically. It is mechanically a marked advantage that since the apertures are small the focal depth is large; focussing is still critical but the photographic film does not need to be at the level of the screen. On the other hand, long depth of focus carries with it the misfortune that all levels in a specimen are focussed at the same time so that differential focussing, so marked a facility with light microscopes, is not possible. This difficulty can to some extent be alleviated by taking, and projecting, stereo pairs of micrographs.

The mean free path of electrons in air is, of course, small. In order to allow electrons to pass from the gun to the screen, of the order of a metre distant, the microscope tube must be sealed and constantly evacuated to about 10^{-6} torr. The specimen is introduced through an air lock but can nevertheless be then manipulated very delicately. Apart from a visualization of the object on the screen and film, the corresponding diffraction pattern can also be recorded. As with optical microscopy, the diffraction pattern — a Fourier transform of the object — occurs in the back focal plane of the objective. This can be focussed by the intermediate lens on to the image plane of the projector lens and thence, enlarged, on to the screen or film. It is necessary to remove the objective aperture since this screens off most of the diffraction pattern. A field aperture inserted before the intermediate lens (sometimes adjustable) limits the field of view; apertures as small as $25\,\mu m$ in diameter can be used corresponding to $1\,\mu m$ in the object plane of the objective so that areas of this diameter can be explored. The object can then be visualized though at low resolution. By analogy with the corresponding X-ray case (Chapter 5) and remembering that the effective camera length D is large so that θ is small, spacings can be calculated from

$$\lambda = 2d\theta$$

$$r/D = 2\theta$$

where the symbols have the meanings given in Chapter 5. D is obtained by calibration from the diffraction rings of, say,

152

aluminium, vacuum deposited on the specimen.

It is to be noticed that when the microscope is to be used in the diffraction mode it is necessary to reduce the tube current (and therefore the electron beam intensity) below the level at which an image is visible on the screen. This is because, in the visual mode, the beam intensity is sufficient to destroy the internal structure of the specimen under bombardment even though it does not change the external morphology. The procedure is, therefore, to use low beam current, switch to the diffraction mode, scan the specimen for diffraction patterns and photograph immediately any diffraction pattern observed. Even at low tube currents the pattern persists for only a few seconds. Having recorded the pattern the microscope can then be switched to visual observation and the area giving the pattern recorded photographically if it is of the type required.

6.3 The specimen

The specimen is therefore inserted into an electron beam in high vacuum and needs to be in a stable condition in the microscope and transparent to electrons. It therefore needs to be prepared in special ways dealt with below. For the moment we concentrate on specimen features necessary for the recording of high quality electron micrographs.

Firstly, the specimen and the necessary supporting film must be thin because electrons must pass freely through in order to produce an image. Beyond this, however, electrons must not be absorbed because this not only transfers energy to the specimen, producing undesirable side effects, but induces changes in velocity of the transmitted electron leading to chromatic aberrations with a loss of resolving power. This reduces the allowable thickness of the specimen to a few 10's of nms. Losses of this kind are, however, inevitable (though by scattering rather than absorption) and the loss is the greater the thicker the section. It is, indeed, a useful rule of thumb that the *practical* resolving power is one-tenth of the specimen thickness. With sections 50 nm thick it is not to be anticipated that the resolving power can be much better than 5 nm.

This raises the matter of *contrast*. It will be clear that contrast cannot be obtained, as it is in the light microscope, by absorption and it has been understood from the first that contrast is achieved only by electron scattering. Scattering arises from several effects which, again by analogy with the X-ray case, we can think of here as both elastic and inelastic. Either form of scattering from an atom is proportional to the atomic number and, for an extended object, to the number of atoms present. Contrast is then achieved if the scattered electrons are not allowed to reach the screen. This is the main function of the objective aperture; to screen off these unwanted electrons and achieve contrast thereby.

It will be apparent that contrast with most biological specimens is of necessity low, since all parts of the specimen consist in the main of C, H, O and N. Contrast is therefore achieved by the deposition, upon or within the specimen, of heavy metals or their compounds in such a way as to produce regional separations, corresponding to regional differences in the specimen, of these highly scattering materials. Corresponding methods will be dealt with below, but first we need to consider the methods available for preparing the thin specimens required for treatment.

6.3.1 Specimen preparation

The methods available for preparation of cell walls are:

(a) manual dissection in special cases
(b) surface replication
(c) ultra-thin sectioning

(a) *Manual dissection*
This cannot normally be used with higher plants since the cells are too small and the walls too thin. It is, however, very effective with the larger cells of some algae (Fig. 8.12) and is the only way in which walls may be examined with no chemical treatment. With cells as large as *Valonia* vesicles, a small square may be cut from the wall or with filamentous cells such as *Cladophora* or *Nitella* a part of the cell can be removed as a cylinder which is then slit open and laid flat. The piece of wall is then placed in water and lamellae stripped off either with fine needles or fine forceps with needle points. The fresh wall is often too gelatinous to allow the separation of intact undistorted lamellae in this way but with practice some success is normally achieved. Brief immersion in 5% formaldehyde helps enormously. This reagent appears to cross link within the

wall lamellae more rapidly than it does between them so that after a short time the lamellae are firm enough to be withdrawn smoothly. Immersion for too long a period, however, prevents entirely the separation of the lamellae; one half hour's immersion is often sufficient though the optimum time interval depends, of course, on the species.

The older method whereby a piece of wall is fixed between two pieces of sellotape which are then separated, tearing the wall into two leaves which can further be fined down by repetition of the process, can be tried if the above method fails. It is, however, somewhat to be deprecated since the lamella has to be removed from the sellotape by means of a solvent and this is inclined to leave insoluble particles of the adhesive on the lamella surface. This can, however, be obviated by fixing a strip of wall over a hole in a piece of adhesive paper, stripping off the bulk of the wall by repeated application of sellotape and mounting the lamella finally left over the hole; this has never come into contact with an adhesive.

With walls which are thin enough — such as the primary walls of coleoptiles (Fig. 12.1) — it is sufficient to macerate the tissue by some suitable means and cut individual cells in two.

In either case the wall needs then to be treated to induce contrast (see below).

(b) *Surface replication*
This is a method whereby the surface of a wall is coated with a layer of carbon sufficiently thin as to be transparent to electrons — 20–30 nm (scc below). This is achieved by sputtering carbon from a carbon arc in high vacuum. The specimen is then removed and only the replica examined. It needs, of course, to be treated to induce contrast (see below). Tissues may be treated differently according to whether they are firm or not in producing straight replicas. It is convenient here, however, to examine also a method which is coming to be exceptionally valuable in examining walls, particularly in relation to the underlying cytoplasmic organelles. These methods will be dealt with in turn.

Extensive firm surfaces
The method is one devised by R.K. White in my laboratory. A block of wood is a suitable example. The block is prepared showing

an area of the face of interest about 3 cm x 1.5 cm (a smaller area suffices for other tissues). This may be prepared either by splitting or by planing on a sledge microtome. Debris and small slivers of wood are then removed by pressing firmly on sellotape and removing the tape. The surface is finally treated with acetone and ether (in this order) and allowed to dry. One side of a sheet of cellulose acetate is then flooded with acetone and pressed firmly onto the surface to be replicated *with no shear*. After a 10—15 min drying time the acetate sheet is peeled off firmly and fixed on a microscope slide, replica uppermost, by sellotape. This is a *negative* replica in which protrusions in the specimen are represented by depressions; it may at this stage be examined under a light microscope for flaws. In order to produce a positive replica, the replica itself is coated with carbon in high vacuum by sputtering from a carbon arc. The film thickness should again be 20—30 nm. (A simple method of achieving an acceptable thickness is to place near the replica a small square of opal glass carrying a drop of silicone oil; the glass darkens as C deposition proceeds and the degree of darkening required may be checked against standards produced by trial and error on an earlier specimen). After the C film is deposited, contrast is added by oblique metal shadowing (see below), the degree of shadowing required being again adjusted by trial and error.

The replica is then scribed (not cut) into 2 mm^2 pieces by means of a razor blade, cut from the sellotape frame and immersed, *replica side uppermost* in acetone. When the acetate film is dissolved, the squares of carbon (which float) are picked up on electron microscope grids, dried and washed in acetone.

Smaller delicate surfaces
This method is useful with filamentous algae and pieces of higher plant tissue which are not seriously affected by drying, as used by Dr. E. Frei of this department (Fig. 9.8). The tissue, face of interest uppermost, is fixed centrally on a glass slide. It is often sufficient to allow the tissue to dry on the slide, but if this does not induce adhesion albumin fixative may be used. When dry, the tissue is framed with sellotape strips and replicated with carbon and shadowed as above. The sellotape is then pulled away, the replica is scored as above, and the sides and ends of the tissue cut away, leaving a central area of tissue still attached to the glass and bearing

a replica. The tissue may then be removed in some suitable solution, but this normally causes swelling which disrupts the replicas. An effective precaution is as follows. A few drops of concentrated HCl are pipetted round the edges of the tissue and in contact with it. The slide is covered with a petri dish and left for 12 h. The tissue is then floated into 40 per cent H_2SO_4 replica side uppermost and left for 1h. The replicas normally float off and may be transferred to 70 per cent H_2SO_4 through a series of graded concentrations. This normally removes the tissue from the replicas quantitatively.

Freeze-etching

This method is of much more general significance and it applies to the contents of cells as well as their surfaces. It was devised primarily as a method of cytological fixation which avoids the hazards of chemical fixation and embedding; an operational advantage of the method is that it readily allows surface views of walls and of the membranes of all organelles. The method may be said to have been initiated by Steere (1957), based on a method devised by Hall (1950) for the examination of crystals of silver halide. The use of the method for critical examination begins however with the meticulously careful work of Moor (1964). It depends upon the freezing of cells to very low temperature in such a way that ice-crystal and other damage is avoided. The effect of freezing rate (to $-150°C$ from room temperature) on yeast cells is shown, after Moor (1964) in Fig. 6.1. When the rate of freezing is either low or high (say $1000°s^{-1}$) virtually all cells are recovered alive on thawing. At medium rates, between say $0.1°C s^{-1}$ and $10°C s^{-1}$, virtually all of them are killed in the absence of an anti-freeze agent; electron micrographs then show large ice crystals in the cells, disrupting all the membranous structures present. Survival at high and low rates may be explained as follows.

All liquids have temperatures on their cooling curves at which (a) crystals begin to form (b) solidification continues, but in the form of a glass rather than crystals. In cytoplasm these temperatures are $-5°C$ and roughly $-120°C$. There is therefore a range of about $115°C$ over which ice crystals can form, and if 0.5 s, say, is taken over this range of cooling, ice damage will occur. At very much slower rates of cooling, however, the water surrounding the cells, freezing before the cell water, causes exosmosis of water,

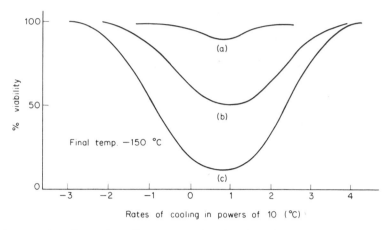

Fig. 6.1 **The effect of cooling rate on the viability of yeast cells (a) in 20 per cent glycerol (c) in water only (b) an intermediate glycerol concentration.**

concentrating the cell contents and lowering their freezing point. When only 'bound' water is left in the cell no large ice crystals can form. Cells frozen in this way, however, appear shrunken, although their structures remain intact. They are quite inadequate for electron microscope examination.

When the freezing rate approaches $1000°C\,s^{-1}$ so that the dangerous range is passed over in 0.1s or less, the molecules of water lose energy too quickly to allow movement into a crystalline lattice. No ice crystals are formed and no shrinkage occurs. The water is said to be vitrified and it is this condition which is aimed at. The specimen, not more than $1mm^3$ in size on account of the high cooling rate desired, is therefore cooled rapidly to the temperature of liquid N_2. Liquid N_2 cannot, however, be used directly because its boiling point is too close to the liquefying point; the specimen becomes surrounded by a layer of gaseous nitrogen which is not a good heat conductor. The specimen is therefore cooled in liquid Freon 22 (a fluorohydrocarbon) which has a boiling point sufficiently high as to prohibit the formation of a gas layer, itself cooled in liquid N_2.

In many cells and tissues the rate of cooling is then still not sufficiently rapid to prevent ice crystal damage. Recourse is then made to an anti-freeze agent, normally glycerine. Treatment of cells with 20 per cent glycerol causes the freezing point to become lowered to $-30°C$ and the vitrification point to be raised to about

−55°C (probably by hydrogen bonding to water). This reduces the 'dangerous' range in the cooling curve to 25°C which is easily passed through in 0.1s (Fig. 6.1). Providing therefore that 20 per cent glycerol itself causes no artefacts, the structure of the frozen cell can be regarded as held in the natural state. Some cells can be cultured in 20 per cent glycerol and confidence can therefore be placed in the structures then revealed. With other cells it is rarely that serious recognizable artefacts are observed and these can be checked against the occasional good preparations achieved in the absence of glycerol.

Otherwise, extremely fast rates of cooling can be achieved by plunging directly into liquid helium II at −272°C with no anti-freeze. This has superfluid properties; it wets the specimen completely giving cooling rates between $1000°C\,s^{-1}$ and $10\,000°C\,s^{-1}$. This is, however, far too costly, apart from its inconvenience, to be usable except by the occasional fortunate.

The specimen is cooled as a drop placed upon a thin copper or cardboard disc about 3 mm diameter. This is then secured, under high vacuum, on a copper finger cooled by liquid nitrogen and its temperature is monitored at, say, −140°C. The ice drop is then 'chipped' by means of a knife also cooled by liquid nitrogen. This fractures cells in the specimen at levels dependent upon their levels with respect to the knife (Fig. 6.2). The knife is then held over the specimen (whose temperature is raised to −100°C) so that ice may sublime to the knife, leaving the chipped surface 'etched' in that the cells lie slightly proud. The surface, still in high vacuum, is then coated with carbon to produce a replica and obliquely shadowed with metal to produce contrast. The temperature is then allowed to rise and the replica is released from the specimen by a method dependent upon the nature of the cells. It carries replicas of the cells at various levels and reveals structures which should be closely like those of living cells.

Observations made in this laboratory, chiefly by Dr. D. G. Robinson whose electron micrographs are used in subsequent pages, show that glycerol is better avoided with algae, fungi and bacteria, whose cells tend to plasmolyse in glycerol. Successful freeze-etching can usually be achieved without glycerol.

(c) *Ultra-thin sectioning*

This is the standard method in electron microscopy and details of

159

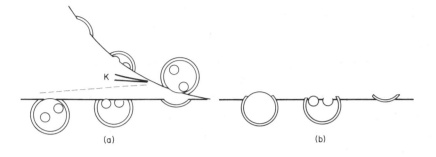

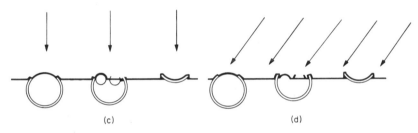

Fig. 6.2 Freeze etching
(a) frozen cells are fractured by the cooled knife K
(b) the ice is sublimed, leaving cells 'proud' of the surface and revealing surfaces determined in part by the levels of the cells in the frozen drop relative to the fractured surface
(c) the surface is replicated by carbon
(d) the replica is metal shadowed and the temperature is raised to free the replica.

various procedures can be found in the standard texts. The difficulty with thick cell walls — a difficulty encountered, of course, whether the wall itself or the cytoplasmic structure is to be revealed — is the hindrance to diffusion which they offer. In order to cut sections as thin as a few 10's of nms the material must be embedded through and through with the monomer of a hard plastic which can then be polymerized inside the wall. It is therefore an advantage to use a plastic of which the monomer, or its solution, is of low viscosity.

In preparing cells and tissues for sectioning, the first step is to apply a fixative which will halt the activity of the cell in such a way as to retain the structure in a state as close as possible to that

160

in the living conditions. There is no guarantee that this can be done and demonstrably all the fixatives in use today change, in some way or other, the structure of, for instance, cytoplasmic membranes. This is, however, of little importance for the cell wall itself and fixation may be omitted unless the relation between the wall and the cytoplasm is to be examined. It is always useful, however, to check by phase or interference microscopy that fixation has not produced gross changes.

Following fixation, if used, the material then needs to be dehydrated through a series of alcohol solutions of increasing strength up to 100 per cent, taken through a solvent of the monomer, to the monomer alone and finally to monomer plus catalyst. Polymerization is then achieved at 60°C. The solvent for the monomer is usually a lipid solvent so that if lipids are of interest it is useful to fix in a fixative which preserves lipids, namely osmium tetroxide or glutaraldehyde and to avoid potassium permanganate, which does not.

A useful procedure for thick cell walls is the following, developed by Dr. Brenda J. Nelmes of this laboratory for the cambium and sapwood of apple trees. Short lengths of twig are immersed in 3 per cent glutaraldehyde in 0.02M phosphate buffer (pH 7.0) at 4°C and the bark removed while in the fixative. After two hours immersion, thin slices are removed, containing both cambium and xylem, to fresh fixative and held in fixative for 2 days at 4°C. The slices are then washed at least three times in buffer and transferred to 1 per cent osmium tetroxide, again in 0.02M phosphate buffer. They are then transferred in stages to absolute alcohol, after which the temperature is allowed to rise to room temperature. Embedding is then achieved in TAAB embedding resin, carried out over a period of 5 days and the slices are finally transferred to fresh resin in flat polythene troughs. Polymerization is achieved by holding at 60°C for 2 days. Sections are cut at approximately 90 nm and post-stained in uranyl acetate and lead citrate.

Sections of this kind are not, however, very informative as far as cell walls are concerned and are normally used in special contexts such as will appear in ensuing pages in this book.

(d) *Contrast*
The recipe recounted immediately above anticipates one particular need, the need for contrast in an electron microscope object, a

need already expressed also in the matter of replication. Contrast may be achieved by the application of materials, usually metals, which have high electron scattering power, in order to enhance some regions in a specimen on account either of their geometry or their chemical constitution. In principle two methods are available.

Shadow-casting

When the surface of a specimen is not smooth, contrast may be enhanced by sputtering metal obliquely onto the specimen in high vacuum. The sputtering metal usually takes the form of a few mm of the appropriate fine wire looped over a tungsten filament subsequently heated white hot by an electric current. The sputtered metal should possess a structure finer than that to be revealed in the specimen, and alloys of Pd/Au or mixtures of Pt and C are useful. The principle of the method is illustrated in Fig. 6.2d. It may be used with dissected lamellae, with replicas, and with sections from which some at least of the embedding material has been dissolved away. The method has the disadvantage that it enhances the size of detail by the unknown (though always small) amount of metal deposited. It has uniquely, however, the advantages that (a) the thickness as well as the width of a detail of structure can be determined (Preston, 1951) (b) *surface* detail only is revealed with no confusion due to overlapping images.

Staining

There are no specific stains for electron microscopy such as are available to light microscopists. The closest approach to such a stain is given by osmium tetroxide which becomes attached at double bonds; broadly speaking osmium delineates therefore the distribution of lipids, and removal of lipids removes the staining. Proteins do not stain with osmium.

Two 'stains' which have come into general use are uranyl acetate and lead citrate. These give contrast without, however, being specific for any particular molecular species. With cell walls they form useful 'negative' stains in that the heavy metal is adsorbed by the non-crystalline components so that the non-adsorbing crystalline components stand out as unstained. Deposition of silver by impregnation with $AgNO_3$ followed by a reducing agent gives the same effect (Bailey and Preston, 1969; Chou and Preston, 1971). These methods are not, however, without the danger of yielding gross misrepresentations (p. 176).

162

CHAPTER 7

General principles of wall architecture

7.1 Introduction

Application of the principles and techniques described in the last five chapters — and others which will be taken up in context as appropriate — have led to a rather thorough general understanding of wall structure. There are still many details of importance to be worked out but sufficient is known with certainty to allow general statements which apply to all plant cell walls. This chapter deals with these generalities.

All plant cell walls are multiphase systems which for simplicity we may nevertheless regard as two-phase. All walls known contain a crystalline polysaccharide of specific composition embedded in a matrix consisting usually of the wide variety of polysaccharide and of other compounds already dealt with in Chapter 3. As already mentioned, the crystalline polysaccharide of most plants is cellulose and it is with these plants we shall be mostly concerned here. Many of the features to be described are, however, shared also by those plants in which the crystalline polysaccharide is a β-1,3-linked xylan or a β-1,4-linked mannan. The details of these plants will be found in Chapter 9.

It was recognized in the last century by Carl von Nägeli, working in Zürich, that the 'crystals' of cellulose are thin relative to the wavelength of visible light and are very much longer than broad (p. 144). He based his belief that the material is crystalline on its appearance under the polarizing microscope and that the crystals are separate long thin entities on the swelling behaviour. This was a triumph of pure reasoning achieved long before techniques were available for its proof, an outstanding example of the heights which a meticulously careful observer can reach. Once the crystallinity had been confirmed by X-ray diffraction analysis and the crystallites shown undoubtedly to be long and thin, the 'micelles' of Nägeli, which he considered as separate and separable entities, became the basis of all work during the next twenty years. They did not indeed disappear without trace even with the onset of electron microscopy.

In the intervening period, the nature of the putative 'micelle' became modified. The X-ray indication that these units are long and thin did not prove their existence as separable bodies; indeed all that it could imply is that in cellulose there is a crystalline lattice which is uniform only over limited volumes. Consideration of the physical and chemical properties of cellulose and its derivatives led, indeed, to two variations on the micellar theme which are not, however, formally distinguishable in structural terms. One of these, the 'fringed micelle', imagined the chain molecules within the classical micelle as fraying out from the sides and ends and passing over into other micelles (Fig. 7.1a). This was favoured by Kratky and Frey-Wyssling. The other, a continuous deformed structure, envisaged a lattice in extent broader than the micelle but interrupted by lattice distortions so that again only in small volumes were the molecular chains precisely spaced (Fig. 7.1b). This was the view held by Meyer.

It was, of course, recognized that the micelles are in any case embedded in and attached to the matrix of intermicellar polysaccharides and it is instructive to recall that even in 1948 there seemed no reason to believe that micelles could be visualized in untreated walls. Attention had been repeatedly drawn to the possible existence of supermolecular units of structure ranging from fibrils some 0.4 µm in diameter to spheres or ellipsoids up to 1.5 µm in diameter (see references in Frey-Wyssling 1948, 1959; Preston 1952), both on the grounds of the physical properties of walls and in the light of microscopic observation, yet it could never be established that these bodies, when seen, were more than mechanically induced artefacts. In 1948, however, first Preston, Nicolai, Reed and Millard (1948) then Frey-Wyssling, Mühlethaler and Wyckoff (1948) revealed by electron microscopy the existence in cell walls of rodlets 20–30 nm wide and of indefinite length. One of the illustrations presented by Preston *et al.* is reproduced in Fig. 7.2 as historically the first published demonstration of the

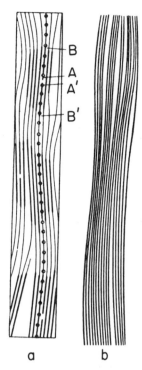

a   b

Fig. 7.1 Earlier concepts of micellar structure; lines represent molecular chains.
(a) The 'fringed' micelle of Kratky and Frey-Wyssling
(b) The continuous, deformed structure proposed by K.H. Meyer.

existence of what are now called microfibrils. This represents two
lamellae of the wall of the green alga *Valonia ventricosa* and should
be compared with Fig. 8.13 illustrating similar material visualized
following the advances since then both in instrumentation and
technique. The close parallel with the run of the two sets of
crystallites already known to exist (Preston and Astbury, 1937)
(see p. 193 *et seq*), made it immediately acceptable that these fibrillar bodies were cellulosic. This was eventually confirmed by the
use of electron diffraction methods in the electron microscope
(Preston and Ripley, 1954). The microfibrils consist, in part at
least, of bundles of cellulose chains and, accordingly, their birefringence is positive.

The advantage of *Valonia* in this regard (and this was found later
to apply also to the Cladophorales) is that lamellae can be stripped
from the wall, with no treatment save a brief submersion in a weak

165

SCIENCE

Fig. 7.2 **One of the first three electron micrographs of a plant cell wall. This is a pseudo-replica of lamellae from the wall of** *Valonia ventricosa*, **metal-shadowed. This is historically the first electron micrograph showing cellulose microfibrils; the microfibrils average about 25 nm wide.**

formaldehyde solution, thin enough to provide excellent electron microscope preparations. At that time thin sectioning techniques had scarcely begun to be developed so that for higher plants, with smaller cells, recourse had to be made to disintegration either in a blender or by ultrasonics. Nevertheless, confirmation was soon forthcoming that microfibrils of around the size already seen were present in the secondary wall of cotton hairs and the primary wall of flax fibres (Frey-Wyssling and Mühlethaler, 1949 and 1950) again with the orientations anticipated by the earlier work. About the same time the extracellular cellulose of *Acetobacter xylinum* and the dispersed celluloses of seed mucilages were shown to be microfibrillar with microfibrils of about the same size,

166

i.e. 25−30 nm diameter. It began therefore to seem likely that all celluloses might prove to be composed of microfibrils uniformly about 30 nm diameter. The first dissent concerning the size of microfibrils in higher plants seems to have come from Hodge and Wardrop (1950) and Ränby and Ribi (1950). They claimed that with the cellulose of conifer tracheids the microfibrils are much narrower, of the order of 10 nm wide. This has turned out to be correct over a wide range of higher plant celluloses examined since that time, and of most lower plants. Exceptionally, the c. 25 nm microfibrils of *Valonia*, the Cladophorales, *Glaucocystis* and *Oocystis*, together with a few other plants also algae, have been confirmed. The width is not invariate and the original observation (Preston 1951) that in *Valonia* it can vary from about 15 nm to a somewhat ill-defined upper limit of about 35 nm with an average in the order of 25 nm, still stands. In considering developments since these early days it is important to keep these plants, in the present context, separate.

Events have shown that even with higher plants the uniformity of microfibril width is not as complete as might have been expected. Betrabet and Rollins (1970) appear to have demonstrated conclusively that in a particular cotton, *Gossypium herbaceum* the microfibrils measure 23.0 x 8.5 nm in section as against 12.0 x 3.0 nm and 11.0 x 2.5 nm in Indian and American *G. hirsutum* respectively. In another context, Hunsley and Burnett (1968) have found significant differences in microfibril size (cellulose or chitin) between apical and distal regions of some fungi, and Frei and Preston (1964a) have recorded, for marine algae in which β-1,3-xylan is the skeletal polysaccharide, microfibril widths of 10 nm in inner wall lamella but widths two or three times as great in outer lamellae of the same wall. Hunsley and Burnett (1968) tentatively suggest that these differences might be explained by secondary intussusception of molecular chains but are commendably cautious. As already observed by Colvin (1972), an understanding of these sophisticated variations may come only when the mechanism of microfibril synthesis is understood.

Experience has shown that the c. 10 nm microfibrils of higher plants and most lower plants can readily be observed even in walls with no pre-treatment though with greater clarity after the mild treatments required partially to remove the non-cellulosic substances which surround them. The original claim that the

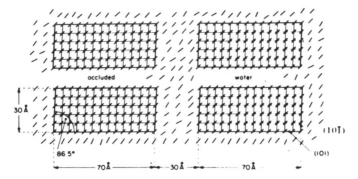

Fig. 7.3 The elementary fibril concept as visualized by Frey-Wyssling (1954); by courtesy of Prof. A. Frey-Wyssling; fibrils viewed in transverse section.

microfibrils were wider than this was an error understandable in the first preparations of such difficult material at a time when reliable methods had not yet been worked out. It led, however, to what seems to have been an unfortunate development in terminology which has persisted, in a different form, to this day. The proposal was made (Frey-Wyssling, 1954) that the c. 25 nm microfibril (as it had become) of these plants is composed of *elementary fibrils* each c. 10 nm wide (Fig. 7.3). This concept can be traced back to a schematic presentation of wall architecture made at a much earlier date by Frey-Wyssling (1936), in which the term microfibril was introduced as a unit lying in size between the micelle (roughly the present elementary fibril) and the optically visible fibril (Fig. 7.4). This can be classed only as a truly remarkable forecast, based as it was upon a culling of evidence from widely disparate fields. It does not lose in elegance by being an attempt diagrammatically to unite various lines of evidence and to present a formalized view rather than the absolute truth.

It had been recognized already that the microfibrils are flattish ribbons (Preston 1951), and that the flatter faces lie parallel to the surface of the walls in which they occur. Since, moreover, in the walls as a whole the (101) planes of cellulose also lie more or less parallel to wall surface (in those walls in which the appropriate observations can be made, see p. 185) then these planes must lie within the microfibril parallel to its flatter faces. The microfibrils were therefore considered by Frey-Wyssling to consist each of four subunits — elementary fibrils — as depicted in Fig. 7.3. Each fibril was supposed, correctly and for reasons which will appear below,

168

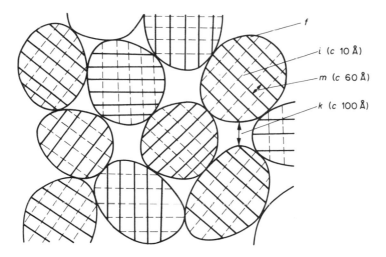

Fig. 7.4 The micellar structure of plant cell walls as visualized by Frey-Wyssling (1936); by courtesy of Prof. A. Frey-Wyssling.

to contain a central crystallite or micelle surrounded by a matrix of molecular chains in paracrystalline array. The error lay in assuming that a fixed small number of elementary fibrils — the true microfibrils — lie disposed in such a close and regular manner as to constitute a formally acceptable superunit.

Experimental investigation of the internal architecture of these microfibrils began with the work of Ränby and Ribi (1950), Ribi (1950) and Ränby (1951). They found that when cellulose is boiled for 1h in 2.5N H_2SO_4 and washed, the wash liquids suddenly became turbid at the point when the pH had reached 3—3.5. The suspended particles appeared in the electron microscope as rodlets some 10—50 nm long and about 7.5 nm wide, i.e. about as wide as the microfibrils from which they came. The rodlets gave the diagram of cellulose I and were hydrophobic. The authors concluded that the hydrolysis had cut the microfibrils into short pieces which formally represent the micelles of Nägeli and the microfibrils could therefore be regarded as micelle strings. The later demonstration that in wood cellulose (though not in *Valonia* cellulose) the microfibrils take up Ag at points irregularly disposed along them, indicating probably locations with unusually high contamination with hemicelluloses (Preston, 1962), is clearly in conformity with this idea and is incorporated in the model of

169

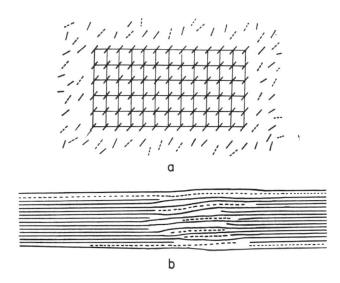

a

b

Fig. 7.5 Diagrammatic representation of the structure of a cellulose microfibril (a) in transverse, (b) in longitudinal view. Thick solid lines represent glucan chains, broken lines chains of other sugars. The central area in a represents the crystallite. As drawn, the microfibril is some 10 nm wide and 5 nm thick; the rectangles in the crystalline regions in a are the basal planes of the Sponsler and Dore unit cell, so that planes spaced 6.1 Å apart lie parallel to the longer edge of the crystallite. Note that as drawn, the crystallite encompasses 60 one-chain unit cells but only three or four whole eight-chain unit cells.

Fig. 7.5. More recently, Stöckman (1972) has argued on thermo-dynamic grounds that such an alternation of ordered and disordered regions must occur, though arising secondarily. This is a point to consider when we take up the matter of biosynthesis.

Cellulose crystallites in higher plants are about 5−7 nm wide so that it is possible to assume that the rodlets of Ränby are the crystallites. The question remains of the association of these crystallites within the microfibril. It had been shown long before this that the removal from a cell wall of pectic substances, hemi-celluloses and lignin had no effect upon the X-ray diagram of the wall (Astbury, Preston and Norman, 1935; Preston and Allsopp, 1939) (so long as the extraction procedures were not so severe as materially to degrade the cellulose) except for some sharpening of the arcs such as would indicate a small increase in crystallite size (Preston, Wardrop and Nicolai, 1948). Expressed in modern terms, this means that the *extractable* non-cellulosic wall compounds appeared to surround (and in some way be weakly bonded to) the microfibrils. The morphological identification of cellulose made possible by electron microscopy, coupled with the ability which had developed to identify the sugars of wall hydrolysates by chromatographic methods, has led to some changes in this simple view. When plant cell walls are extracted by the method of Jermyn and Isherwood (1956) (p. 40) the so-called α-cellulose remaining as the final residue, exclusively microfibrillar, still contains some hemicelluloses judging by the non-glucose sugars in hydrolysates (Cronshaw, Myers and Preston, 1958) (p. 42). These range from 15 per cent in the xylem of some higher plants to 50 per cent in the red alga *Rhodymenia* and in some green algae. These hemicelluloses cannot be removed without degradation of the microfibrils and are therefore integral with them. When, however, the microfibrils are subsequently treated by the method of Rånby, the rodlets into which the microfibrils then disperse still give the X-ray diagram of cellulose but now hydrolyse to give glucose only. This was demonstrated by Dennis and Preston (1961) for *Rhodymenia*, *Ulva* and *Laminaria* among the algae and for elm and pine wood among higher plants. Since the X-ray diagram is that of cellulose I, as found also by Rånby, the molecular chains of cellulose cannot have been widely separated at any stage of the extraction and it seems highly likely that the rodlets represent, in only slightly modified form, an original central core of the microfibril from which the tightly-bound hemicellulose have been stripped. This lends strong support to the view which had by then already been expressed (Rånby 1958; Preston 1959), since that time supported by Ellefsen, Kringstad and Tönnesen (1963), Colvin (1972) and others, and illustrated in Fig. 7.5. Each microfibril at any point along its length contains one central crystalline core the dimensions of which vary with, but are somewhat smaller than, microfibril width. This is surrounded by a paracrystalline 'cortex' of molecular chains which lie parallel to microfibril length but are otherwise not stacked in crystalline array. Taking the microfibril size as 10 nm x 5 nm in higher plants and the crystal as 5 nm x 3 nm, this implies the presence of about 48 chains ($\equiv$ six, 8-chain unit cells) in the crystalline core and about 100 in the 'cortex'. This implies further that not all chains in the cortex are hemicellulose chains and these must be mixed with cellulose chains; and it is presumably this admixture which prevents the chains from regular packing. It

171

might be that the proportion of non-cellulosic chains increases from the crystallite surface outwards and that the 'surface' of the microfibril comes at the plane in which the proportion is almost 100 per cent. The corresponding figures for a microfibril of the *Valonia* type (20 nm x 12 nm say; crystallite size 17 nm x 11 nm (p. 145)) are about 500 chains in the crystallite and about 160 in the cortex; these latter are, however, almost all glucose chains since the microfibrils yield only glucose on hydrolysis (p. 42). The number of chains in the cortex is an overestimate, since the microfibril sizes stated are those measured after shadow-casting, over-emphasizing the width of the microfibril by the amount of metal coating. A correction for microfibril size can be roughly calculated. In a typical shadow-casting exercise, about 3 mm of wire 11/1000th of an inch thick is evaporated on to the specimen some 10 cm away at an angle of 1:3. This will coat a side of the microfibril, considered flat and vertical, to a thickness of about 1.5 nm and the horizontal, flat surface 0.5 nm. A better estimate for microfibril size in higher plants is therefore 8.5 nm x 4.5 nm and of the *Valonia* type 18.5 nm x 11.5 nm; the number of chains in the cortex would then be 70 and 80 respectively.

7.2 Microfibril size and structure in fresh tissue

The X-ray derivation of the structure of cellulose, and therefore the demonstration of crystallinity, applies only to air dried material; and the demonstration of the existence of microfibrils and the determinations of microfibril size stems almost exclusively from specimens bone-dry in high vacuum. Yet all celluloses, and all other skeletal polysaccharides, are synthesized in an aqueous medium and are operative in the plant mostly fully wetted and never dried, and it is in this condition that structure has meaning as far as this book is in the main concerned. Considerable importance must therefore be attached to an extrapolation of the dry structure to the previously undried condition.

To begin with, fresh, undried cell walls are birefringent and the changes in birefringence on drying can be understood by the withdrawal of water without any need to invoke a change in crystallinity. This means that the molecular chains already lie at least parallel to each other in fresh walls. Whether they also lie in a crystalline lattice, that is whether they are spaced regularly side by side, was first examined by Berkeley and Kerr (1946). They showed that fresh, undried cotton hairs yielded an X-ray diagram showing only a water halo, with little sign of reflections due to crystalline cellulose. The same bundle of hairs when dried, how-ever, yielded a good cellulose diagram as did also the same bundle on rewetting. With flax fibres, on the other hand, fresh material yielded good X-ray diagrams. It seemed, therefore, for a time possible that cotton cellulose, at least, is laid down in the fresh wall in only paracrystalline array. In observations of this kind, however, the technicalities of the methods involved are of para-mount importance and Berkeley and Kerr had not taken these into account. Preston, Wardrop and Nicolai (1948) soon showed that never-dried *Chaetomorpha* walls give identically the same diagram as do dried walls except for the overlying water halo. They pointed out, moreover, that with cotton hairs the water-filled lumen is so large in volume with respect to the wall that the water halo may completely mask the cellulose diagram. Moreover, on drying, the hair collapses and, on wetting, the lumen does not fill out again and the water halo is then less pronounced. The situation in dried, re-wetted cotton is not therefore the same as that in fresh material, and the cellulose diagram could be masked only in the latter. The conclusions drawn from cotton are therefore suspect. More recently Caulfield (1971) has shown that the crystallite size in never-dried *Valonia* microfibrils is the same as that in dried material and is so large that a substantial part at least of the cellulose in never-dried *Valonia* must be fully crystalline.

It is clear that observations of this kind are valid only when the wall is so thick that a characteristic cellulose diagram could appear against the masking water halo, and it seems on these grounds reasonable to conclude that the evidence against crystallinity in any fresh cellulose is weak. If it does eventually turn out that in some fresh walls the cellulose chains are not arranged in a detect-able lattice, little water can lie between them; for on drying with well-separated chains the lattice would be the cellulose II lattice, not the cellulose I lattice observed. The 'sheets' of chains con-sidered by Warwicker and Wright (1967) to be the reactive unit rather than single chains would then be especially significant.

There can be, on the other hand, absolutely no doubt about the existence of *microfibrils* in fresh cell walls; they may readily be seen on carbon replicas of fresh material.

7.3 Elementary microfibrils

There remains the claims frequently made of the existence of units of structure narrower than the microfibril which together make up the microfibril and replace it as the true structural entity. It is necessary here to distinguish the view that microfibrils may be broken down to subunits from that which claims that the subunits pre-exist. The latter follows the chimaera that units must have subunits and stem from the erroneous belief that the 20 nm wide units first seen with higher plants in early electron micrographs were the true microfibril (p. 168). In following the argument that microfibrils contain, as subunits, bodies 3.5 nm wide, variously called *elementary fibrils* or *protofibrils*, it is essential to keep apart the microfibrils common to higher plants and the 25 or so nm microfibrils of some algae.

The observational basis for the existence of the elementary fibril as a real, intrinsic pre-existing subdivision of the microfibril is entirely electron microscopic. It began in isolated observations of celluloses which had been treated by ultrasonics, oxidation or hydrolysis, all disruptive techniques likely to produce fission artefacts. The first definitive statement on material which had not been treated in these ways referred to the extracellular cellulose of *Acetobacter xylinum* (Frey-Wyssling and Mühlethaler, 1962) and to the primary walls of root meristems in *Allium cepa* (Mühlethaler, 1960). Negative staining in phosphotungstic acid revealed unstained threads, some 3.5 nm diameter against the black background of the stain. Similar results with other celluloses have, since that time, been quoted in support. This type of observation must clearly be regarded with some scepticism (Colvin 1963), if only because the heavy metal stain is likely to penetrate the (partly hemicellulose) cortex of the microfibrils and reveal therefore only the impenetrable central crystallite. Similarly, all observations made on material which had been prepared by treatment with ultrasonics, even when subsequently metal-shadowed, must be regarded with suspicion. As shown by Sullivan (1968), the diameter of the 'protofibrils' then depends upon the time the material has been subjected to the ultrasonic field. Even the demonstration

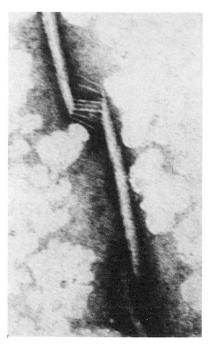

Fig. 7.6 Electron micrograph of 'kinked' microfibril of *Valonia* cellulose, negatively stained (Frey-Wyssling *et al.*, 1966); by courtesy of Prof. A. Frey-Wyssling. Magnification 100 000 X. The strands linking the two parts of the microfibril are about 3.5 nm wide.

(Frey-Wyssling and Mühlethaler, 1962) that 3.5 nm fibrils can be observed both after negative staining and positive staining is not convincing; for, as the authors admit, measurement of positively stained microfibrils is difficult and subjective. A further line of evidence, that frayed ends of microfibrils or sharply bent microfibrils (Frey-Wyssling, Mühlethaler and Muggli, 1966; Manley, 1971) show strands some of which may be 3.5 nm in diameter (Fig. 7.6) is evidence only that microfibrils can be split, as one might expect of an assembly of parallel polysaccharide chains.

With these criticisms in mind, search for proof of the existence of elementary fibrils has more recently turned in another direction. Clearly, if a set of parallel, closely packed microfibrils is negatively stained, a measure of fibril width can be obtained as the distance between neighbouring dark bands (Fig. 7.9). The appropriate measurements have been made on wall sections in higher plants (Heyn, 1966 and 1969) and again a width of 3.5 nm has been derived. Unfortunately, apart from dangers arising from phase-contrast effects consequent upon even a slight departure from exact focus, a simple geometric circumstance has been overlooked in all such interpretations. This effect has been explored in detail (Preston 1971) and will be dealt with only briefly here.

Let us assume that the microfibrils are 10 nm diameter and that the microfibrils lie parallel to the surface of the section. An ultra-thin section of a wall will be about 50 nm thick and will encompass five microfibrils. These must be arranged in some relationship with each other; let us assume in the first instance that they are hexagonally packed. The metal stain will penetrate the interstices (Fig. 7.7) so that for *any one layer* of microfibrils the centre-to-centre distance will correctly measure microfibril width (cp. optical simulation, Fig. 7.8a). When two or more superposed layers are

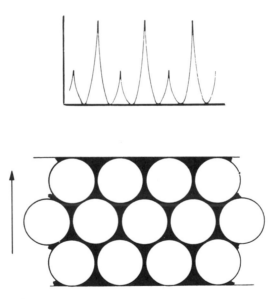

Fig. 7.7 End view of parallel microfibrils (white circles) with the interstices filled with an absorbing substance (negatively stained) (below) with the intensity distribution (above) when viewed along the arrow.

imaged, it has to be remembered that the long focal depth of electron lenses condenses the image of a three-dimensional object into one two-dimensional plane (the film) so that with, say, three layers of microfibrils (Fig. 7.7) the intensity profile is as shown and the centre-to-centre distance is 5 nm. The corresponding optical

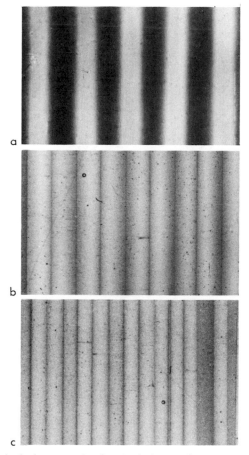

Fig. 7.8 (a) Optical photograph of a single layer of perspex rods 1 cm diameter in an absorbing medium. The centre-to-centre distance measures the rod diameter.
(b) Photograph of perspex rods, 1 cm diameter, arranged as in Fig. 7.7.
(c) Photograph of an array of perspex rods, 1 cm diameter as in Fig. 7.7 but tilted 12.5° about an axis parallel to the length of the rods.

simulation is presented in Fig. 7.8b. This sets the highest limit to the measurable repeat distance. If the section is not precisely set at 90° to the line of sight, the image will be more confused, but if the error happens to be 12.5° a regular repeat will reappear at (100/3 =) 3.3 nm (optical simulation Fig. 7.8c). In a slightly uneven section, therefore, distances of 5 nm, 3.3 nm and smaller will be measurable, but the full 10 nm will never appear. This is, of course, a consequence of hexagonal packing and it may be claimed that

Fig. 7.9 Electron micrograph of a layer of packed microfibrils in the wall of *Glaucocystis nostochinearum*, negatively stained, with a single overlying microfibril. Magnification 171 000 X, shadowed Pt/Au. (Preparation by Dr. D. Robinson). Note that the overlying microfibril appears broader than do the close-packed ones below.

the packing in real cell walls in unlikely to be so precise. Irregular packing, however, can only reduce both the average size and the minimum size observable. The reality of the effect is shown in the electron micrograph of Fig. 7.9. Here, one single microfibril overlies a section of parallel microfibrils, all of the same diameter. The overlying microfibril is *c*. 20 nm diameter. Most repeat distances measurable on the layer are 10 nm (20/2) and some are *c*. 7 nm (20/3) or even smaller. Reverting to the situation with 10 nm microfibrils, the observation of repeat distances at 5 nm or 3.5 nm appear therefore to demand a conclusion which is the reverse of that made by proponents of the elementary fibril; observation of 3.5 nm repeats in negatively stained sections *proves that the units composing the section are 10 nm wide.* Cox and Juniper (1973) have recently repeated this error in claiming to establish the 3.5 nm elementary microfibril. They unfortunately compound the error by failing to realise the significance of X-ray line broadening and low angle scattering discussed below.

There seems therefore no reason to doubt but that the elementary fibril is an illusion based in part upon a confusion between microfibril size and crystallite size and in part upon a misinterpretation of electron microscope images.

On the whole, therefore, the evidence for the existence of elementary fibrils vanishes. It has been occasionally claimed that celluloses exist in which the elementary microfibril is even smaller in width than 3.5 nm, in the range 1–2 nm. One such claim is made for the tunica of the Chrysomonad *Ochromonas malhamensis* by Kramer (1970). Such claims are based either on negatively-stained arrays of microfibrils, with the consequent errors due to overlapping of images mentioned above, or to a misunderstanding of the nature of negative staining. Correct measurement of the individual fibrils presented by Kramer (*loc. cit.*) gives a width of about 10 nm, not 1–2 nm.

The evidence *against* the elementary fibril concept, on the other hand, is weighty. Firstly, ends of the cellulose microfibrils of the algae *Glaucocystis* have been observed (Schnepf, 1965) to taper smoothly down to diameters much less than 3.5 nm. Secondly, the crystallite diameter, measured as already mentioned (p. 145) by accepted reliable physical methods in such a way as to give *minimal* values are such as to allow only one central crystallite per microfibril. The broader crystallites of *Chaetomorpha* and *Valonia* celluloses are the most revealing in this context. They measure on average some 17.0 nm x 11.4 nm (Table 5.2) with a standard deviation of 5.0. Sizes of the same order have been found for *Valonia* cellulose by Caulfield (1971), using methods similar to those of Nieduszynski and Preston (1970) and by Bourret, Chanzy and Lazaro (1972). The latter authors used line broadening in electron micrographs, but supported their findings by dark-field electron microscopy on which the image is recorded by use of the coherent reflections from the $10\bar{1}$, 101 and 002 planes of the Meyer and Misch unit cell. In none of these more recent investigations was any trace found of a 3.5 nm subunit in undisturbed microfibrils. Bourret *et al.* (1972) record occasional ends of *Valonia* microfibrils splayed into *c.* 4 nm threads. They conclude that these represent mechanical splitting which has no relevance for the unsplit microfibril. This leaves open no possibility of the existence of 3.5 nm

179

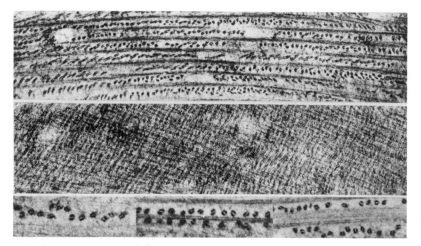

Fig. 7.10 Electron micrographs of wall layers in *Oocystis* positively stained (1.75 h) in uracryl acetate and lead citrate (Preston, White and Robinson, 1972).
(a) Section apparently almost paradermal
(b) Section (a) tilted 45° about an axis parallel to shorter edge of page
(c) Sample of electron micrographs of microfibrils in section.

entities for the vast bulk of the microfibrils. The standard deviation of 5.0 nm does, however, allow the chance that in a very few microfibrils the crystallite(s) may be as small as 3.5 nm. If these exist, they might by chance have been seen in the electron microscope but should not mislead into generalizations which are not valid. Observation of microfibrils in section leaves no room for doubt as to the unitary nature of the microfibril. In wall sections, microfibrils are hardly ever seen in accurate cross-section for sections cannot be cut deliberately with the necessary accuracy. With a tilting stage, however, a section far away from the transverse (Fig. 7.10a) may be tilted to reveal the required section (Fig. 7.10b) (Preston, White and Robinson, 1972). Examples of electron micrographs of microfibril sections prepared in this way (all *positively* stained) are given in Fig. 7.10c. The microfibrils are about 17 nm wide, with a clear centre of about 10 nm wide, presumably the central crystallite) and tend to be ovoid in shape as anticipated. Finally, the vast majority of microfibrils measured in a large number of laboratories range around 10 nm for higher plants and 25 nm for some algae. These are sometimes identified without any chemical pre-treatment of the wall, as with *Valonia, Cladophora, Chaetomorpha, Glaucocystis, Oocystis,* and sometimes after fairly drastic chemical treatment to remove non-cellulosic compounds, as with most higher plants. Since these treatments remove hemicelluloses some of which remain, however, as integral parts of the microfibrils, one is inclined to the view that the bonding within the microfibril is different at least in degree from that outside it. On this ground alone, the microfibril would be the valid unit. Only occasional observations of internal subunits have been recorded which are free from one or other of the errors discussed above. The earlier interpretation of the structure of a microfibril remains the valid one and is as depicted in Fig. 7.5.

There is, however, one concept of the elementary microfibrils which remains of value not because it maintains the elementary fibril concept, since this must clearly be abandoned, but because it raises the validity of an assumption implicit in the structure of cellulose discussed so far – the assumption that the molecular chains are continuously and uniformly straight within the crystal-lite. The challenging papers on this topic by Manley (1964, 1971), have arguably been of greater service to the study of cellulose structure than almost any others in the past ten years, in opening up the possibility that the molecular chains of cellulose may be folded just as the chains of many polymers are now known to be. The conceptual basis of this possibility lies in the known features of crystalline plates formed by some cellulose derivatives (Manley, 1964; Bittiger and Huseman, 1964; 1966). Briefly, the molecular chains in these plates stand normal to the surface and are so very much longer than the plate is thick that they must be folded repeatedly back on themselves; and the same situation arises with some other polymers. Only with extreme caution, however, may this be taken as significant for natural cellulose itself, since the derivatives are precipitated from solution, whereas cellulose microfibrils are not (p. 439), and since the added side groups modify considerably the physical properties of the chain. The case for cellulose must rest on observations made upon this substance alone. Apart from the standard observations concerning 3.5 nm filaments discussed above, these comprise (a) photographs showing sharp kinks in microfibrils (b) photographs of 'beaded' microfibrils. Neither of these categories of observation is conclusive, as shown by Colvin (1972), and even the author of the idea writes about (b) 'the phenomenon must be regarded as suggestive but not conclusive' (Manley, 1971).

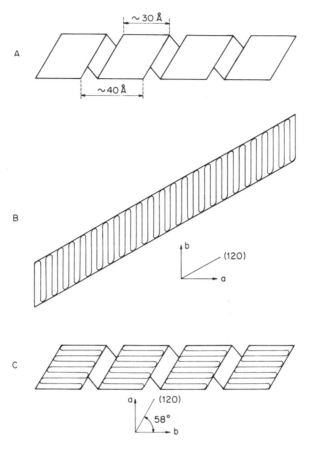

Fig. 7.11 **The folded chain concept (Manley, 1971); by courtesy of Dr. R. St. J. Manley.**

As first formulated (Manley, 1964), a long molecular chain of cellulose was supposed to be folded repeatedly upon itself to form a ribbon 35–40 Å wide (Fig. 7.11B) which in turn was coiled helically to form a hollow elementary fibril, several of which together formed the *Valonia* microfibril. This gave completely the wrong density for *Valonia* cellulose and did not accommodate the known uniplanar orientation in which the 101 (6.1 Å) planes lie parallel within the microfibril. In the newer modification (Manley, 1971) the helix is in effect flattened to occlude the central cavity (Fig. 7.11A,C). In either case, of course, the model meets the requirements, except at the folds, that the chains be parallel to microfibril length. In the newer model the two inner faces of the

182

elementary fibril (*protofibril* in Manley's terminology), and the neighbouring faces of two opposed fibrils, are supposed to be so tightly in contact that continuity of structure is achieved and the total effect is of one single large central crystallite. In that case it is surely simpler to apply Occam's razor and take it that the proto-fibrils have no separate existence.

This apart, Manley's original publication led to a flurry of investigation aimed at testing the validity of the folded chain concept. The resulting evidence tells strongly against the idea. Muggli and his co-workers (1969) have shown that when ramie fibres (in which the microfibrils lie almost longitudinally) are cut transversely in thin sections the degree of polymerization falls from 3900 to 1600. This is to be expected if the constituent chains are straight but would not happen if the chains were folded in the way Manley suggests. This is clear compelling evidence for extended chain conformation. Otherwise, Mark and his associates have shown that a model of a protofibril with helically wound chains and circular cross-section would show an elastic modulus three orders of magnitude lower than that observed (Mark, Kaloni, Tang and Gillis, 1969). This helical model differs only in detail from the later flattened model of Manley and the results will be equally valid for this newer model or, indeed, in general, for any folded model. Finally, if it is correct as Gardner & Blackwell (1974) claim (p. 140) that in cellulose the chains do not lie anti-parallel, there remains no possibility of a fold.

There are thus four strong lines of evidence against any model of the microfibril with folded chains and it seems fair to conclude that although the concept has provoked a good deal of very useful investigation the idea of chain folding in native cellulose must once and for all be abandoned.

7.4 Disposition of the wall substances

Accepting it, then, that the microfibrils are the unit of structure and that the molecular chains within them are straight, there remains to be considered the dispositions of the microfibrils with respect to each other and to the matrix. We shall here deal with these matters only in the most general way in preparation for the details to be discussed later with individual cell types.

It has always been taken for granted with any one cellulose – and indeed for cellulose generally – that there is (except in

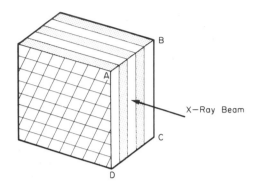

Fig. 7.12 (a) A block of wall with 5 lamellae cut with section ABCD at right angles to one of two sets of microfibrils in the wall (seen on face of block as two sets of crossed lines). Alternate layers carry alternate microfibril directions. The microfibrils lying parallel to the beam are called *transverse* in the text and the others *longitudinal*.

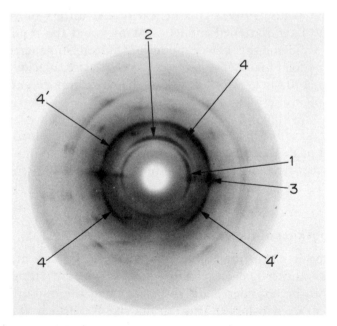

Fig. 7.12 (b) X-ray photograph of (a) with beam normal to ABCD (arrow in (a)).

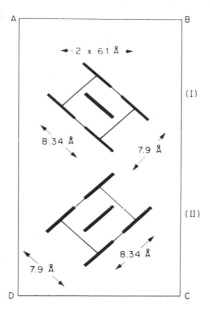

Fig. 7.12 (c) **Disposition of the basal planes of the two unit cells necessary to explain (b).**

size) only one type of microfibril, as illustrated in Fig. 7.5. One piece of evidence that this may not precisely be so comes from those cell walls showing uniplanar orientation. Fig. 7.12b shows the X-ray diagram of such a wall when the beam lies parallel to wall surface and along a direction parallel to one of two sets of microfibrils constituting the wall and therefore approximately at right angles to the other (Fig. 7.12a). Though a complex one, the diagram can be easily interpreted. The arc at 5.4 Å (arrowed, 1), corresponding to $10\bar{1}$ planes lying parallel to the b axis, can come only from the transversely oriented microfibrils; while the arc at 6.1 Å (arrowed 2) (101 planes) must come from both sets of microfibrils. This is the evidence for uniplanar orientation. Note, however, the complex disposition of the 3.9 Å arcs. The section labelled 3 can have come only from the longitudinal microfibrils and labelled 4 and 4′ from the transverse. Now the two arcs labelled 4 come from microfibrils within which the disposition of the unit cell is as in Fig. 7.12c (i). Equally, the arcs 4′ cannot be derived from this unit cell; instead they must be derived from microfibrils with a unit cell as in Fig. 7.12c (ii). In this sense there must be two types of microfibril; as far as the basal plane is concerned, it is as

185

though one kind is upside down with reference to the other. In other words, whether or not the molecular chains in a single crystallite lie antiparallel, the molecular chains of the whole wall might do so.

All walls which are extensive enough to allow this type of observation reveal precisely this type of uniplanar orientation and it may well be that walls in general are similarly organized. In that case there are always present two kinds of microfibrils with two different unit cells or, if only one unit cell, with one upside down with reference to the other.

Most walls are laminated so that the whole cell wall consists of superposed hollow shells — co-axial cylinders in cylindrical cells, for instance, or concentric spheres in spherical cells — more or less tightly bonded together. The lamellae may be so coarse as to be visible in the light microscope or so fine as to be detectable only in the electron microscope, or both types may occur, and often do so, in the same wall. Coarse lamellation appears sometimes to be associated with an alternation of sheets of material which are cellulose-rich and cellulose-poor or cellulose-free as with the green algae *Chaetomorpha* (Nicolai and Frey-Wyssling, 1938) and *Clado-phora* (Hanic and Craigie, 1969; Frei and Preston, 1961) (see p. 205) or with rather abrupt net changes in general microfibril orientation coupled, however, also with changes in the type and proportion of non-cellulosic substances as with xylem elements (see p. 288) and some collenchyma cells (see p. 312). In the absence of quantitative data, it is therefore not possible to decide with any precision how the microfibrils are distributed within the amorphous or paracrystalline matrix in which they lie. Added to this is the difficulty that while the types of inter-chain bonding possible in the wall are known it is not yet certain which, if any, predominate. In other words, no precision can as yet be reached concerning the amount of matrix substance with which a single microfibril may be associated or about the nature of the bonding involved.

Nevertheless it seems desirable to have some sort of mental image of the texture of the wall, particularly of the fresh wall. Electron micrographs, necessarily of intensely dehydrated material, certainly on the one hand, present the wall as too condensed. On the other hand, diagrams of wall structure indicating only the general run of the microfibrils, and even that only by widely spaced lines, tend to implant the subconscious image of a loose network which is the looser the lower the cellulose content; when it is easy to slip into the misconception that the spaces in the mesh are holes as in, for instance, some past and even current researches involving water movement through cell walls.

Sufficient information is available to allow a guess, perhaps even an informed guess, at the texture of cell walls at the molecular level within the limits defined by certain assumptions and this guess is attempted here and presented diagrammatically. Attention is focussed in turn on (a) walls with a low cellulose content symbolizing primary walls (b) walls with a medium cellulose content which symbolize the secondary walls of higher plants and (c) the extreme case of some algae with high cellulose content. The assumptions which need to be made throughout are as follows:

1. In a real cell wall the microfibrils are probably oval in section (on account of surface forces), and take up specific orientations ranging from the purely random to the almost perfectly parallel. In the models only short lengths of them are drawn and these are made rectangular in section and parallel to each other.

2. The microfibrils are assumed to be uniformly distributed throughout the wall so that the separation between them is the mean separation, or 50 per cent of the matrix substance is assumed to occur in non-microfibrillar lamellae and the microfibrils assumed to be uniformly distributed among the remaining 50 per cent. These two assumptions possibly give the two extremes of mean distribution.

3. The separation between the microfibrils is assumed to be the same at right-angles to the broader faces as at right-angles to the narrower faces. This has no justification other than its simplicity.

4. The molecular chains of the matrix substances are assumed to have a DP (degree of polymerization) which is not very high but is nevertheless sufficient to encompass the thickness of the section drawn, namely greater than about 100. This seems rather safe (Chapter 5).

5. There remains the question of the degree of orientation of the matrix chains. The matrix is often written off as amorphous, but this cannot always, be the case. The 'hemicellulose'

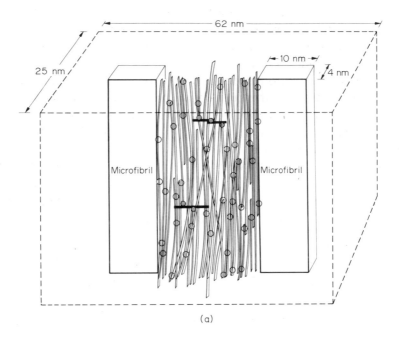

62 nm

25 nm

10 nm

4 nm

Microfibril

Microfibril

(a)

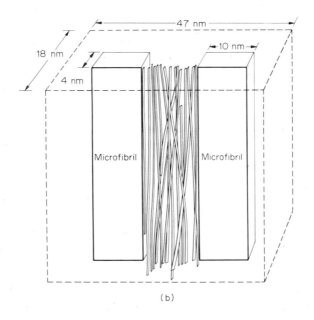

47 nm

18 nm

4 nm

10 nm

Microfibril

Microfibril

(b)

chains which form part of the cortex of the microfibrils
certainly lie parallel to microfibril length (p. 170) and there
is no reason to believe otherwise of molecular chains in the
vicinity of the microfibril surface. There is, indeed, evidence
which suggests a tendency for the molecular chains of the
matrix to lie parallel in this way. When a beam of polarized
infrared radiation impinges on a bond, the stretching frequency is excited more strongly when the bond lies parallel
to the direction of vibration than it is when the two directions
lie at right-angles; the bond shows dichroism. Marchessault
and Liang (1962) used this phenomenon to show that in the
xylem of a conifer the molecular chains of xylan tend to lie
parallel to the molecular chains of cellulose. Again, Roelofsen
and Kreger (1951) have shown conclusively that the X-ray
diagrams of some collenchyma cells show arcs characteristic
of pectin which demonstrate that the bulk of the galacturonic
acid chains lie more or less longitudinal, parallel to the cellulose microfibrils, and a similar situation has been described
for the alginic acids of brown seaweeds by Frei and Preston
(1962). Although it may be dangerous to generalize, these
observations, made upon cells which can yield them, form the
basis for the assumption to be made, namely, that in most
cells, whether or not the appropriate observation can be
made, the molecular chains of the matrix tend to adopt the
orientation of the microfibrils, though the agreement need
not be perfect.

6. For the sake of simplicity side groups and side chains are
   ignored except in so far as departures from parallel packing
   will allow for them.

Fig. 7.13 Diagrammatic and speculative view of the distribution of microfibrils
in fresh cell walls (to scale). The circumscribed dotted lines mark out the
volume of the wall 'occupied' by the microfibrils shown and their associated
hemicelluloses. Only one 'layer' of hemicelluloses is shown, and that only
between the microfibril. The whole volume outside the microfibrils should be
imagined packed with similar ribbons.
(a) A wall containing 5 per cent cellulose, 33 per cent hemicelluloses and pectin, 2 per cent protein and 60 per cent water. The vertical ribbons represent
hemicellulose chain molecules (and pectin), the black bars protein connected
at serine (o) to galactans. The circles not attached to protein represent H bonds.
(b) As above, but with one half of the hemicelluloses deposited in separate,
non-cellulosic lamellae (not shown) (protein not included).

7. Various types of bonding are possible, depending upon the relative composition of the matrix. The observation that the cation atmosphere in the wall, and particularly the balance between monovalent and divalent cations, can influence both the mechanical properties of the wall and cell growth (p. 402) points to the involvement of salt links. These are likely to be affected through —COOH groups and would implicate especially the pectic compounds. This has been visualized by many workers including Henglein (1958) who proposed a model of 'protopectin' in the wall as a union of polygalacturonic acid chains through Ca or Mg phosphate groups. On the other hand, evidence has accumulated over recent years that the bonds involved in the mechanical properties of cellulose of the cell walls of growing cells (Chap. 13) and of timber (Kauman, 1966) are hydrogen bonds. As a corollary it is to be expected that these are the bonds contributing most to the stability of the wall. The models presented are therefore in the main hydrogen bond models but it is not to be overlooked that other bonds are anticipated also, to a smaller extent; with wall proteins, covalent bonds must be involved (p. 63) and these are included where appropriate.

Taking first, therefore, a wall symbolizing a primary wall. Let the wall contain 5 per cent cellulose, 33 per cent matrix substances, 2 per cent protein and 60 per cent water (w/w) and assume that the microfibrils measure 10 nm x 4 nm in section. Then it is simple to show that each microfibril must lie centrally in a sheath of matrix plus water 105 Å wide. Each hemicellulose chain, measuring about 4 x 2 Å in section occupies a cross-sectional area about 8.5 Å x 8.5 Å. The resulting model is presented in Fig. 7.13a. Since the molecular weight of the protein is not known, it is represented here by chains only a few amino acids long with, however, the serine-galactose groups at the correct distance given by Lamport (1972) (p. 63) and in the orientation to be expected if the galactose is part of a galactan. This makes it visually clear that the wall is rather solid even when the cellulose content is low and it is not to be thought that there are huge gaps through which water and enzymes can move with complete freedom.

It makes remarkably little difference if the wall is thought to be lamellated in such a way that one half of the matrix substance is separated into non-microfibrillar lamellae (Fig. 7.13b) or if the cellulose content is increased, since the microfibrils move closer together to compensate for the fewer chains between them. The wall is, of course, equally solid and the degree of penetrability the same, whatever the cellulose content.

CHAPTER 8

Detailed structure – cellulosic algae

The walls of most of the green algae so far investigated are based upon cellulose as the skeletal polysaccharide, the microfibrils varying in orientation from completely random to the most perfect alignment found anywhere in the plant kingdom. The microfibrils range in width from about 10 nm to 25 nm according to species and it seems to be the rule that the more perfectly aligned micro-fibrils are the broader ones. On account of the regular beauty of their structure, and because the microfibrillar orientation − and therefore changes in orientation − can be quantified, the species with the broader microfibrils have been the more thoroughly investigated and these will be described first. Because *Valonia* was historically the first to be examined, these will be referred to as *Valonia*-type walls.

8.1 *Valonia*-type walls

The species to be considered include *Valonia, Siphonocladus, Dictyosphaeria Apjohnia* and the Cladophoraceae, together with two unicellular algae *Glaucocystis* and *Oocystis*, presenting a series of remarkable structures which have proved central to questions of biosynthesis. The walls are comparatively thick, finely lamellated and the wall organization is remarkably uniform over the species. In all cases the wall increases in thickness as the cell increases in size and there is no obvious structural difference between the wall of a growing cell and that of the fully mature cell; in the Clado-phorales, there is no obvious distinction between the wall at that part of a cell which is growing and another part of the same cell which is not. There is therefore no distinction here between a primary wall layer and a secondary wall layer and these terms, useful as they are with higher plants, do not apply. For this reason, and although the phenomena associated with wall growth are considered later in a separate chapter, the structure of the growing wall must be included here. For the morphology, histology and reproduction of these algae the reader is referred to Fritsch (1956) and only brief details will be given here. *Valonia* itself will be dealt with first as the archetype of the group.

8.1.1 *Valonia*

Valonia is in the main an alga of tropical and subtropical seas though a few species are found in the Mediterranean. The young plants consist of a macroscopic multinucleate vesicle anchored basally by rhizoids. The cytoplasm is parietal, containing a network of lobed chloroplasts and enclosing a large central vacuole. At a later stage, the cytoplasm and chloroplasts accumulate in masses which are roughly circular in surface view and become cut off from the main vesicle by a curved wall. These then form small lenticular cells, called watch-glass cells, and always appear at the basal end of the vesicle which, indeed, they define by growing out to form secondary rhizoids (Fig. 8.1). In *Valonia ventricosa* occasional watch-glass cells are found elsewhere in the vesicle but these do not develop and the vesicle continues to grow as effectively a single undivided vesicle to the size of a hen's egg. In other species some of the upper peripheral cells are rather large and grow out into branches repeating the structure (e.g. *V. utricularis*, Fig. 8.2). In this way some members of the Valoniaceae attain a habit strikingly similar to that of the Cladophoraceae. Reproduction by swarmers has occasionally been described but there is no record of swarmer formation by vesicles in culture. On the other hand, reproduction by the formation of aplanospores can readily be induced and the development of these bodies has been followed in some detail. When a mature vesicle is punctured by a glass rod 1.5 to 2.0 mm in diameter the cytoplasm begins to segregate into numerous roughly spherical masses around which a wall is then formed (if the rod is less than 1 mm diameter the hole is re-sealed and no aplanospore formation occurs). The resulting aplanospores may be sown on marble chips in seawater, when rhizoids develop

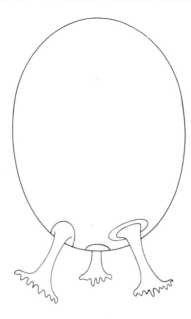

Fig. 8.1 Drawing of a vesicle of *Valonia ventricosa* with its basal rhizoids (about 2 × life size).

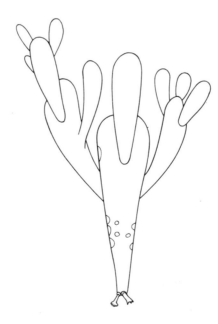

Fig. 8.2 Drawing of a vesicle of *V. utricularis* (about 3 × life size).

as above and the spore enlarges.

All species of *Valonia* show the same wall characteristics though only three — *V. ventricosa, V. macrophysa* and *V. utricularis* — have been examined in any detail. Attention here will be confined to the species about which most is known, namely *V. ventricosa*.

The wall is exceptionally rich in cellulose, to the extent of 70 per cent (Cronshaw, 1957; Cronshaw and Preston, 1958) and the cellulose yields glucose only on hydrolysis. The material soluble in hot water (21 per cent) yields on hydrolysis, galacturonic acid, galactose, glucose, arabinose and xylose. The sugars of the alkali-soluble fraction (9 per cent) have not been recorded; indeed, sugars have not been detected in this fraction.

Valonia ventricosa has figured largely in wall studies. It was used by Sponsler in the earliest days of the X-ray investigation of cellulose to demonstrate that the 101 and 10$\bar{1}$ planes lie approximately at right-angles and to derive his unit cell (p. 134); it was the first organism in which the now commonly recognized 'crossed fibrillar' wall structure was proved (Preston and Astbury, 1937); and it was the organism used in the first demonstration of cellulose microfibrils (Preston *et al.* 1948) (Fig. 7.2).

Critical examination of the wall may be said to have begun in the demonstration by Correns (1893) of two sets of striations in the wall, lying commonly at about 90° to each other but sometimes at angles as low as 60°, together with an occasional third, fainter striation, though these had also been seen some thirty years earlier by Nägeli. He concluded, correctly, that the two striations lay in different wall lamellae but was at that time unable to offer any interpretation. Striations were then, and subsequently, observed on the walls of many cell types and were variously interpreted as discontinuities, or folds, and so forth. It was not until the underlying molecular structure was known that Preston and Astbury (1937) were able to prove by X-ray methods that the striations lie exactly parallel to two sets of crystallites in the wall. Before this, Preston (1931b) had shown that, as seen in surface view under the polarizing microscope, the wall is birefringent but that the m.e.p. is constant in direction only over very small areas so that at any one setting the wall presents a mosaic of light and dark patches (see Plate III, Fig. 3, Preston, 1952); and Astbury, Marwicke and Bernal (1932) had shown, by the methods described in Chapter 4, that the birefringence is in harmony with the presence of two crossed crystal plates of cellulose.

The first major investigation by modern crystallographic methods was made by Preston and Astbury (1937). The X-ray diagram of the wall with the beam normal to wall surface (Fig. 8.3) shows two linear arrays of intense arcs at 5.4 Å ($10\bar{1}$) and 3.9 Å (002) (the arc at 6.1 Å (101) is missing since the 101 planes tend to lie parallel to wall surface), conclusive proof of the presence in the wall of two parallel arrays of cellulose crystallites with the crystal axes crossed at somewhat less than 90° though again angles as low as 60° were recorded. The two sets of crystallites lie in separate lamellae and the angle at any one point in the wall is constant throughout the

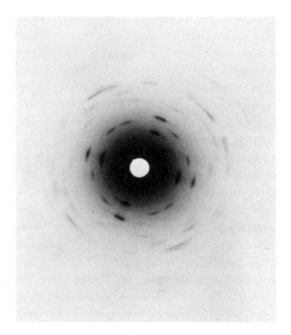

Fig. 8.3 X-ray diagram of a wall of *Valonia ventricosa*, beam normal to wall surface.

wall thickness. Since fibrils can be pulled from the wall, each lamella must be fibrillar and the whole structure was called the 'crossed fibrillar structure'. In section the wall is lamellate at the level of the light microscope, but these lamellae individually still possess a crossed fibrillar structure. The lamellae containing only one fibrillar direction must therefore be submicroscopic in thickness and this was much later confirmed by electron microscopy. Occasionally, a third orientation was detected on the X-ray diagram, although this was never detected as a third microscopically visible striation. The photomicrographs of Sisson (1941), however, do show a third striation direction though this is not mentioned by him. Subsequently the structure was confirmed visually in the electron microscope (Fig. 7.2) as two crossed sets of cellulose microfibrils, some 25 nm wide, with an occasional third orientation (Preston *et al.* 1948). All these orientations have been recorded also in the electron micrographs of Wilson (1955) and of Steward and Mühlethaler (1953) and it appeared that the third orientation turns up in electron micrographs more frequently than the X-ray diagram would suggest. This could be because the microfibrils lying in the third orientation are either less abundant or less crystalline than are those in the two major directions. A careful examination of numerous electron micrographs later showed that the lamella with the third orientation occurs with about two-thirds of the frequency of the other two lamellae, and that within it the microfibrils are more sparse (Cronshaw and Preston, 1958). These authors also showed that, while the planes of 6.1 Å spacing (101) in the two major sets of microfibrils tend to lie parallel to wall surface with a possible 'error' of ±60° (a figure also recorded by Preston and Astbury (1937)), the 'error' for the third set is only ±34° at most; accordingly, in the normal X-ray photographs the 3.9 Å (002) spacing for the third set does not appear.

Valonia lends itself readily to observation in the electron microscope since the wall can be stripped into exceedingly thin lamellae by any of the stripping techniques described in Chapter 6. The illustration presented in Fig. 7.2, the first view ever obtained of microfibrils, is a pseudoreplica and is consequently of low definition; the effect of advances in technique may be appreciated by comparison with Figs. 8.12 and 8.13, which, although referring to other algae, are indistinguishable from some preparations of *Valonia*. With care it is possible to strip away a single lamella, carrying only one striation direction (Fig. 8.4) (Preston and Kuyper, 1951).

The problem early arose as to the ways in which two (or three) sets of parallel microfibrils could be so arranged as completely to cover a globular vesicle. This was solved in three ways (Preston and Astbury, 1937; Cronshaw and Preston, 1958). Use was made in

Fig. 8.4 Electron micrograph of a single lamella of the wall of *Valonia ventricosa*, magnification 36 000, shadowed Pd/Au.

each case of the observation that the striation direction runs sensibly straight over several millimetres. The contents of a vesicle were squeezed out, incrustations of $CaCO_3$ removed in dilute hydrochloric acid, and the vesicle reinflated with air.

1. Taking the axis of a vesicle as the polar axis, a series of circles of latitude were drawn on the vesicle and on an enlarged scale model. The inflated vesicle was clamped in an X-ray spectrometer in such a way that the X-ray diagram of the wall in contact with the slit, and the nearest fiducial lines, were recorded on the plate (the latter as shadows of two wires arranged parallel to them). The directions of the cellulose chains were then drawn in the correct orientation on the corresponding (small) area of wall and on the model. One direction was continued over a length of 1.5 mm and another

photograph recorded; and so on. In this way, and at each point, the second orientation was recorded. The third orientation was ignored.

2. The vesicle was marked as in 1. with fiducial lines 4 mm apart, vertical lines drawn between neighbouring circles dividing the wall into 4 mm squares, and an enlarged model prepared. The squares were numbered, cut out, and the X-ray diagram of each recorded, using long exposures so that the third orientation could be identified. The directions were transferred to a model.

3. A slow helix was drawn round a vesicle, with turns 4 mm apart and the wall again marked in 4 mm squares by lines drawn at right-angles to the winding and a model prepared. The wall was then cut along the helical winding into a long ribbon which was then threaded into an X-ray spectrometer so that the diagram could be recorded at the centre of each square.

All three methods yielded the same mode, except that method 1., used by Preston and Astbury, did not yield the third orientation. Photographs of the models will be found in Cronshaw and Preston (1958), who, incidentally, proved, by careful checking of the microfibril orientation in the electron microscope against the electron diagram of the same specimen, that the arcs in the diagram reflect accurately the run of the microfibrils. A schematic diagram is given in Fig. 8.5. One major set of microfibrils lies in a slow left-hand helix, the other cuts across this in a steep left-handed helix and the third set lies in a right-hand helix the winding of which approximately bisects the angle between the other two. The intrafibrillar angles are remarkably constant at somewhat less than 90° between the two major sets. All three helices converge to a 'pole' at each end of the cell, defining the cell axis. The X-ray diagram of the wall of this region is, of course, a ring diagram (Fig. 8.6). As the point of observation moves away from a pole, the ring breaks up into long arcs (Fig. 8.7) which then progressively shorten to the normal diagram (Fig. 8.3).

Although the X-ray diagram alone left no room for doubt, this proposed model of the envelope of the *Valonia* vesicle has not gone without criticism. Steward and Mühlethaler (1953) concluded from an electron microscope examination of ultra-thin section

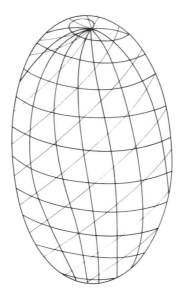

Fig. 8.5 A schematic diagram of the run of the cellulose crystallites (i.e. of the microfibrils) in a vesicle of *Valonia*. Thick lines represent the run of the two major sets; thin lines the run of the set in the third, minor orientation.

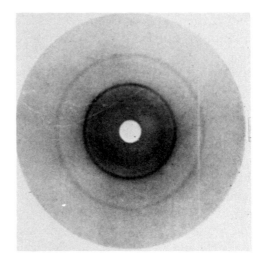

Fig. 8.6 The X-ray ring diagram at a 'pole'. Beam normal to wall surface.

(which it is now realized were paradermal sections) gave the view that the three sets of microfibrils are crossed at 120°. Since it is impossible for three sets of microfibrils crossing so symmetrically

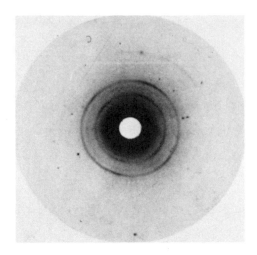

Fig. 8.7 **The X-ray diagram slightly away from the pole. Beam normal to wall surface.**

to converge to two poles, they further attributed the existence of poles to the method used in finding them. These views were, unfortunately, founded upon both an incorrect interpretation of paradermal sections and upon demonstrably incorrect measurement of interfibrillar angles (Cronshaw and Preston 1958). They ignored, moreover, the by then fully documented situation of the closely similar structure in the walls of *Cladophora* and *Chaetomorpha* (p. 203). In fact, the 'poles' were subsequently discovered under the light microscope (Wilson 1955) and have since that time been seen again and again in globular cells with walls structured precisely as recorded by Preston and Astbury. The work of Cronshaw and Preston (1958) and of all subsequent work on *Valonia* and related algae have fully validated the model of Fig. 8.5, if this were necessary. The structure is basically a two-lamella repeat, interrupted occasionally by a lamella carrying the third orientation (Fig. 8.8).

Steward and Mühlethaler (1953) were correct, however, in drawing attention to the first lamella laid down around a previously naked aplanospore. This is a thin lamella with sparsely distributed microfibrils deposited at random. At this early stage, therefore, the cytoplasmic surface has already developed the full mechanism for cellulose synthesis but has not yet the authority to parallelize the

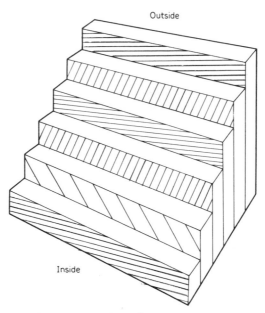

Fig. 8.8 Diagrammatic representation of the periodic repeat of structure through the wall of *Valonia*. The oblique lines show the run of the microfibrils in adjacent (fine) lamellae.

microfibrils. This comes dramatically in the next lamella to be laid down, in an abrupt change from complete randomness to almost perfect alignment. This has been amply confirmed by Cronshaw and Preston (1958) and is paralleled by later work on the swarmers of the Cladophoraceae. From this stage onwards, the cytoplasm retains the ability to lay down microfibrils in three specific directions and in no other. Taking the wall of a mature vesicle as 0.04 mm (= 40 000 nm) thick and containing 80 per cent cellulose, and assuming each individual lamella to be two microfibrils thick (= 20 nm), there must be some 1600 lamellae or about 600 each bearing the major microfibril directions and about 400 with the third orientation. The cytoplasm therefore 'remembers' the orientations over many cycles of change in orientation.

Apjohnia is not known in as much detail as is *Valonia*, but *A. laetevirens* has been investigated. This is a member of the Siphonocladales, endemic to the south coast of Australia and resembling somewhat in habit the branched *Valonias*. The mature plant is tufted and about 15 cm in height. The thallus consists of a

202

basal club-shaped cell some 5 cm long by 4 mm diameter attached to the substrate by rhizoids. Above, the segments are 1–2 cm long and 1–2 mm diameter, and branched. The wall is lamellated and has a wall structure involving crossed microfibrils closely resembling the condition in *Valonia* (Dawes and Wardrop, *unpublished*). The chemical constitution of the wall has been reported to resemble closely that of *Chaetomorpha melagonium* (see below) but containing 61 per cent cellulose, identified both by chemical means and by X-ray diffraction (Stewart, Dawes, Dickins and Nicholls, 1969). The peaks on the X-ray diagram are narrow, indicating broad crystallites and wide microfibrils as in *Valonia*.

8.1.2 *The Cladophoraceae*

Members of the Cladophoraceae are found in fresh water and in the sea where they inhabit mainly the rocks or sandy stretches in the littoral zone. With the exception of *Spongomorpha* (which will be treated separately below) the wall structure is remarkably uniform. The plants are all filamentous either unbranched (*Chaetomorpha, Rhizoclonium*) or richly branched (*Cladophora*), the filaments lying between about 0.1 and 0.6 mm diameter and many centimetres in length (Fig. 8.9). The wall is thick and finely lamellated. The apical cell continues to grow and divide throughout the life of the plant in *Cladophora* but not in *Chaetomorpha*, but in all forms almost all cells in the filaments elongate and divide continually. This means that only the innermost wall lamellae are common to an individual cell; lamellae toward the outside envelope groups of cells (derived from one individual by division) and eventually the whole filament. In *Cladophora* growth of a cell is confined to the apical half and it is from this region that the cell 'blows out' to produce a 'branch cell' which continues to divide, repeating the morphology of the parent filament. With *Chaetomorpha*, growth in each cell is confined to the basal part. The plants are again anchored by rhizoids.

The cells are multinucleate and the nuclei are usually numerous, interior to or embedded in, the chloroplasts, and sometimes arranged with great regularity; though with some Cladophoras the number may be reduced to one or two. Wall deposition continues throughout most at least of the life of the plant so that the walls of the basal cells (the older cells) are much thicker than are those

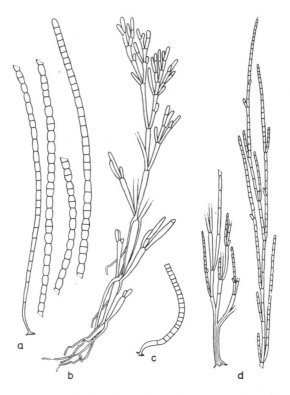

Fig. 8.9 *Camera lucida* drawings of (a) *Chaetomorpha melagonium* (b) *Cladophora prolifera* (c) *Ch. princeps* (d) *Cl. rupestris.* (a), (b) and (c) × 3; (d) × 7.

of the apical cells.

The walls are not as rich in cellulose as are those of *Valonia*, ranging from about 29 per cent (*Cl. rupestris*) to about 41 per cent (*Ch. melagonium*) on a dry weight basis (Cronshaw, Myers and Preston, 1958; Frei and Preston, 1961a). As with *Valonia* this cellulose on hydrolysis yields glucose only. The hemicellulose content ranges from about 40 per cent to about 17 per cent and contains polysaccharides which on hydrolysis yield mainly galactose, arabinose and xylose. About 42 per cent to 32 per cent of the wall is soluble in hot water and this fraction contains glucose-, galactose-, arabinose- and xylose-containing polymers. *Chaetomorpha* has not so far been shown to contain pectic compounds and it is doubtful if *Cladophora* contains this polysaccharide derivative either.

As with *Valonia* the walls are striated when seen in surface view (Correns, 1893) the striations normally lying very oblique to the cell axis in two helices which are continuous over the septum between two cells. The first X-ray diagram published referred to *Chaetomorpha* (Nicolai and Frey-Wyssling 1938), and demonstrated a structure almost exactly the same as that of *Valonia*; two sets of arcs corresponding to two sets of crystallites lying at rather less than 90° to each other and in uniplanar orientation (101 planes parallel to wall surface). Again, the lamellation visible in the light microscope is due to an alternation of cellulose-rich and cellulose-poor lamellae (Nicolai and Frey-Wyssling, 1938); Hanic and Craigie, 1969) (Fig. 8.10) and not to an alternation of microfibril orientation.

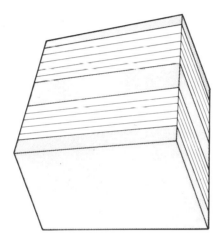

Fig. 8.10 **Wall lamellation in the Cladophoraceae. Stippled areas indicate non-cellulosic lamellae thick enough to be detectable by light microscopy. The lamellation in the unstippled, cellulosic lamella can be seen only in the electron microscope. (The number of lamellae in the unstippled areas in much reduced, for clearness of figure.)**

The first thorough examination of this group of algae was made by Astbury and Preston (1940) on *Cl. rupestris* (at that time wrongly named *Cl. arcta*), *Cl. gracilis* and *Cl. prolifera*. The cells of *Cl. prolifera* are just large enough to allow a single piece of wall to be cut out and examined in an X-ray spectrometer with the beam normal to the surface. The diagram then contained two sets of equatorial arcs demonstrating two major crystallite directions

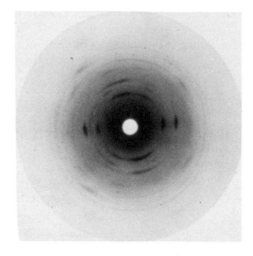

Fig. 8.11 X-ray diagram of a single wall of *Cl. prolifera*, beam normal to wall surface. Two rows of equatorial arcs present, inner 5.4 Å, outer 3.9 Å (the arc at 6.1 Å is missing on account of uniplanar orientation), crossing at less than 90°. A third set, running from top right to bottom left, is represented by 5.4 arcs only, the uniplanar orientation being sufficiently perfect that the planes at 3.9 Å do not reflect.

lying at rather less than 90° to each other. It was not until 21 years later (Frei and Preston 1961a) that a third orientation was demonstrated in this species (Fig. 8.11), again approximately bisecting the angle between the other two and with the 3.9 Å arc missing. Again the wall striations lie in the same directions as those of the two major sets of crystallites. With *Cl. rupestris* and *Cl. gracilis* the cells are too narrow for this simple approach but the diagram of whole filaments, though complex, nevertheless reveals the presence of two sets of crystallites with axes crossed at the customary angle (Astbury and Preston, 1940; Preston, 1952). The striations, and therefore the crystallites, run in two helices round each cell in the filament, one becoming steeper toward the end of the cell nearer the filament apex and the other flatter. This is a matter to be taken up again later in another context. It was therefore by 1940 established that cylindrical cells which are elongating can possess a crossed fibrillar structure, in simulation of a somewhat similar structure already considered for elongating cells of higher plants.

The first electron micrographs published did not reveal a third

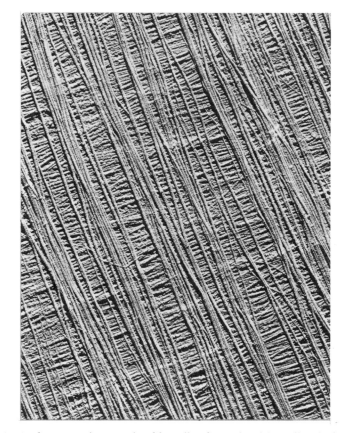

Fig. 8.12 Electron micrograph of lamellae from the side walls of *Ch. mela-gonium* viewed from outside the cell, magnification 24 000 X, shadowed Pd/Au. Two microfibrillar orientations. Cell axis ↕.

orientation either (Cronshaw, Myers and Preston, 1958; Nicolai and Preston, 1959) but they did confirm that the walls contain two sets of parallel microfibrils lying exactly in the orientations predicted by the earlier X-ray results. It was left to Frei and Preston (1961a) to lay down precisely the wall organization of the various species, by electron microscopy and in terms of the X-ray diagram. Up to this point the X-ray diagrams had referred to whole walls and the electron micrographs reflected the appearance of lamellae stripped at random from the walls. The intention now was to examine lamellae from known locations within the wall and particularly the lamellae just deposited lying next to the cyto-plasm.

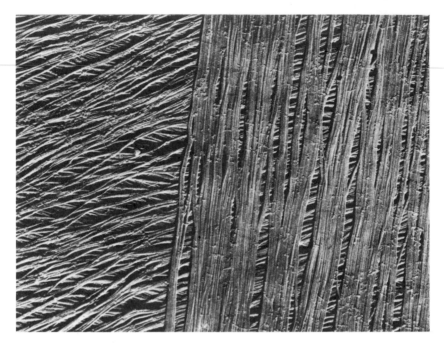

Fig. 8.13 Electron micrograph of inner side-wall lamellae of *Cl. prolifera* viewed from outside the cell; magnification 21 000 X, shadowed Pd/Au. Three microfibrillar orientations. Cell axis ↕.

All five species examined — *Ch. melagonium, Ch. princeps, Ch. aerea, Cl. rupestris, Cl. prolifera* — revealed in their side walls two sets of parallel microfibrils lying approximately at right-angles to each other in separate lamellae, each orientation running helically around the cell one a slow helix and the other fast (Fig. 8.12). In *Ch. melagonium, Ch. aerea* and *Cl. rupestris* no third orientation is detectable either on the X-ray diagram or in the electron microscope. A third orientation does occur, however, in *Cl. prolifera* (Fig. 8.13) and *Ch. princeps* (Fig. 8.14). The third orientation is not characteristic of a genus, therefore, and is not necessary for the integrity of the wall. Its situation differs from that in *Valonia* in that the lamellae with the third orientation alternate regularly with the other two — the wall is a three-lamella repeat. The microfibrils of the third set are often very sparse but even on inner lamellae are sometimes grouped into broad anastomosing bands or even into broad sheets indistinguishable in microfibril aggregation from the lamellae of the other two kinds. In each of the species

208

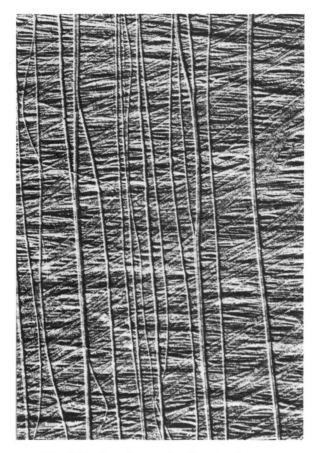

Fig. 8.14 *Ch. princeps* details as in Fig. 8.13.

showing the third orientation the order in which the lamellae are deposited is the same – slow set, followed by the steep set, followed by the third orientation and so on. As judged either by the X-ray diagram or by electron microscopy the signs of the three helices are constant within a species and are therefore genetically controlled. These signs are (in the order slow, fast, third) left, right for *Ch. melagonium* and *Cl. rupestris*; left, right left for *Ch. princeps*; left, right for *Ch. aerea* and right, left, left for *Cl. prolifera*, compared with left, left, right for *Valonia ventricosa*. It is intriguing that all five plants nevertheless conform to a general pattern. If a grid is set up defining the run of the microfibrils (Fig. 8.15) then the run of the microfibrils in any of the five plants may be

209

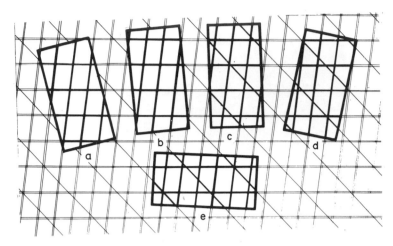

Fig. 8.15 Diagram representing the relationships between the structure of the wall in *Chaetomorpha*, *Cladophora* and *Valonia*. The two major sets of microfibrils are represented by two sets of double lines and the third orientation by single lines. The longer edge of each rectangle corresponds to the axis of the cell; a, *Ch. melagonium*. b, *Cl. rupestris*. c, *Ch. princeps*. d, *Cl. prolifera*. e, *V. ventricosa*.

correctly figured by suitable orientation of a superposed rectangle the larger side of which corresponds to the cell axis. It is then possible to move from one plant to another by an angular shift in the grid which, except for *Valonia*, is small. Angular displacement of one and the same orientating mechanism serves to account for all five structures. Perhaps it is arrogant to conclude that the large shift required for *Valonia* is sufficient to place this genus in a separate order.

The rhythm of deposition persists through several cell divisions without disturbance. This is evidenced by the observation that, at any one time, neighbouring cells in a filament tend to show innermost lamellae, butting on the cytoplasm, with microfibrils in the same direction. For example, in a filament quoted by Frei and Preston (1961a) with 75 cells, the innermost microfibrils of 36 cells were transversely oriented and 39 longitudinally oriented. The succession of orientation (s = slow, f = fast) from cell to cell was:

Base s ff s ff t f s ff s ff sssss fffff ss fffff ss ffffff ss fff ss fff ss ff sss fff ssssssss ff ss f s Tip.

The rhythm can therefore persist unchanged through three

210

Fig. 8.16 Electron micrograph of lamellae from the side walls of *Ch. princeps*; magnification 21 000 X, shadowed Pd/Au. Note two microfibrils curved through about 90°, incorporated in wall at each end.

successive divisions. The descendants of any one cell do, however, eventually fall out of step; if this were not so the above sequence would be all *s* or all *f*.

The rhythm of deposition carries the implication that lamellae are laid down one after the other and that one lamella is completed before another is begun. The possibility of stripping off single lamella with microfibrils lying in only one direction would point in the same direction were it not that many preparations are covered

211

Fig. 8.17 Electron micrograph of side-wall lamellae of *Cb. melagonium*; magnification 30 000 X, shadowed Pd/Au. Note interwoven microfibrils.

by apparently random microfibrils which might, at first sight, have been pulled away from the superposed lamella which has been stripped off (Fig. 8.16). Preparations such as that illustrated in Fig. 8.17 argue convincingly against this. Here the microfibrils are interwoven. With this clue it is then obvious that in Fig. 8.16 all the 'loose' microfibrils belong to the lamellae observed in the specimen; the microfibrils marked by an arrow, for instance, run at one end with the fast set and at the other end with the slow set and must have been deposited exactly as imaged. In other words, more than one lamella must have been deposited at one and the same time. This, together with the observations common to all electron micrographs of plant cell walls, that in any one lamella the microfibrils are occasionally twisted round each other (e.g. Fig. 8.4), means that the microfibrils must be produced by end synthesis by a mechanism not confined to a monomolecular surface layer.

One final point should be noted before we leave the side walls. The orientations of the microfibrils on the innermost lamellae depicted here are by no means perfect – are not, indeed as perfect as the X-ray diagram would suggest. Since the X-ray diagram reflects the whole wall it is pertinent to note that with lamellae only a little way in from the cytoplasmic surface the more nearly longitudinal microfibrils have become more nearly, and those almost transversely oriented less nearly, parallel to each other. This is a consequence of growth which will be considered further later in this chapter (p. 215) and again in Chapter 12.

In these filaments each cell is capped by a plane transverse wall at each end. The lamellae visible in the light microscope pass smoothly from the side walls to the end walls but this does not imply a similar smooth transfer of structures. When a beam of X-rays is directed normally through a cross wall the resulting diagram is a simple ring diagram with only one intense ring, at 5.4 Å; this contrasts strongly with a powder diagram of the corresponding cellulose in which all the rings characteristic of cellulose appear. If, on the other hand, the beam is directed through the wall parallel to its surface the diagram (now a fibre diagram) carries arcs at 6.1 Å and 3.9 Å. The microfibrils must therefore lie in the plane of the wall almost, if not quite, at random, and in a uniplanar orientation more perfect than that found in the side walls. As seen in face view under a polarizing microscope, the cross-walls mostly show a mosaic of patches as with *Valonia* walls but occasionally a maltese cross suggesting a tendency for microfibrils to sweep toward the centre, as to a 'pole'. Lamellae stripped from the cross-wall show in the electron microscope that the microfibrils tend on the whole to lie at random. At the edges, however, the microfibrils of the side walls, both of the slow and fast set, sweep over to the cross-wall and pass round close to the edge of the wall forming sometimes a densely packed rim. In the large central area of the wall, however, the microfibrils are always loosely arranged, strongly interwoven and on the whole lie at random. The very central region of the wall – which, be it noted, does not expand in area as do the side walls – appears to be a point of weakness since a hole sometimes develops there in specimen preparation.

The apical cells of these plants will be considered later when problems of wall growth and synthesis come under review. It is necessary to return here, however, to the side walls and deal with further aspects of their fine structure which have implications for both synthesis and growth. Firstly, there are two situations in which the beautiful organization of the wall is lost. When cells of *Ch. melagonium* are plasmolysed and held in that condition the plants remain healthy for long periods even though the bathing solution is then 0.9M glucose (or mannitol) in sea water. The cytoplasm pulls away from the lower cross-wall (Fig. 8.18) (but not from the upper) and proceeds to lay a wall over the free, dome-shaped end.

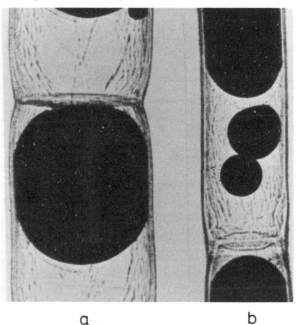

a   b

Fig. 8.18 Photomicrographs of plasmolysed cells of *Ch. melagonium*, magnification 64 X.
(a) Note the cap wall over the lower face of the protoplast and the absence of a wall over the face of the upper parts of the protoplast where is has drawn away from the wall.
(b) As in (a) after transfer to a slightly weaker sucrose solution. The protoplast has exuded globules through the cap wall which then become surrounded by a wall.

The microfibrils in this new wall lie at random; disturbance of the cytoplasmic surface has involved loss of the facility for parallel orientation though not for cellulose synthesis. Correspondingly, if the filaments are allowed to recover from plasmolysis, the normal sequence of deposition is restored soon after the cytoplasm again fills the cell. Again, during the months of February and March the plants can be persuaded to sporulate by storing overnight at 5°C and bringing back to room temperature in light the next morning. Swarmers are then liberated in about half an hour. If the filaments are plasmolysed during this half-hour so that the innermost lamella can be stripped away for electron microscope examination, then this lamella, deposited during a time when the cytoplasm is about to segregate, contains only microfibrils lying at random. This marks again a loss of order in microfibrils synthesized at a time of disturbance of the cytoplasmic surface.

Secondly, and as already remarked, the features of the microfibril arrays are not constant throughout the wall. In an examination of the difference (Frei and Preston 1961b), it has turned out that not only does the degree of order change as the lamella under observation is stripped from locations progressively deeper in the wall, but the net orientations change. This is best quantified by splitting the wall into three layers, inner (I), middle (M) and outer (O), setting them up in the correct orientation in an X-ray spectrometer and recording the average inclination of the microfibrils from that of the arcs in the diagram. Some average values are given in Table 8.1. It is clear that as the point of observation moves out through the wall the flatter helix becomes steeper and the steeper helix flatter. The steepening of the flatter helix is understandable since the cell has been elongating while the lamellae have been sequentially deposited one upon the other by apposition. The further a lamella is from the cytoplasmic surface the further has it been stretched. This is an example of the effect of 'multinet growth' first advanced, for higher plant cells, by Roelofsen (p. 388). Wall extension alone does not explain, however, the flattening of the steeper helix. This can be explained only if, while the cell extends by ΔL, one end twists through an angle $\Delta \phi$ in the direction of twist of the steeper helix. Simple geometry then shows (Frei and Preston 1961b) that:

$$\Delta\phi/\Delta L = \frac{(K \cot \theta_2{}^r - M \cot \theta_2{}^l)}{a(K + M)}$$

where
$$K = \cot \theta_2{}^l - \cot \theta_1{}^l$$
$$M = \cot \theta_2{}^r - \cot \theta_1{}^r$$

TABLE 8.1

Average inclination, θ, of the two major microfibrillar directions to the transverse in inner (I), middle (M) and outer (O) layers in the wall.

Specimen	No. of cells examined	$\theta_r^{\circ}{}^*$	$\theta_1^{\circ}{}^*$	Sign of predicted growth spiral	$\Delta\phi/\Delta L$ (calc.) (deg mm^{-1})
Ch. melagonium					
Young plants (spring)	7	I 69	10	right-hand	430
		O 66	22		
Older plants (summer)	10	I 72	11		
		M 65	15	right-hand	458
		O$_i$ 60	18		
		O$_0$ 60	33		
Cl. rupestris					
Cells near base	8	I 67	14		
		M 49	19	right-hand	4600
		O 37	19		
Cl. prolifera					
Cells near base	6	I 8	82	left-hand	−710
		O 17	76		

* r = right-hand helix; l = left-hand helix.

if there is no increase in cell breadth or

$$ K = \cot\theta_2^{l} - \cot\theta_1^{l}\left(1 + \frac{\Delta b}{b}\right) $$

$$ M = \cot\theta_2^{r} - \cot\theta_2^{r}\left(1 + \frac{\Delta b}{b}\right) $$

if the breadth b also increases by an amount Δb,
in which the superscripts r and l refer to the sign of the microfibrillar helix concerned, suffix 1 refers to inner lamellae and 2 to outer lamellae and a is the cell radius. The corresponding calculated rotations are given in Table 8.2. These calculations were made at a time when it was not known whether or not these filaments show this so-called 'spiral growth'. Laboratory tests soon showed that they in fact did, and the observed rotations agree remarkably well with those calculated for the same filaments (Table 8.2) particularly remembering that the calculated values refer to times past. During the period of the experiments recorded, the breadth b

TABLE 8.2

Calculated and observed values for $\Delta\phi/\Delta L$

| | | | $\Delta\phi/\Delta L$ | | |
Species	Sign of growth spiral predicted observed		Predicted (b const.)	Predicted (b increase by 30%)	Observed
Ch. melagonium	r	r	360	–	350
Ch. princeps	r	r	115	–	small, not measured
Cl. rupestris	r	r	4600	1040	1050
Cl. prolifera	l	l	−170	−120	−200

increased by about 30 per cent. It was further shown that cessation of length growth involves cessation of spiralling. The spiral growth of this kind clearly explains why in the sea filaments of *Chaeto-morpha* growing in clusters are often twisted together into a rope.

(a) *Swarmers*

All these plants can be induced to produce swarmers and these present an excellent opportunity to examine the development of a new wall at a naked cytoplasmic surface. The nature of the plasma-lemma and the events leading to microfibril synthesis will be postponed to Chapter 13 dealing with biosynthesis. Attention here will be confined to the new wall itself.

The life history of these plants is complex. All we need note is that one of two kinds of swarmers may be produced. Those with four flagella are zoospores and those with only two are demon-strably gametes. Either type of swarmer can settle, however, and deposit a wall in precisely the same way. The first true wall lamella is thin, with sparsely distributed microfibrils in random array, resembling the first wall of aplanospores of *Valonia* (Nicolai, 1957; Nicolai and Preston, 1959; Frei and Preston, 1960). The next lamella contains closely packed microfibrils in a slow helix con-verging at each end to a 'pole' (Fig. 8.19). This initiates the rhythm which then continues – slow, fast, third – throughout the life of the plant. Once, therefore, the first oriented wall lamella is laid down, the axis of the cell, and of the filament into which it develops, is defined. There is reason to believe, however, that the axis is already defined at an even earlier stage. The rhizoid grows

217

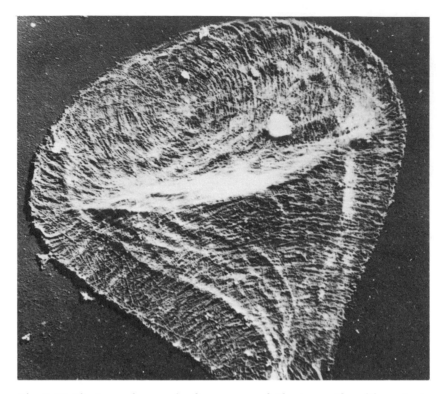

Fig. 8.19 Electron micrograph of a swarmer of *Chaetomorpha* with contents removed, shadowed Pd/Au, showing the development of the first 'transversely' oriented microfibrils sweeping in to a pole at one end. The other end of the swarmer has grown out into a rhizoid.

out from the region of one pole (with no cell division), while subsequently cell extension proceeds also at the other pole where, however, cell division also occurs in the production of the filament. This electron microscopic visualization of a 'pole' in the structure of globular cells can surely, incidentally, leave no doubt as to the reality of the pole in *Valonia*.

8.1.3 *Glaucocystis and Oocystis*

The swarmers of the Cladophorales are available only for a short period each year and search for another organism suitable for wall initiation studies which can be cultured all year round has led to these two unicellular organisms. The relationship between them

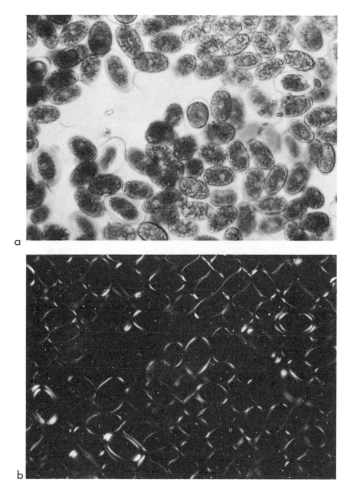

Fig. 8.20 Photomicrographs of *Glaucocystis nostochinearum*, magnification 315 X, (a) in normal light (b) in a polarizing microscope with crossed polaroids showing the birefringent wall. The same cells are visualized in (a) and (b).

has always been confused and neither modern biochemical examinations nor structural observations by electron microscopy have cleared the air to any extent. *Glaucocystis nostochinearum* Itzigs is blue-green in colour with chromatophores which are called *cyanelles* in terms of an interpretation which was long supported. This was that *Glaucocystis* represents a symbiosis between a host cell, probably *Oocystis*, and a blue-green alga reduced to its chromatophores. A brief history of this idea will be found in a paper by Robinson and Preston (1971c) in which the claim is made

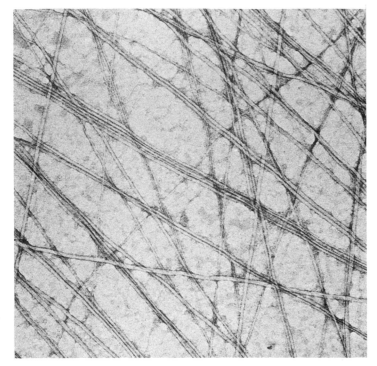

Fig. 8.21 Electron micrograph of microfibrils of *Glaucocystis* negatively stained with uranyl acetate. Magnification 73 200 X.

that the organism cannot be understood on this basis. These authors see three possibilities: *Glaucocystis* could be (1) a red alga (with rudimentary flagella) (2) a primitive dinoflagellate (3) a representative of a different and perhaps now unrecognized taxonomic group. *Oocystis*, represented by many species, is, on the other hand, without doubt one of the colonial Chlorococcales.

The cells of each genus are microscopic in size and ovoid, measuring some $50\,\mu$m long by $30\,\mu$m in diameter with *Glaucocystis* (Fig. 8.20) and some $30-40\,\mu$m by $18-30\,\mu$m in *Oocystis* with a thin wall (*c*. $1\,\mu$m in Oocystis) which is birefringent (Fig. 8.20b) and lamellated. The cells divide by fission producing daughter cells within the old mother cell wall which is eventually discarded. Sometimes the cytoplasm withdraws from the wall in preparation for division and a new wall may then be deposited round it, making the whole wall morphologically complex; two examples are shown in Fig. 8.20b. The X-ray diagram demonstrates

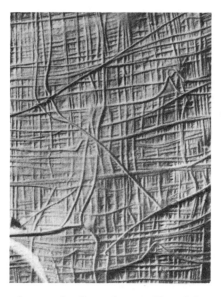

Fig. 8.22 Electron micrograph of a carbon replica of the surface of a cell of *Oocystis* produced by freeze-etching. This represents a fracture through the wall of an autospore. Magnification 30 000 X, shadowed Pd/Au.

clearly the presence of cellulose in uniplanar orientation (Robinson and Preston, 1971b; Robinson and White, 1972).

The wall architecture is in each case strikingly like that of *Valonia*. The microfibrils have been visualized by shadow-casting, by negative staining and in freeze-etched preparations. They are of the order of 25 nm wide (Schnepf, 1965; Robinson and Preston, 1971b; Robinson and White, 1972) and 8–10 nm thick (Robinson and Preston, 1971b) (Fig. 8.21; Fig. 8.22). According to Schnepf they consist of at least two elementary fibrils 10 nm wide but this is probably a misinterpretation of the kind discussed in Chapter 6. They certainly, however, taper at their ends at least down to 5 nm. In the intact wall two major sets of microfibrils occur, crossing at somewhat less than a right-angle (Fig. 8.22) coiled round the cell in helices (Fig. 8.23) and curling in to two poles, one at each end of the long axis of the cell. An occasional third orientation occurs, precisely as with *Valonia*. The sign of the helices is, however, (for slow, fast, third) right-hand, left-hand, right-hand, a sequence which does not occur in any of the other algae studied here (note that in Fig. 8.23 the signs appear to be (slow, fast) left, right; this is due to a reversal consequent on the preparation of the print).

Fig. 8.23 Electron micrograph of freeze-etch replica of *Glaucocystis* viewed from outside the cell, magnification 6300 X, shadowed Pd/Au. Note that the two sets of microfibrils are sweeping in toward a pole near the crack in the upper part of the replica.

This difference apart, it must be of significance that two algae so completely unrelated either to *Valonia* or the Cladophorales should have so strikingly the same structure. Significantly, these cells also fit on the grid of Fig. 8.15, closer to *Valonia* than to any of the Cladophoraceae.

8.2 Other algal walls

The group of the algae, therefore, contains organisms of exceptional interest and importance and a good deal of attention has for

222

this reason been paid to other forms. Attempts were made some time ago to define the skeletal polysaccharides of a range of green algae (Nicolai and Preston, 1952), brown algae (Cronshaw *et al*. 1958) and red algae (Myers, Preston and Ripley, 1955; Myers and Preston, 1959), using in the main the method of X-ray diffraction analysis. In each group a large number of genera were found which appeared not to contain cellulose in the form of cellulose I and in a considerable number of them it appeared that cellulose might still be present but either in the crystallographic form of cellulose II *or* as a derivative. It has turned out since that time that this conclusion arose from an unsuspected contaminant completely masking the diagram of cellulose (which is mostly present only in low proportion).

During a re-examination of these intriguing genera (Frei and Preston, 1963) it turned out that many arcs remained on the X-ray diagram after all polysaccharides had been removed from the wall or even after the material had been ashed. It was then found that the outer surfaces of all the algae concerned were covered in microscopic platelets of clay minerals, all therefore fairly accurately parallel to each other and giving beautiful orientation diagrams. The clay is so firmly adherent that it cannot be removed by vigorous scrubbing and is present in such amounts as to mask the diagram due to the polysaccharide and, indeed, seriously to affect the colour of the algae. Since some clay minerals have planes of spacing 7.2 Å (e.g. kaolinite) lying parallel to the surface of the platelets, or 14.7 Å (montmorillonite, chlorite, vermiculite) (second order reflection at 7.35 Å), fairly close to the 101 spacing of cellulose II, this undoubtedly accounted both for the view that cellulose II might be present in uniplanar orientation and the doubts that it could be. Some of these algae are dealt with further below; it turns out that most of them contain perfectly ordinary cellulose.

Most of the algae to be dealt with have finely lamellated walls with, commonly, an outer thin layer and an inner, thicker layer either of different chemical constitutions or of different micro-fibril orientation. In cut sections, or in optical section, the walls are usually positively birefringent, i.e. the direction of light vibration with the greater refractive index lies parallel to wall surface. This is to be expected for cellulosic walls though it is, of course, not proof that cellulose is present.

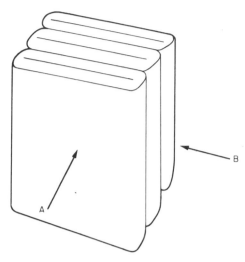

Fig. 8.24 Sector of a bundle of three flattened filamentous cells. An X-ray beam may be passed through this either in direction A or in direction B.

8.2.1 *Green algae*

Nicolai and Preston (1952) did, however, correctly add to the forms dealt with already containing cellulose as the skeletal polysaccharide; *Trentepohlia* in the Chaetophorales and *Botrydium granulatum* in the Heterosiphonales. With *Vaucheria* and *Spirogyra* they were doubtful. A pellet of wall material gave in each case a ring X-ray diagram with all the spacings of cellulose. When, however, the filaments are flattened and the beam passed through the wall parallel to wall surface (direction B, Fig. 8.24) (a common technique for the separation of arcs) the disposition of the 6.1 Å, 5.4 Å and 3.9 Å arcs was not that expected for cellulose. There can, however, be no doubt about the presence of cellulose; in 1950, Vogel (published by Frey-Wyssling 1953) had demonstrated the presence of microfibrils and Frey (1950) had shown for *Spirogyra* and *Mougeotia* (Conjugales), *Tribonema* (Heterotrichales) and *Vaucheria* (Siphonales) that the precipitate from a solution of the walls in cuprammonium contains cellulose II. Subsequently, Kreger (1957) has supported Nicolai and Preston by showing that *Spirogyra* is exceptional in that either the 5.4 Å (10$\bar{1}$ of cellulose) planes or the 3.9 Å (002 of cellulose) planes lie parallel to the wall surface. More recently Limaye, Roelofsen and Spit (1962), Buer (1964) and Dawes (1965) have demonstrated a tendency to axial alignment of microfibrils in the outer layers of the wall with a tendency for transverse alignment in inner layers, in harmony with the multinet growth hypothesis (p. 388).

Among the Charales, internodal cells of *Nitella* have been studied intensively, particularly with reference to cell growth and wall biosynthesis (p. 397). The wall was for a long time thought to be cellulosic without experimental proof. This was eventually provided by Hough, Jones and Wadman (1952) using chemical methods and by Probine and Preston (1962) (using *Nitella opaca*) by X-ray diffraction methods. Green (1958) and the latter authors also demonstrated the presence in the wall of microfibrils about 10 nm wide, with changes in microfibril orientation across the wall much as recorded above for *Spirogyra*. This last point has been confirmed by Limaye *et al.* (1962) for *Nitella mucronata. Chara*, on the other hand, remains as uncertain as Nicolai and Preston (1952) found it. According to Krishna Murti (1965) material pre-treated following the methods of Cronshaw *et al.* (1958) shows X-ray spacings at: 12.03, 8.63, 3.33, 3.89, 3.56, 3.04, 16.36, 7.52, 6.44 and 5.6 Å. These do not fit either cellulose I or cellulose II. If his X-ray diagram does reflect the skeletal polysaccharide, this cannot be cellulose. The findings in this laboratory are that the central cell of the internodes (which cannot carry a contaminant such as clay) yields a perfectly good cellulose I diagram (Frei, 1962 *unpublished*).

With the Cladophorales, Nicolai and Preston (1952) added *Rhizoclonium* as a cellulose-containing form and showed, indeed, that the whole wall structure resembles that of *Cladophora*. Similarly, Kreger (1962) has added *Spongomorpha arcta* and in this he is supported by Frei and Preston (unpublished) though the microfibrils in the wall differ from those of *Cladophora* in being thinner (*c.* 10 nm) and in random arrangement.

Similarly *Ulva lactuca* and *Enteromorpha* sp. in the Ulotrichales have been shown by Dennis and Preston (1961) and by Kreger (1962) to give a cellulose I X-ray diagram after the walls have been physically cleaned and treated with dilute mineral acid.

Several members of the Chlorococcales and the Volvocales have been examined still more recently, and some have been found still recalcitrant. *Chlorella pyrenoidosa* clearly contains cellulose to the extent of 15 per cent on a dry weight basis (Northcote, Goulding and Horne, 1958) with microfibrils, in a loose network, said to be 50 Å wide and yielding glucose and galactose on hydrolysis. On the other hand *Hydrodictyon* in spite of much investigation remains obscure. According to Kreger (1960) the untreated wall of *H. reticulatum* (which is lamellated and about 5 μm thick) gives only a diffuse X-ray diagram which is, however, improved after the material has been boiled for 1 hour in 2% HCl. Nicolai and Preston (1952) had observed a similar diagram and found it uninterpretable. Kreger concluded, however, that this is a diagram of mannan of the ivory nut type (p. 239) and, in this sense, *Hydrodictyon* leads directly to the algae in the next chapter. Precipitates from walls dissolved in cuprammonium nevertheless contain cellulose in the form of cellulose II, though the native wall diagram does not contain arcs relating to this substance. In Kreger's view, the polysaccharide is an association of glucan and mannan chains, each of which prevents the crystallization of the other, though some mannan is crystalline. Only glucose and mannose are identifiable in any quantity in wall hydrolysates (in a ratio varying from 3:4 to 1:1), and Kreger believes that the mannan and glucan may be linked through glucose and/or mannose in bonds susceptible to mild acid hydrolysis. Northcote *et al.* (1960) reached much the same view with *H. africanum*; they add that the microfibrils are thin and that the 'α-cellulose' amounts to about 70 per cent of wall weight, containing about equal proportions of glucose and mannose.

The X-ray diagram and the chemical analysis of *Pediastrum* are said to resemble those of *Hydrodictyon* (Parker, 1964) though it is claimed that the electron microscope reveals a peculiar structure of an interwoven chain of rings 18 nm in diameter. An equally possible interpretation of the published electron micrographs would be a straight forward interweaving of microfibrils. A most interesting case has recently come to light with *Eremosiphaerea viridis* (Mix, 1972). The autospore mother cell walls appear to consist of alternating lamellae of parallel microfibrils crossing in direction from one lamella to the next. Although there is yet no proof that these are cellulosic walls, and though no systematic investigation has yet been made, it could well be that they should be classed as *Valonia*-type.

A situation somewhat similar to that of *Hydrodictyon* obtains for *Halicystis* (Siphonales), which again leads to the algae of Chapter 9. Van Iterson (1937) had looked at this alga as early as 1936 and had found no signs of striations though in surface view under the polarizing microscope he found the same mosaic of m.e.p. found by Preston (1931) for *Valonia* with which the habit of *Halicystis* had suggested it might be compared. He already saw the wall as in three layers and thought them chemically different. Some two years later, Sisson (1938) obtained X-ray diagrams. These revealed no orientation, but the diagrams were compatible with the presence of cellulose II in uniplanar orientation (101 planes parallel to wall surface) though with an unexplained spacing at 12.5 Å. Roelofsen, Dalitz and Wijman (1953) subsequently confirmed these points. They found, however, in the hydrolysate of the whole wall glucose and xylose in about equal amount and almost no other sugars. Since the diagram was increased in intensity when the wall material had been extracted in hot alkali (in which the xylan was soluble), giving a residue which still contains equal amounts of glucose and xylose, they concluded that the polysaccharide is a xyloglucan and not cellulose II. Roelofsen *et al.* also confirmed the layered structure described by van Iterson and added that, while the inner and outer layers are positively birefringent, the central layer is negatively birefringent; more will be said about this later. In a re-examination of *Halicystis*, Frei and Preston (1961c) confirmed these findings and agreed further that the arc on the X-ray diagram at 12.5 Å is to be attributed to the xylan. According to them, however, treatment of the holocellulose of *Halicystis* with increasing strengths of alkali solution progressively reduces the strength of this arc so that finally it has disappeared at 4N KOH leaving only the arcs attributable to cellulose II which have persisted throughout. Hydrolysates of the small amount of material remaining to give this diagram, still contain, however, in addition to glucose some 40 per cent of xylose. Moreover, the glucan and xylan can be separated by treating the whole with cuprammonium. The glucan dissolves and can be precipitated to give a cellulose II diagram. The residue yields a diagram identical with those to be described for the xylan algae in the next chapter; it hydrolyses to give xylose predominantly and is undoubtedly 1,3 linked. In these terms it is possible to interpret the birefringence of the wall layers described by Roelofsen *et al.* The outer and inner layers must contain, predominantly, the glucan and the central layer, predominantly the 1,3 linked xylan. It is still a moot point whether the glucan is cellulose II; the evidence is now marginally in favour.

The situation of *Halicystis* is complicated by the circumstance that this alga is the gametophytic stage of the sporophyte *Derbesia* which contains only mannan as the wall skeletal polysaccharide This is therefore another alga which leads directly into the next chapter.

A number of investigators have given preliminary accounts of some members of the Volvocales. Lange (1963) has reported that *Gonium* and several other Volvocales lack a compact microfibrillar wall, differing in this respect therefore from *Chlamydomonas* (Sager and Palade, 1957). Parker (1964) has found that walls of *Gonium sociale* var. *Sacculum* do not stain in I_2/H_2SO_4, give no X-ray diagram and yield no detectable sugars in the hydrolysate, confirming in this way the work of Lange. He has also reported the intriguing finding that *Oedogonium* (Oedogoniales) possesses a finely lamellated wall containing microfibrils 4—5 nm wide in two sets crossing at about 90° giving a clear cellulose I diagram. The arc at 5.4 Å is missing from the diagram so that the microfibrils may lie in uniplanar orientation. The microfibril size is almost certainly an underestimate and it could be that *Oedogonium* should take its place with *Valonia*.

Finally, Mix (1966 and 1968) has closed a gap by giving some attention to the Desmids. She has reported the presence of microfibrils, which have the appearance of cellulose microfibrils, in *Tetramonas laevis*, *Pleurotenium trabecula* and *Cosmarium* sp. and in inner wall layers of *Euastrum oblongum*, *Microsterias rotata* and *M. denticulatum*. In the outer layers of the wall the microfibrils are only sparsely distributed, and in *Pleurotaenium* show a change in orientation through the layer consistent with multinet growth. The inner wall layer here displays helically oriented ribbons, 8—10 microfibrils wide, of different orientation to each other and occurring in different lamellae. In this they resemble closely the structure of *Spirogyra*. In a more recent publication, Mix (1972) has demonstrated the same peculiar structure in *Netrium digitus* and *Gonatozygon monotaenium* (though in this plant the bands are only 3—4 microfibrils wide).

8.2.2 Brown algae

All the brown algae so far examined contain cellulose I as the skeletal polysaccharide. The cellulose content is usually low (Black,

1950, 1954; Cronshaw *et al.* 1958) and the X-ray diagram is revealed only after the non-cellulosic polysaccharides have been at least in part removed by alkali extraction (Schurz, 1953; Percival and Ross, 1948; Frei and Preston, 1961c). The clarity of the X-ray diagrams and the sharpness of the arcs distinguishes these algae sharply from the red algae. The cellulose becomes mercerized in KOH solutions greater than 14 per cent in concentration, another difference from red algae. In the holocellulose of most of the Fucales a reflection appears on the diagram corresponding to a spacing of 14.1 Å which is not attributed either to cellulose in any form or to a clay contaminant since this arc is found even when *inner* layers only of the thalli are used. It probably represents a second crystalline component in the wall but it remains uncertain whether the treatment necessary to allow the diagram to be revealed at all has induced crystallization of a substance not crystalline in the native wall.

Exceptionally, untreated material from the stipe of *Alaria* (but not from the lamina) already contain a second crystalline substance additional to the cellulose.

In these algae the X-ray spacings of cellulose vary, as they do in all plants, not only between species but within species. Representative values obtained during the work presented by Frei and Preston (1961c) but not hitherto published, are given in Table 8.3, giving four major spacings of cellulose. The 040 spacing remains constant throughout, so that the molecular chain must maintain the same conformation. The hol spacings, however, differ, implying differences in the dimensions of the basal plane probably indicating different inter-chain hydrogen bonding. With those species which allow the thallus to be flattened so that the X-ray beam could be passed through the wall parallel to wall surface (B, Fig. 8.24) the diagram revealed uniplanar orientation of the crystallites (101 planes parallel to wall surface).

The walls are throughout microfibrillar (Cronshaw and Preston, 1958) with microfibrils about 10 nm broad; so far as the writer is aware the question of preferred orientation has not been examined, except in the cortex of *Chorda* where some cells have been shown, in terms both of the microscopic appearance of helical fissures and of the X-ray diagram (Fig. 8.25), to resemble somewhat the tracheids of conifers in that the microfibrils lie helically round the cell.

TABLE 8.3

Species	Formal cellulose spacing				
	2.58(040)	3.91(002)	4.32(021)	5.4(10$\overline{1}$)	6.1(101)
Brown algae					
Himanthalia	2.58	3.90	4.32	5.23	6.10
Fucus serratus*	2.58	3.89	4.37	5.24	6.09
Dictyota					
dichotoma*	2.58	3.92	4.36	5.26	6.13
Chorda filum*	2.58	3.85	4.38	5.24	6.12
Laminaria					
saccharina*	2.58	3.94	4.35	5.25	6.12
Alaria esculenta	2.58	3.91	4.42	5.25	6.12
Cladostephus					
spongiosus	2.58	3.91	4.34	5.26	6.12
Scytosiphon					
lomantarius	2.58	3.98	4.24	5.28	6.15
Chordaria					
flagelliformis*	2.58	3.92	—	5.31	6.10
Elachistea sp.	2.58	3.92	4.43	5.25	6.15
Pilaiyella					
littoralis*	2.58	3.94	4.47	5.24	6.13
Cutleria					
multifida*	2.58	3.89	4.40	5.21	6.14
Red algae					
Griffithsia					
flosculosa*	2.58	3.95	4.38	5.33	5.94
Polysiphonia					
fastigiata	2.58	3.92	4.42	5.31	6.10
Gelidium sp.	2.58	3.89	4.38	5.35	5.90
Gigartina					
stellate*	2.58	3.99	4.47	5.31	5.82
Dumontia					
incressata*	2.58	3.90	—	5.30	5.72
Rhodymenia					not
palmata	2.58	3.94	—	5.46	present

* Uniplanar orientation

(a) *Alginic acids*

The characteristic of the brown algae is that many if not all of them possess in their walls polyuronides in quantity which are not pectic compounds and are called alginic acids. These compounds

230

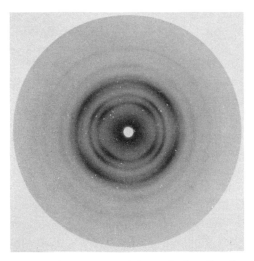

Fig. 8.25 X-ray diagram of the holocellulose of a bundle of cortical cells of *Chorda filum* after washing in cold 2 per cent HCl. The crossed sets of arcs refer mostly to cellulose I (though traces of arcs due to alginates are present), showing that the microfibrils coil helically round the cells.

are not, however, unique to these plants since it is now known that some bacteria also contain them. The alginates and alginic acid have been subjects of intense investigation and only the details necessary for elucidating structure will be considered here. They occur naturally in the form of calcium alginates, often far outweighing the cellulose in amount, which are crystalline and yield X-ray diagrams often obscuring the cellulose diagram until the alginate has been extracted by, for instance, sodium carbonate solutions.

At the time when Astbury (1945) was obtaining the first crystallographic data, the main constituent was thought to be β-1,4-poly-D-mannuronic acid (Hirst, Jones and Jones 1939; Lucas and Stewart 1940) (Fig. 3.4) and the X-ray diagrams were referred to this substance. Ten years elapsed before Fischer and Dörfel (1955) demonstrated the presence, often in large quantity, of another polyuronic acid, poly-L-guluronic acid (Fig. 3.4), together with small amounts of polyglucuronic acid. They found the proportions of the two major acids to vary from 2:1 to 1:2 in the 22 species they examined. Poly-L-guluronic acid is also known to be 1,4-linked (Drummond, Hirst and Percival, 1958). The presence of a second acid changed the situation considerably. Astbury had

231

correctly deduced that for his sample b (fibre axis) = 8.7 Å, not the 10.3 Å common to 1,4-linked polysaccharides. He therefore considered that neighbouring mannuronic residues in the chain were tilted with respect to each other by rotation about the glycosidic $-C-O-C'$ (Fig. 3.7) bonds so that the chain was kinked. It proved later that it was then impossible to fit poly-mannuronic acid into the unit cell. Frei and Preston (1962) resolved this problem by showing conclusively that Astbury's diagram referred to polyguluronic acid, not polymannuronic acid, and this has been supported by all later workers. They achieved this by demonstrating that the fraction of the alginic acid of *Himanthalia elongata* soluble in hot water contained a preponderance of poly-mannuronic acid while that subsequently extracted by a sodium carbonate solution is enriched in polyguluronic acid. The diagrams of these acids are given in Fig. 8.26; the axial repeat is 8.7 Å for polyguluronic acid and 10.3 Å for polymannuronic acid so that the former, not the latter, has kinked chains. When alginic acid is extracted from the seaweeds and precipitated, crystallization of polyguluronic acid inhibits crystallization of polymannuronic acid; this is presumably why Astbury's diagram was of polyguluronic acid although his specimen was largely polymannuronic acid.

The inter-relationships between the two major acids has been examined for *Laminaria* and *Ascophyllum* by Haug, Larsen and Smidsrød (1966, 1967). They have concluded, following electro-phoretic separation after partial acid hydrolysis, that alginic acid is a block polymer with long sequences of mannuronic acid and long sequences of guluronic acid. Following this, Haug and his co-workers have found it possible to produce fractions which are very much enriched in either polymannuronic acid or polyguluronic acid and Atkins, Mackie and Smolko (1970) have used this material to carry the observations of Frei and Preston (1962) a little further. Their photographs are identical with those of Fig. 8.26. The unit cell of polymannuronic acid is given as $a = 7.58$ Å, b (fibre axis) = 10.35 Å, $c = 8.58$ Å (orthorhombic), containing two dissacharide chain segments with probable space groups $P2_12_12_1$. With polyguluronic acid the unit cell is $a = 8.6$ Å, b (fibre axis) = 8.72 Å, $c = 10.74$ Å (orthorhombic) with the same space group. These agree with the unit cells found by Frei and Preston (unpublished) with which they did not proceed owing to their inability to pack the chains satisfactorily in the unit cells of

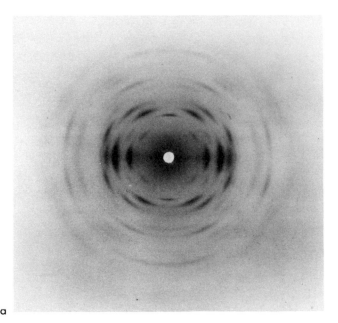

a

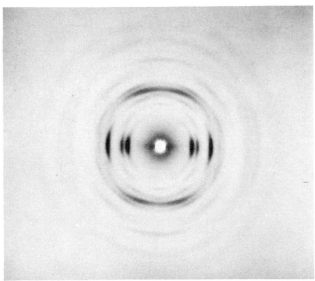

b

Fig. 8.26 Fibre X-ray diagrams of alginic acids
(a) polymannuronic acid
(b) polyguluronic acid
(diagrams by Dr. Eva Frei).

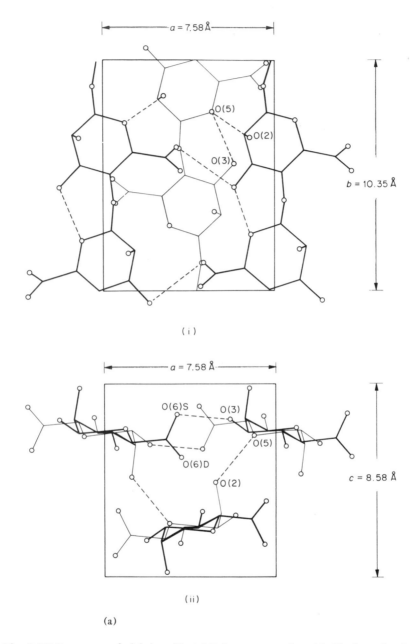

(a)

Fig. 8.27 Structure of alginic acids (a) Polymannuronic acid. (i) *ab* projection of unit cell. H—bonds dotted O(3)—H to O(5) is intrachain H—bond. O(2)—H to O(5) is inter-sheet H—bond. (ii) Equatorial projection. O(2)—H to O(5) is inter-sheet H—bond. O(6)S—O(3) is intra-sheet H—bond.

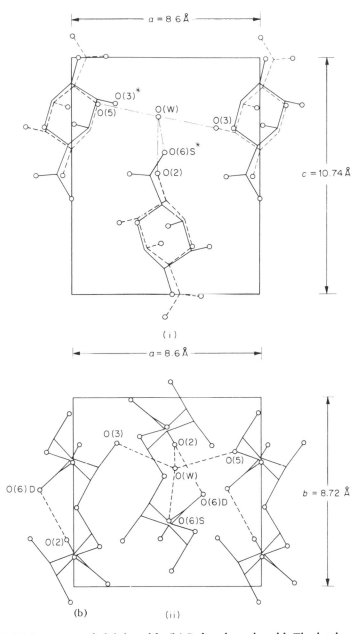

Fig. 8.27 Structure of alginic acids (b) Polyguluronic acid. The lattice contains water. (i) *ab* projection. H–bonds dashed. Only one water molecule is shown (O(W)) to avoid confusion. (ii) Equatorial projection. * denotes two atoms which may also be H–bonded. The dotted lines represent the next residue in the chain. (by courtesy of Atkins *et al.*, 1973).

these high density (1.62) compounds. Atkins *et al.* have found it possible to position the chains in the unit cell of polymannuronic acid, using the usual β-1,4-conformation, with each residue in the energetically favourable Cl form and with the intra-residue H bond within the chain at 2.7–2.8 Å. The basic structure (Atkins, Mackie, Nieduszinski, Parker and Smolko, 1973) is presented in Fig. 8.27a. With polyguluronic acid, it was necessary to assume that the residues are in the 1C form. The putative structure is given in Fig. 8.27b.

As far as the writer is aware, Frei and Preston (1962) have been the only authors seriously to consider the distribution of alginic acid within the wall and between the wall and the intercellular mucilages. It had earlier been concluded for *Laminaria* and *Fucus*, on somewhat slender evidence, that none (Thiele and Anderson, 1955) or very little (Moss, 1948) is contained within the wall. The structural evidence speaks to the contrary view. With few exceptions the X-ray diagrams of the wall examined by Frei and Preston (13 genera, 7 orders) showed clear arcs referring to alginate as well as cellulose and that the molecular chains lie parallel to wall surface. With *Chorda*, indeed, the molecular chains (of polyguluronate) in the walls of cortical chains are demonstrably aligned with the cellulose microfibrils (Fig. 8.25). This constitutes strong evidence that polyguluronate, at least, occurs in the wall. There is, however, further evidence (Frei and Preston, 1962); (1) the cell walls swell appreciably during extraction of alginic acid in sodium carbonate solutions, (2) when two contact X-ray microradiographs are compared of transverse sections of a *Laminaria* stipe, one treated with acid only and one treated with acid followed by copper acetate to form copper alginate, the cell wall as well as the mucilage is much more opaque to X-rays in the latter than in the former, (3) when the copper-treated section is washed to form a spodogram, the section appears unchanged with the cell wall distinct (a section merely washed in acid leaves no residue). Finally the holocelluloses of brown algae (the residue after boiling in hot water and chlorination) still contain alginic acid in some quantity, chiefly polyguluronic acid, although the tissue has become macerated.

All this evidence taken together implies that polyguluronic acid may be characteristic of the wall, and polymannuronic acid of the intercellular material, at least in the more complex brown algae.

If it is correct, as Haug and his co-workers assert, that alginic acids are block polymers, then the polymer in the wall must contain more polyguluronic acid than does the intercellular polymer.

8.2.3 *Red algae*

The situation in the red algae was also cleared up once the confusing presence of clay minerals was realized. With many of these algae, and especially with the Corallinaceae, heavy incrustations of calcium carbonate, in the form of calcite or aragonite need to be dissolved away before examination. All members of the red algae examined (Frei and Preston 1961c) then give X-ray diagrams of cellulose I except the members of the Bangiales which are considered on p. 000. The cellulose content of the walls is low (Cronshaw *et al.*, 1958) as is also the degree of crystallinity; indeed, the lack of clarity in red algal diagrams distinguishes them markedly from those of brown algae. The X-ray spacings vary somewhat from species to species, as in the brown algae (Table 8.3) though on the whole the (101) spacing is greater and the (10ī) spacing is less in red algae than in brown algae. The cellulose is exceptionally easily mercerized, changing from cellulose I to cellulose II in KOH solutions as low in concentration as 10 per cent.

The walls are invariably finely lamellated and the lamellae have been said to contain microfibrils in random array (Myers *et al.*, 1955; Myers and Preston, 1959). There is no doubt about the presence of microfibrils but the random orientation must be doubted in some cases at least, since the relevant observations were made on material homogenized in a blender. The walls of *Rhodymenia palmata* are somewhat peculiar (Myers and Preston, 1959) in allowing the thallus to be stretched by at least 100 per cent reversibly; the cellulose content, as in all red algae, is of course low (2–7 per cent).

CHAPTER 9

Detailed structure – non-cellulosic algae

Until comparatively recent times, the crystalline skeletal materials of all plants other than fungi were thought of as constructed uniquely of cellulose, and the many other sugar polymers known to be present in cell walls were considered always to form the cementing materials binding the cellulose microfibrils together. Both in the concept of the microfibrils as aggregates of straight unbranched chains of 1, 4-linked β-D-glucose residues and the uniformity of its application over a major part of the plant kingdom, this gave a picture of beautiful simplicity which was very persuasive. There were always, however, a few plants for which the evidence had to be stretched in order to preserve this uniformity. These are all algae, mostly tropical, and all members of the old order, the Siphonales (Fritsch, 1935). Examination of these plants over the past twelve years or so has shown that they are, in fact, highly unusual plants of very great interest through their quite exceptional cell walls.

As already mentioned, these plants mostly fall into two groups on the basis of wall architecture, one group in which the wall is based upon mannan instead of cellulose and another in which a 1, 3-linked xylan is the sole skeletal polysaccharide. These will be dealt with in turn as *mannan weeds* and *xylan weeds* respectively. There is a third variation upon the skeletal polysaccharide theme in the two genera of the Bangiales in the red algae and this will be taken up after the two major groups have been considered. It will be clear from what has been said in the last chapter that all these plants are, in terms of wall architecture, linked to the cellulosic algae through *Hydrodictyon* and *Halicystis/Derbesia*.

9.1 The mannan weeds

These fall into two families, the Codiaceae and the Dasycladaceae, and therefore include the genera *Codium, Dasycladus, Batophora, Halicoryne, Neomeris, Cymopolia* and *Acetabularia*; and take in *Derbesia* (classed as the sole genus in the family Derbesiaceae by Fritsch, 1935). It has already been shown that three groups of workers, Miwa and his colleagues, Percival and her colleagues, and Frei and Preston, almost simultaneously and quite independently showed that the walls of these plants (though not all groups worked with the same plants) contain large quantities of β-1, 4-linked mannan, which is the skeletal polysaccharide (Frei and Preston, 1961c, 1968).

The Dasycladaceae are all plants of the warmer seas. They are all, moreover, built upon a common plan consisting of a long axial cell – the central siphon – bearing regular whorls of simple or forked thin-walled branchlets of limited growth with no septa. The wall of the main axis consists of a thin outer layer with an inner layer which is finely lamellated and becomes progressively thicker as the plant ages, until in mature plants the lumen is narrow and there is only a restricted pore leading into each branchlet. Characteristically the plants are incrusted, sometimes heavily, with calcium carbonate (aragonite) and as a consequence this family can be traced far back into the fossil record. This incrustation needs to be removed (e.g. by immersion in cold 0.1 N HC1) before attempting either chemical or physical determinations of wall structure. Members of the Codiaceae, however, range into northern waters. They are filamentous in construction, the filaments being thin-walled, branched and aggregated to form a conspicuous pseudoparenchymatous plant body.

9.1.1 Chemical constitution

The chemical composition of the walls is remarkably uniform and simple. When walls freed of cytoplasm and incrustations are fractionated by the standard hot water-chlorination-alkali sequence only hot water treatment leaves a residue markedly different from the extract. Subsequent treatment by alkali solutions 'dissolves' part of the residue but cannot achieve further fractionation since the residue is overwhelmingly mannan. Some 10 per cent of the whole wall is soluble in hot water, giving a soluble fraction which on hydrolysis yields a variety of sugars, mostly galactose and glucose. The hydrolysate of the residue, and even of the residue after further treatment with 24 per cent KOH, always contains a little glucose which appears therefore to be attached to the mannan. Love and Percival (1964) have indeed suggested for *Codium* that the glucose residues are linked to mannan by a primary valence.

These walls are therefore quite unique in their chemical simplicity. The physical state of the mannan is, however, demonstrably not uniform even within one species. The residue remaining when any of the walls are treated with alkali solutions (6 per cent LiOH, 10 per cent NaOH or 14 per cent KOH at room temperature) retains, intact, the membrane structure. When subsequently placed in water, the inner wall layers of the central axis in *Dasycladus*, *Batophora* and *Cymopolia* differ from the outer layers and from the walls of *Codium* and *Acetabularia*, in swelling and rapidly dissolving. This suggests strongly a spatial differentiation of mannan chain length such as recorded for bulked samples in Chapter 2. If this is the case, then these plants produce larger molecular chains when young and shorter chains when older, the reverse of the situation claimed with some cellulosic plants (Chapter 13).

Under no circumstances have the walls of the mannan weeds been found to stain blue with I_2/H_2SO_4 or $I_2/ZnCl_2$. On the other hand, the walls do take up metal ions from metal salts, recalling the behaviour of hemicelluloses in higher plants. Indeed the mannan of these walls is chemically indistinguishable from the equivalent hemicellulose of higher plants so that here we have a structural polysaccharide in an alga which has continued into land plants where, however, it plays a different role.

9.1.2 Physical organization

(a) Polarization microscopy

The native walls are as a rule birefringent, not only in section, but also in face view (Fig. 9.1 Frontispiece) when the walls of ivory nut endosperm (containing 90 per cent mannan (Meier, 1958)) are isotropic. These are therefore the first organisms in which an orientation other than random, of mannan chains, could be envisaged. In optical section the m.e.p. lies parallel to wall surface (Fig. 9.2) as with cellulosic walls. Since the chains of mannan presumably lie parallel to wall surface as do cellulosic chains (and as they are proved to do below) this means that mannan, like cellulose, is positively birefringent and the m.e.p. gives the net preferred orientation. The direction of the m.e.p. of walls seen in face view therefore has meaning. This lies parallel to filament length in *Codium* and in the branchlets of *Cymopolia* and *Neomeris*, where the walls are thin and the birefringence weak. By far the most revealing birefringence is found, however, in the large siphons which make up the central axis of *Dasycladus*, *Batophora* and *Cymopolia*. The m.e.p. in the thin-walled apical region lies transversely to the axis and, though the birefringence at the apex is weak, it increases lower down the siphon and can become fairly high. Further down, as the wall increases in thickness, the birefringence (of the whole wall) decreases again until the wall becomes briefly isotropic and then positive (i.e. the m.e.p. now parallel to the axis), the positive birefringence then increasing to the base of the plant. Here the wall can be split into an inner part with the m.e.p. parallel to the axis and an outer, thinner part with the m.e.p. transverse to the axis (Fig. 9.1 Frontispiece). In this figure the outer layer (left) and the inner layer (right) of the same wall are displayed for (a) *B. oerstedi* and (b) *D. vermicularis* as seen between crossed polaroids with a Red I plate (55x magnification). The m.e.p. of the plate, and the axis of the siphon, lie parallel to the longer edge of the page; blue and green are addition colours, yellow and orange subtraction colours. Reference back to Chapter IV will serve to verify the stated run of the m.e.p. of the two wall layers. The behaviour of the birefringence on passing down the siphon can now be understood by reference to Fig. 9.2. At the apex A, the mannan chains lie on the whole transversely. At point B, however, a new wall lamella begins to be deposited, with chains lying axially. The whole wall is now a crossed crystal plate with crystal axes at right angles and the negative birefringence is in part compensated. Full compensation is achieved at point C, where the wall is isotropic and below this the positive birefringence of the inner layer becomes overriding. In *Dasycladus*, at least, the birefringence of the outer layer itself continues to increase down to the base of the plant, a feature possibly connected with the peculiar growth habit of this plant which will be discussed below. The outer wall layer in these plants clearly resembles the primary walls of higher plants, and the inner layers the secondary walls, in both form and function.

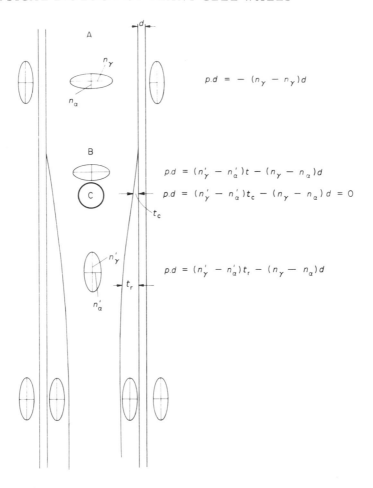

$$p.d = - (n_\gamma - n_\gamma)d$$

$$p.d = (n'_\gamma - n'_\alpha)t - (n_\gamma - n_\alpha)d$$

$$p.d = (n'_\gamma - n'_\alpha)t_c - (n_\gamma - n_\alpha)d = 0$$

$$p.d = (n'_\gamma - n'_\alpha)t_r - (n_\gamma - n_\alpha)d$$

Fig. 9.2 Optical properties of the walls of central siphons in Dasycladaceae showing outer layer, thickness d, and inner layer, thickness t. The ellipses represent the trace in the plane of the wall (or of the section of the side walls) of the effective index ellipsoid.

Both inner and outer faces of walls with 'secondary' thickening show markings in various degrees. On the inner face these are striations lying axially, in conformity with the putative chain direction (Fig. 9.1 a). On the outer face they are not striations; instead they represent corrugations of the outer layer lying transversely though sinuously (Fig. 9.1). Accordingly, each wall can be torn cleanly in the direction of the markings. The interruptions of the wall shown in Fig. 9.1 represent the points of insertion of branchlets. The 'scars' are roughly circular in outline on the outer face of the outer layer and continue as pores into the inner layer which, as the layer becomes thicker, progressively become slit-like with the long axis of the slit lying axially. Between crossed polaroids the scars stand out as Maltese crosses with the m.e.p. running round each scar parallel to its edge. The physical form and properties of these walls are therefore in close harmony with the proposed chain orientations.

(b) *X-ray diffraction*

Pellets of wall material of all members of the Dasycladaceae examined by Frei and Preston (1961c and 1968) give good powder diagrams charactersitic of β-1, 4-linked mannan. It had been reported that the mannan of ivory nut endosperm occurs in two crystalline forms (Meier, 1958) and this they confirmed. In the algae, however, only one of these forms is found, yielding only one X-ray diagram (Fig. 9.3). The background is variable, though high, indicating the presence of much non- or para-crystalline material, a good deal of which must also be mannan. In this family no treatment of the wall is ever necessary in order to elicit the diagram; even with highly calcified plants treatment to remove calcification can always be avoided by using dissected-out inner layers which always contain crystalline mannan. *Codium* is distinctly different since untreated walls yield a diagram consisting of only two broad rings centred upon 6.4 Å and 4.9 Å with no apparent relationship with any rings attributable to mannan. When material of *Codium* is treated with boiling water and re-dried, however, it then gives the diagram of Fig. 9.3, typical of mannan. The evidence is that this change in the diagram represents a real change in the material and not simply an unmasking of an existing diagram by the removal of material giving overriding background scatter; that hot water induces a crystallization in material which before treatment was at most para-crystalline. We apparently have here for once, a plant which does not lay down its wall in strict crystalline condition. With the Dasycladaceae, on the other hand, extraction merely sharpens an already existing diagram by removal of material contributing to background scatter.

Inner wall layers from the central siphon of all members of the Dasycladaceae, examined either singly or as carefully aligned stacks, yield a beautiful fibre diagram (Fig. 9.4) when the beam lies at right-angles to the cell axis. The diagram is the same whether

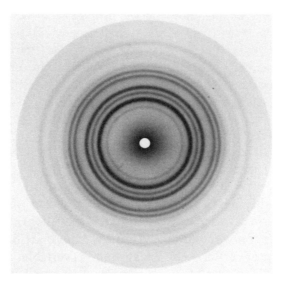

Fig. 9.3 X-ray powder diagram of a pellet of the walls of *Acetabularia crenulata* after treatment with hot 1 per cent HCl.

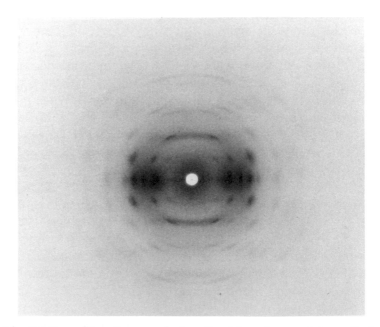

Fig. 9.4 X-ray fibre diagram of a bundle of inner layers of the wall of *Batophora oerstedi* after water extraction and gentle chlorination. Siphon axis ↕, beam normal to axis.

244

the beam also lies normal to wall surface or parallel to wall surface so that, unlike cellulose walls, mannan walls do not show uniplanar orientation. These are the only plants known from which such a diagram can be obtained and therefore provide the only basis from which the structure of mannan may be deduced. With outer wall layers, the same diagram is found, but much more dispersed and with the meridian lying transversely. This confirms the different orientations of the mannan chains between outer and inner wall layers deduced above. Even with the best of these fibre diagrams the arcs, particularly the stronger equatorial arcs, are merged into a surrounding penumbra which must be taken into account in discussing the possible structure.

The X-ray spacings derivable from the fibre photograph are given in Table 9.1 together with the Miller indices corresponding to an orthorhombic cell with dimensions $a = 7.21$ Å, b (fibre axis) = 10.27 Å, $c = 8.82$ Å (Frei and Preston, 1968). This is the smallest unit cell consistent with the diagram. There are indications that, as with cellulose, the true unit cell may be either twice or four times as large as this (involving a doubling of one or both of the a and c parameters) so that the mannobiose residues shown in Fig. 9.5 may need to be slightly adjusted by, for instance, relative translation shifts. The model of Fig. 9.5, presented by Frei and Preston (1968) is based on the following considerations:

(i) The 10.27 Å repeat along the fibre axis means that the chains must be in the Herman's configuration and the sugar residues be in the C1 form, as for the parallel case of cellulose. There is therefore probably a hydrogen bond between the —OH on carbon 3 of each residue and the ring oxygen of the next in the same chain.

(ii) The density of the mannan (a good deal of which is not crystalline) is 1.57. Two mannobiose residues per cell gives a crystalline density of 1.63 which is acceptably close enough to 1.57.

(iii) The a and b parameters of the unit cell, and the strong 101 and 102 reflections, demonstrate that the sugar rings cannot lie either in the ab or the bc plane.

(iv) The meridional reflections are either weak (010, 040 (missing)) or diffuse (020).

The corner chains are accordingly positioned with the rings

SCIENCE

TABLE 9.1

X-ray spacings of algal mannan

mannan I			mannan II		
	observed	calculated		observed	calculated
E 1	5.58 *VS*	5.58 (101)	E 1	8.26 *VS*	8.26 (202)
E 2	4.41 *S*	4.41 (002)	E 2	4.50 *VS*	4.50 (20$\bar{2}$)
E 3	3.76 *S*	3.76 (102)	E 3	3.95 *VS*	3.95 (004)
E 4	3.34 *W*	3.34 (201)	E 4	2.95 *VW*	2.95 (20$\bar{4}$; 506)
E 5	2.73 *W*	2.73 (103)	E 5	2.74 *VW* d.	2.75 (606)
E 6	2.29 *VW* d.	2.28 (203)	E 6	2.67 *VW* d.	2.68 (407); 2.66 (307); 2.63 (006, 507)
			E 7	2.47 *VW* d.	2.50 (20$\bar{5}$)
			E 8	2.35 *VW* d.	2.35 (408); 2.36 (70$\bar{7}$); 2.34 (508); 2.31 (308)
			E 9	2.26 *W*	2.25 (40$\bar{4}$); 2.23 (30$\bar{5}$); 2.22 (208)
I 0	10.20 *VVW*	10.27 (010)			
I 1	6.72 *W*	6.69 (011)	I 1	8.73 *W*	8.73 (111)
I 2	4.89 *M*	4.90 (111)	I 2	3.99 *S*	3.98 (11$\bar{3}$)
I 3	3.53 *M*	3.53 (112)	I 3	3.57 *M*	3.52 (315)
I 4	2.68 *W*	2.69 (212)	I 4	2.74 *W*	2.74 (516); 2.75 (116); 2.72 (11$\bar{5}$)
I 5	2.32 *VW*	2.34 (310)	I 5	2.59 *W*	2.58 (317); 2.60 (417); 2.55 (016)
I 6	2.08 *VW*	2.07 (312)	I 6	2.35 *W* d.	2.33 (11$\bar{6}$); 2.30 (717)
II 10	4.87 *M* d.	5.13 (020)	II 0	5.10 *S*	5.10 (020)
II 1	4.45 *M*	4.44 (021)	II 1	4.70 *W*	4.85 (021); 4.87 (121)
II 2	4.14 *VW*	4.18 (120)	II 2	4.29 *S*	4.28 (022)
II 3	3.78 *VW*	3.78 (121)	II 3	3.48 *MS*	3.45 (224)
II 4	3.38 *VVW*	3.34 (022)	II 4	2.73 *W*	2.77 (525); 2.67 (025)
II 5	3.03 *W*	3.03 (122)	II 5	2.47 *VW*	2.50 (126); 2.42 (626)
II 6	2.45 *VVW* d.	2.45 (222)			
II 7	2.11 *VVW* d.	2.11 (321)			
III 1	3.19 *W*	3.19 (031)	III 1	3.35 *S*	3.33 (031; 131)
III 2	2.92 *W*	2.92 (131)	III 2	3.02 *W*	2.98 (233)
III 3	2.70 *W*	2.70 (032)	III 3	2.69 *W*	2.70 (134); 2.68 (133)
III 4	2.47 *W*	2.48 (230)	III 4	2.55 *MW*	2.58 (034); 2.52 (335)
III 5	2.37 *VW*	2.39 (231)	III 5	2.47 *W*	2.47 (435; 23$\bar{3}$)
III 6	2.15 *VW*	2.16 (232)	III 6	2.10 *W* d.	2.10 (337); 2.14 (636); 2.11 (43$\bar{1}$); 2.08 (036)
			IV 0	2.55 *VW*	2.55 (040)
IV 1	2.47 *W*	2.46 (041)	IV 1	2.53 *VVW*	2.52 (041, 141)
IV 2	2.33 *W*	2.33 (141)	IV 2	2.45 *W*	2.45 (14$\bar{1}$); 2.46 (142); 2.43 (042; 242)
IV 3	2.22 *W*	2.21 (042)	IV 3	2.25 *W*	2.24 (244); 2.23 (344); 2.22 (24$\bar{2}$)
IV 4	2.12 *VW*	2.12 (142)	IV 4	2.16 *W*	2.18 (444); 2.14 (044)
IV 5	2.03 *VVW*	2.03 (241)	IV 5	1.99 *VW*	1.99 (045); 1.98 (346); 1.97 (446)
V 1	2.00 *VW*	2.00 (051)	V 1	1.97 *VVW*	1.97 (052); 1.99 (152)
V2	1.93 *VW*	1.93 (151)	V 2	1.94 *VW*	1.94 (253)

Key *VS* = very strong; *S* = strong; *M* = medium; *VW* = very weak, reflections.

E = equatorial; I = first layer line; II = second layer line; III = third layer line; IV = fourth layer line; V = fifth layer line. The arabic numerals denote the sequence of the reflections measured from the meridian.

d = diffuse.

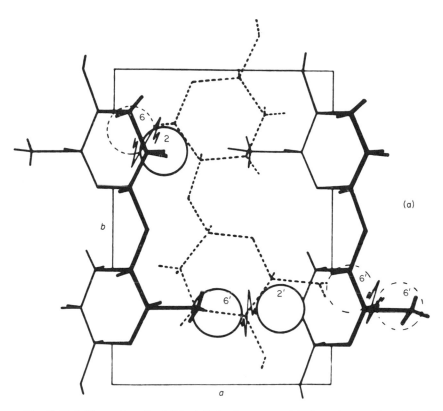

Fig. 9.5(a) The unit cell of β-1, 4-linked mannan (a) viewed normal to the *ab* plane (∿ represents interchain H bonds).

lying in the 102 plane and the central chain inserted in such a way as to allow hydrogen bonding to the corner chains. There is some uncertainty as to the hydrogen bonding to the —OH's on carbon 6' of the corner chains, depending on the orientation of these groups and two possibilities are included in the model. The whole structure recalls that of cellulose (p. 115), with differences due to the different hydrogen bondings following the translation of the —OH on carbon 2 from one side of the sugar ring to the other. Conformational analysis carried out by Nieduszynski and Marchessault (1972) are broadly in support of this model.

The mannan of these walls can be persuaded to adopt a different crystalline structure just as can cellulose. Material treated with alkali solutions above a certain limiting strength (e.g. 12 − 14 per cent KOH, depending on the source of the mannan and its pre-

247

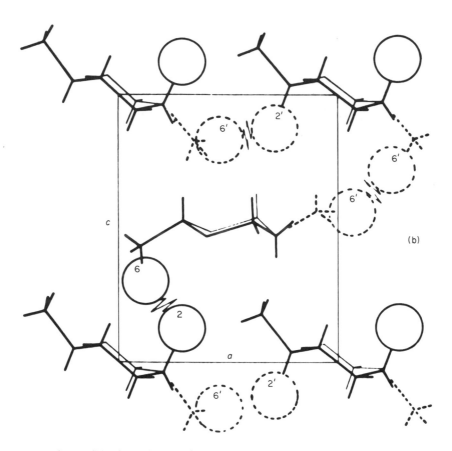

Fig. 9.5(b) The unit cell of mannan viewed normal to the *a c* plane.

treatment) and washed, yield the X-ray diagram of Fig. 9.6. This change in the diagram, and the general appearance of this new diagram, is strongly reminiscent of cellulose. By analogy, the native mannan has therefore been called mannan I and this crystallographically different form mannan II (Frei and Preston, 1968). The diagrams of Fig. 9.6 make it clear that mannan II, unlike mannan I, shows uniplanar orientation, the planes of about 8 Å spacing tending to lie parallel to wall surface. The unit cell is monoclinic with dimensions $a = 18.8$ Å, b (fibre axis) $= 10.2$ Å, $c = 18.7$ Å, $\beta = 57.5°$; the unit cell contains eight mannobiose residues, and must include a number of water molecules. If the central chain in the unit cell of mannan I lies antiparallel, so must one half of the chains in mannan II. A possible arrangement in the

248

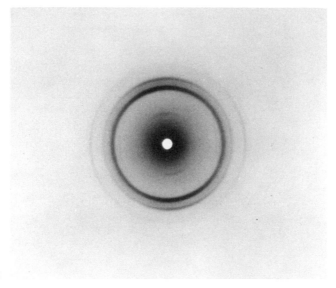

a

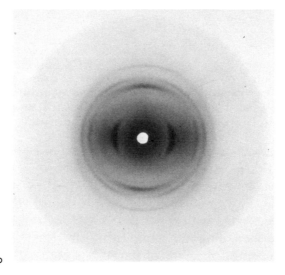

b

Fig. 9.6 Fibre diagram of a stack of aligned outer wall layers of *B. oerstedi* after treatment with hot water and 15 per cent KOH. (a) beam normal to wall surface, siphon axis ↔; (b) beam parallel to wall surface and to siphon axis.

basal plane has been given by Frei and Preston (1968).

There is some difficulty in obtaining diagrams of mannan II, since highly oriented diagrams are to be expected only from inner

249

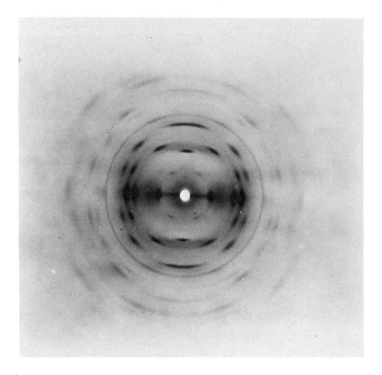

Fig. 9.7 Fibre X-ray diagram of a bundle of inner layers of *Cymopolia barbata* after treatment with 6 per cent lithium hydroxide in 69 per cent alcohol and drying under tension. Siphon axis ↕, beam normal to siphon axis. The faint ring is due to a salt contaminant.

layers of the walls of *Batophora*, *Dasycladus* or *Cymopolia* which unfortunately after alkali treatment become either soluble in water or highly swollen with disoriented crystallites; this is why the diagrams of Fig. 9.6 show so much disorientation. When, however, filaments are treated with 6 per cent LiOH in 30 per cent alcohol and then washed in water while being stretched (about 10 per cent), the beautiful diagram of Fig. 9.7 is obtained, giving X-ray spacings listed in Table 9.1. It will be noted that the uniplanar orientation has been destroyed by stretching. Subsequent boiling of the filaments in water does not induce any change from mannan II to mannan I but treatment with boiling glycerine for 1 hour does cause the material to revert in greater part. This again is analogous to the behaviour of cellulose.

The conversion of mannan I to mannan II is a general phenomenon observed also with ivory nut endosperm. There is, however, one odd and so far unexplained difference between algal mannan and palm-seed mannan. The fraction of the algal mannans dissolved in alkali and reprecipitated, while occasionally giving a sharp mannan II diagram (as to be expected by analogy with cellulose) usually yields only two or three very diffuse rings. Algal mannans do not therefore crystallize well under these conditions – with the corollary that they are not synthesized in the plant as separate chains which then come together, a matter of importance for wall biosynthesis (Chapter 13). Palm-seed mannan, on the other hand, precipitated from alkali solutions still invariably gives the diagram of mannan I.

(c) *Electron microscopy*
The walls of these algae are not readily preparable for electron microscopy. For one thing, the outer regions are too compact and the inner regions too soft to allow lamellae to be stripped away. For another, the amorphous or para-crystalline material which needs to be removed in order to reveal the crystalline mannan is itself mainly mannan. Accordingly, it is never certain that treatments required to clarify any microfibrils, even through carbon replication, may not destroy them. In the upshot, the walls mostly show fine structure which does not change noticeably with any of the mild chemical treatments applied, but with none of the treatments always successful with cellulose is there any sign of the presence of microfibrils.

The general electron microscopic appearance is illustrated in Fig. 9.8 for *Dasycladus*. This represents a carbon replica of lamellae revealed by splitting the inner part of a wall after hot water extraction. The low magnification micrograph (Fig. 9.8a) reveals a 'grain' and lamella tearing which suggests the presence of microfibrils lying almost longitudinally and this would harmonise with the X-ray diagram. At higher magnification (Fig. 9.8b), however, the 'microfibrils' become loosely arranged files of granules, some 10 nm diameter, sometimes fused into short rods not more than 20 nm long. Direct electron micrographs of real wall lamellae show very much the same structure though the rod-like bodies are perhaps more in evidence, but again at higher magnification these tend to fade into the granular background.

The mannan of these algae contrast markedly, therefore, with the mannan of the endosperm of ivory nut and date palms reported by Meier (1956 and 1958). His wall preparations contain

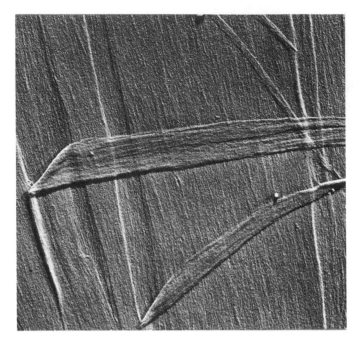

Fig. 9.8 (a) **Electron micrograph of carbon replica of face exposed by tangential splitting of wall of** *Dasycladus vermicularis* **(hot water extracted). Magnification 6900 X, shadowed Pd/Au. Siphon axis ↕.**

mannan B (the fraction with the longer chains in his material) with some 20 per cent cellulose and show clear microfibrils. Although the mannan in these specimens is not crystalline, and while Meier does not rule out the possibility that the microfibrils may be mannan/cellulose complexes, he is strongly of the opinion that the mannan present is segregated into purely mannan microfibrils. This view is, incidentally, accepted by Frey-Wyssling (1959) and Roelofsen (1959) without qualification.

Meanwhile, Mackie and Preston (1968) have re-examined the above findings of Frei and Preston (1961 and 1968) by avoiding hot water treatment of algal cell walls and substituting immersion in 0.3 per cent aqueous sodium lauryl sulphate and 50 per cent aqueous urea for *Codium fragile* and *Acetabularia crenulata*. Some preparations then yield electron microscopically visible microfibrils although the typical appearance is still the rodlet/granule reported by Frei and Preston. The microfibrils have an encrusted

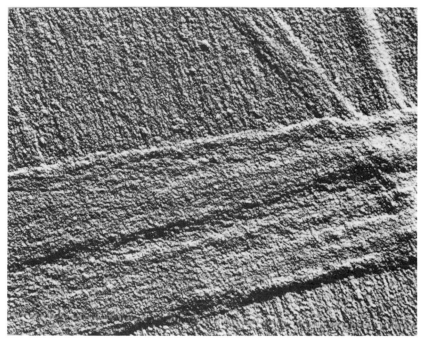

Fig. 9.8 (b) **Part of (a) at magnification 47 000 X.**

appearance and range in diameter from about 10 nm to 25 nm. It is not clear whether only certain parts of the wall are microfibrillar or whether the distribution of microfibrils parallels the distribution of physical properties already mentioned (p. 240).

It is, however, quite certain that the extraction processes normally used with cellulose walls causes any mannan microfibrils in the algae to be altered while not interfering with the coherence of the walls or with the parallel alignment of the crystallites. It remains therefore odd that mannan should be found as microfibrils in higher plants following the cellulose extraction procedure. Perhaps the microfibrils seen by Meier (1956 and 1958) are cellulose microfibrils in which the usual hemicellulose coat (in this case, of mannan) is abnormally but not, as simple calculation shows, excessively thick. This would explain why the so-called mannan microfibrils stain blue with $I_2/ZnCl_2$ which algal mannan does not.

In gross morphology, therefore, the cell walls of mannan weeds, particularly those in the family Dasycladaceae, resemble those of higher plants. The apex of the central siphon is clothed in a thin wall, continuous over the branchlets and down the whole siphon, in which the mannan crystallites tend to lie transverse to the axis. This therefore resembles, if in fact it is not ontogenetically identical with, the primary wall of higher plant cells. Away from the growing apex, the deposition of an inner layer begins in which the mannan crystallites tend to lie longitudinally, becoming progressively thicker toward the base of the plant. This clearly is equivalent to the secondary wall of higher plants and explains the appearance of Fig. 9.1. It represents, moreover, one cycle of a crossed-fibrillar structure which becomes a multi-cycle in *Valonia*-type cellulosic walls.

The 'secondary' layer may well, as in higher plants, lead to cessation of longitudinal growth. There are, however, odd features of growth in the Dasycladaceae. *Dasycladus clavaeformis*, for instance, grows quite normally by tip growth, producing successive whorls of branchlets separated by internodes which proceed to elongate for a short time. As shown by Hämmerling (1944), while the apex grows vigorously in this way, producing a series of new internodes, the whole plant may become shorter by progressive contraction of the older internodes below. Similarly, the 'stipe' of *Acetabularia* contracts in length, with transverse folding of the wall, though only after the umbrella has developed (Hämmerling, 1944). There is no evidence of contraction in *Batophora* or in *Neomeris* and *Cymopolia* where, however, the heavy incrustation with calcium carbonate may prevent it; but the wall morphology, particularly the transverse folding of outer lamella (Frei and Preston, 1968) (Fig. 9.1) suggests that it may occur. While this recalls the contraction of contractile roots of higher plants, the phenomenon seems more far reaching and extensive and would seem to call for further attention. It is always associated with transverse folding of outer lamellae which may extend to inner lamellae. The innermost face of this 'secondary' layer − always present in contracting cells − remains, however, smooth. This could mean that contraction of the cells is due to contraction of the inner wall layer in which the mannan chains lie parallel to the axis of contraction. Since the cells do not become wider, the wall must become thicker or more dense and how this could happen is not clear. Perhaps contraction is associated with a change in the deposited mannan (or some of it) from an amorphous or a para-crystalline condition to a crystalline condition. It could be significant that the mannan is not crystalline in fresh *Codium* walls and at the best shows only limited crystallisation in the highly differentiated walls of the Dasycladaceae.

9.2 The xylan weeds

The xylan weeds fall into four families, the Bryopsidaceae, the Caulerpaceae, the Udotaceae and the Dichotomosiphonaceae, all at one time members of the old order Siphonales and mostly inhabitants of the warmer seas. They are all of filamentous construction with finely lamellated walls and therefore lend themselves admirably to all the physical techniques involved throughout this book. The filaments are coenocytic and sometimes branched (often dichotomously) with septa occurring in rare instances if at all. The degree of complexity reached in the plant body depends partly on the extent to which the filaments are aggregated and partly upon the level to which they are specialised. The simplest expression is found in *Bryopsis* in the form of a single, flattened pinnately branched filament. As already mentioned, a greater morphological differentiation in *Caulerpa* is based on modifications of the branches of the filament, simulating in a remarkable way the habits of diverse higher plants. Complexity of form is reached in the Udotaceae by association and specialisation of filaments to form a pseudoparenchymatous body associated with incrustations, often heavy, of calcium carbonate (as aragonite). Finally, *Dichotomosiphon,* a small freshwater alga, consists of stiff threads with marked dichotomous branching.

9.2.1 Chemical constitution

The walls of all these plants are prominently microfibrillar and the microfibrils contain chains of a xylan which is β-1, 3-linked as shown by Frei and Preston (1961 and 1968) from whose papers the following description is taken. Hydrolysates of the whole wall contain predominantly xylan with much smaller amounts of glucose and some galactose but with little trace of other sugars or uronic acids. With the exception of *Bryopsis,* xylan is in excess over glucose to the extent of $3:1$, a ratio which must be regarded as minimal since the method of hydrolysis (using sulphuric acid) tends to remove xylose. The high glucan content of *Bryopsis* is a genuine characteristic of the walls of this genus and is not a contaminant by, for instance, starch. The glucan can be located in the wall, as will appear below.

The fraction of the wall soluble in hot water ranges from about 7 per cent to 17 per cent, about a mean of 10 per cent. It consists in the main of a glucan. Similarly, the fraction extracted by chlorination (12 — 48 per cent of whole wall weight) is in the main a glucan, and a repetition of the hot water treatment then removes further amounts of a polysaccharide which is again glucan. None of these treatments remove more than trace amounts of xylan so that the residue — the *holoxylan* — is enriched in this polysaccharide and free of all the sugars found in the hydrolysates of whole walls except for some glucan. The ratio xylan : glucan is about 4 : 1 (except again for *Bryopsis*), a figure which agrees well with that of Miwa *et al.* (1961) relating to *Caulerpa* and *Halimeda.*

Holoxylan, unlike holocellulose, is characteristically soluble in alkali, the outer layers of the wall, referred to in the literature as the 'cuticle' being the last to be dissolved. The concentration of alkali necessary for complete solution is low, ranging from 10 per cent to 17 per cent KOH according to genus and species. Treatment with solutions weaker than these removes a quantity of material which increases as the concentration increases, but does not result in any further fractionation; the holoxylan has much the same composition as the residue left after treatment with an alkali solution which just fails to dissolve it. Indeed, this residue is still microfibrillar and crystalline and still contains glucan. By a somewhat loose analogy with cellulose, it could be that the glucan (which is alos 1, 3-linked) forms a cortex in the microfibril surrounding a central xylan crystallite.

Only *Bryopsis* gives a blue colouration with I_2/H_2SO_4 reminiscent of cellulose, i.e. only the genus with a high concentration of glucan in its walls. The holoxylan also gives a positive reaction, like holocellulose. After extraction with 6 — 8 per cent KOH, however, the reaction fails. Removal of the ability to stain can also be achieved by treatment with 2 per cent alkali at room temperature after pretreatment with 1 per cent HCl, and the residue is then xylan with only a trace of glucan. These are not the solubility properties characteristic of normal cellulose and this is in part why *Bryopsis* was not treated earlier with *Halicystis.* Frei and Preston (1968) were unable to substantiate the claim of Parker, Preston and Fogg (1963) that the walls of *Dichotomosiphon* stain blue with I_2/H_2SO_4 using identically the same material.

The holoxylan of all species is to a large extent, but not wholly, soluble in cuprammonium, an outcome which is unexpected in view of the reported insolubility in this reagent of β-1, 3-linked polysaccharides. The material which dissolves can in part be fractionated with the help of Benedict's reagent to give a precipitate, mainly xylan, and a solution containing mainly glucan.

9.2.2 Physical organization

(a) Polarization microscopy

The birefringence of the walls has two components, an intrinsic birefringence due to the internal structural anisotropy of the crystallites, and a form birefringence due to their mutual arrangement. The following statements refer to material in which the latter has been removed by mounting the walls in Canada balsam.

Except for the outer layers in *Bryopsis*, and perhaps *Caulerpa*, the walls are negatively birefringent when viewed in optical section, in complete contrast to the positive birefringence of cellulose or mannan walls. This, of itself, means that these xylan walls are built on a plan in some way materially different from that of these other walls; *in effect* the molecular chains must stand at an angle of less than 45° to the line normal to wall surface. The negative birefringence of *Caulerpa* walls was noticed long ago (Nägeli and Schwendener, 1877) and the contrast with cellulose walls was clearly pointed out (Correns, 1894; Preston, 1931a). In face view also the walls are birefringent though the birefringence is low. With filaments of *Udotea*, *Halimeda* and *Penicillus*, and with the trabeculae of *Caulerpa*, the birefringence is again negative. Exceptionally, the birefringence is patchy with the thick walls of *Caulerpa*, though weak even in the most highly birefringent areas. When the wall is stretched, however, the birefringence in face view becomes strongly negative with respect to the axis of strain. Each area over which a trabecula merges into the wall, shows between crossed polaroids a Maltese cross with the characteristics of that shown by starch, i.e. with the m.e.p. lying radially.

The optical properties are, however, demonstrably not uniform throughout the wall. The walls of the filaments of *P. dumetosus* can be split into an inner and an outer part. In face view the outer part is then isotropic and only the inner part shows the (negative) birefringence characteristic of the whole wall. Moreover, only the inner part shows striations; these lie longitudinally, at right angles to the m.e.p. Probably the walls of most species behave in this way.

Optical heterogeneity of a different kind characterises the wall of *Bryopsis*, however. In untreated walls, both of the main frond and the branches, the birefringence in face view is negative. In optical section, a narrow outer layer can be distinguished by its positive birefringence as against the negative birefringence of the rest of the wall (Fig. 9.9). This had already been noted by Küster (1933). Correspondingly, the scars (C, Fig. 9.9) left on the wall by dehiscence of laterals, as seen vertically, show a Maltese cross when viewed between crossed polaroids, in which the inner bulk of the wall is negatively, and the outer layer positively, birefringent. The corresponding holoxylan shows exactly the same phenomenon although the birefringence is somewhat reduced. When the holoxylan is further treated with dilute KOH solutions, however, two major changes are always observed. The positively birefringent outer layer can no longer be detected, and the whole wall in face view has become *positively* birefringent. No other plant examined shows these phenomena or these changes (except for a fine outer lamella in *Caulerpa* walls which may be very faintly and positively birefringent). Explanation of the phenomena is postponed until the fine structure of the wall has been presented (p. 268).

Since one may suppose that the xylan crystallites, like cellulose and mannan crystallites, are long and thin — and must therefore be deposited parallel to wall surface — it may be concluded that the crystallites themselves are negatively birefringent.

(b) *Electron microscopy*
The walls of all xylan weeds contain abundant microfibrils some 10 nm wide which certainly represent at least the crystalline part of the xylan. The above supposition is therefore fulfilled. Microfibril widths in these plants are a little uncertain because the measurements can be made only on whole lamellae; unlike cellulose microfibrils, xylan microfibrils appear to be strongly adherent, never forming a loose fringe at the edges of a torn lamella and never separating satisfactorily in blended material. In outer wall layers which are so heavily incrusted that alkali treatment is necessary in order to reveal microfibrils, they appear to be of the order of 20 — 30 nm wide, a difference from those of inner lamellae which appears to be real in spite of the difficulties of measurement.

258

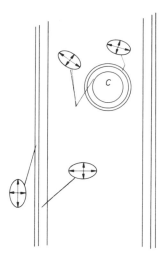

Fig. 9.9 Diagrammatic representation of the optical properties of a flattened, chemically untreated specimen of *Bryopsis*. The vertical lines on either side delineate the outer and inner wall layers and the circles represent a branch scar again with the two wall layers. The ellipses mark the trace in the wall of the optical index-ellipsoid at the points marked.

This is not an effect of alkali treatment since inner layers do not show larger microfibrils after alkali. Either there is a real difference between inside and outside or the microfibrils of outer lamellae are aggregates of narrower fibrils inextricably welded together. The mutual arrangement of the microfibrils varies from complete randomness in the outer wall layers to almost precise alignment parallel to the filament axis in the inner parts of the walls (Fig. 9.10). The microfibrils are often heavily encrusted with amorphous material, much of which can be removed by hot water and chlorination and must therefore be mainly glucan. The amount of this material increases across the wall from inside to outside to the extent that outer lamellae are welded together giving the appearance of a separate entity erroneously called a *cuticle*. Carbon replicas of the outer surface normally appear smooth and microfibrils can be detected only after the material has been treated with KOH solutions. Consequently most of the observations quoted here refer to the holoxylan after KOH treatment.

The simplest microfibrillar arrangement occurs in *Bryopsis*; the microfibrils lie throughout almost at random, with a slight tendency towards transverse orientation on the inner face of the wall and towards longitudinal orientation on the outer face (Fig. 9.11). A

259

Fig. 9.10 **Electron micrograph of the inner wall layer in an untreated filament of *Penicillus dumetosus*. Magnification 34 700 X, shadowed Pd/Au, cell axis ↕.**

somewhat higher degree of order is found in *Caulerpa*. Observation here is complicated by the trabeculae crossing the cell lumen from side to side, preventing any stripping of this beautifully laminated wall. The inner face of the wall can be visualised, however, by slitting a rhizome longitudinally and preparing a carbon replica of one half. This then carries a replica both of the wall and the trabeculae (Fig. 9.12). Overall the microfibrils lie at random, but closely parallel arrays occur of limited extent and at various azimuths, with occasional tendency toward a crossed structure. Within the trabeculae the microfibrils lie in close arrays, parallel to trabecula length except on the outer surfaces where they lie at

Fig. 9.11 Electron micrograph of carbon replica of outer face of cylindrical main axis of *Bryopsis plumosa* (holoxylan after 4 per cent KOH). Magnification 21 700 X, shadowed Pd/Au, cell axis ↕.

random. In section, the inner wall layers at least pass over smoothly into the trabeculae (Fig. 9.13) and it is to be noted that the *inner* lamellae of the wall pass over into the *outer* lamellae of the trabeculae. The microfibrils of the outer surface of the wall are scarcely resolved even after severe treatment of the wall with alkali. The signs are, however, that they also lie at random. From the point of view of the connection between wall architecture and growth already several times referred to, note may be taken of the circumstance that *Caulerpa* grows by tip growth.

These xylan walls are as clearly microfibrillar as are cellulosic walls. The microfibrils cannot, however, be mistaken for cellulose microfibrils especially with the Udotaceae and Dichotomosiphonaceae. When the microfibrils lie parallel, as on the inner face of the wall of filaments of *P. dumetosus* (Fig. 9.10) the parallel array is

261

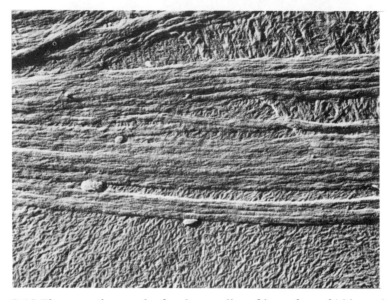

Fig. 9.12 Electron micrograph of carbon replica of inner face of 'rhizome' of *Caulerpa prolifera* (holoxylan after 8 per cent KOH). Magnification 21 700, shadowed Pd/Au. Fragments of trabeculae lie over the wall surface.

always crossed by short fibrillar bodies which are not removed from holoxylan by alkali treatment. These appear to be an integral part of wall architecture and perhaps constitute some form of microfibrillar connection. In wall lamellae progressively further from the inner face the parallel orderliness is progressively reduced, the microfibrils on the outer face are usually disordered and the crossed bodies are no longer evident. In some cases this is accompanied by a marked swing from the longitudinal to a progressively more oblique inclination suggesting that the microfibrils lie in a steep helix.

(c) *X-ray diffraction*
As reported by Frei and Preston (1961 and 1968), air-dried pellets of wall material of any xylan alga, whether untreated or at any

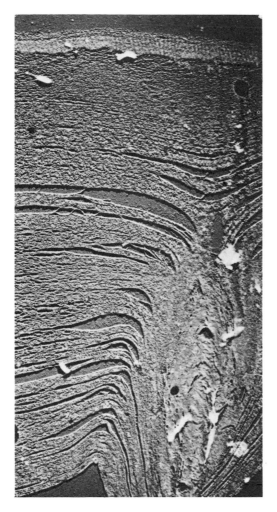

Fig. 9.13 **Electron micrograph of wall section of 'rhizome' of** *Caulerpa* *prolifera* **(holoxylan). Magnification 10 900 X.**

stage of extraction, yield the same distinct X-ray powder diagram. This is characterised by a ring of spacing about 12 Å with a cluster of outer rings, some almost fused, ranging around 4.5 Å and only vaguely defined relfections further out. Since the material after full extraction yields xylan only on hydrolysis, this must be a diagram of xylan. Since the diagram does not resemble that of the β-1, 4-linked xylan of higher plants, the xylan must be linked in some other way and it was realised that the link must be 1,3 some time

263

before Mackie and Percival (1959) and Iriki *et al.* (1960) demonstrated the existence in some of these plants of β-1, 3-linked xylan.
Kreger (1962) reported the finding of the same diagram which he
agreed referred to β-1, 3-linked xylan though he gave no details.
The diagram is much sharper and richer when the material is
conditioned to high relative humidity (Fig. 9.14) and becomes
blurred when the pellet is held dry over P_2O_5. On wetting,
moreover, the lattice swells so that water enters the lattice and is
involved in maintaining crystallinity. This constitutes a distinct
difference from cellulose and mannan. The change in the diagram
is abrupt between 58 per cent R.H. and 65 per cent R.H. and does
not change further with further change in R.H. At 65 per cent R.H.
the xylan contains 30 per cent of its weight as water so that water
in about this amount must partake in the lattice. Even with
Bryopsis in which the wall may contain a glucan which may be
ordered, the diagram is attributable to xylan and xylan only.

When the xylan is reprecipitated from solution, it does not
crystallise well. The diagram is weak and diffuse and hardly improved at high R.H.

A stack of carefully aligned layers taken from the inner part of
the wall of *P. dumetosus* filaments (in which the microfibrils lie
parallel to filament length), set in an X-ray spectrometer with the
beam lying normal to filament length and conditioned to 98 per
cent R.H., yields the beautiful fibre diagram reproduced in Fig.
9.15. The meridian of this diagram lies parallel to microfibril length.
Corresponding X-ray spacings are given in Table 9.2 together with
the indices derived from the monoclinic unit cells given in the
table, each corresponding to an hexagonal lattice. The diagram is
the same when the beam lies normal to wall surface or parallel to
wall surface so that here again there is no uniplanar orientation.
Although most other xylan algae are not so constructed as to allow
the preparation of such a stack, bundles of filaments yield sufficiently good diagrams as to demonstrate that the xylan structure is
the same throughout.

The two lattice structures of Table 9.2 correspond to sets of
circular cylinders in hexagonal close packing, 13.7 Å diameter in
air-dried material and 15.4 Å diameter above 65 per cent R.H., the
repeat distance along the cylinder (fibre) axis being 5.85 Å (air-
dried) or 6.12 Å (moist). The density of purified xylan dried over
P_2O_5 is 1.57 so that the number of xylose residues in a cylinder

TABLE 9.2

X-ray spacings (Å) in fibre diagrams of the holoxylan of Penicillus dumetosus, observed and calculated from hexagonal lattices.

position	air-dried material (unit cell: $a = c = 13.7$, $b = 5.85$, $\beta = 60°$) spacing		material at 72.6 per cent relative humidity (unit cell: $a = c = 15.4$, $b = 6.12$, $\beta = 60°$) spacing	
	observed	calculated	observed	calculated
E 1	11.93 VS	11.90 (100, 101)	13.18 VS	13.2 (100, 101)
E 2	6.75 M	6.75 (201)	7.70 M	7.71 (201)
E 3	6.05 M	6.05 (200, 202)	6.58 S	6.60 (200, 202)
E 4	4.02 W	4.02 (301, 302)	5.05 VW	5.05 (301, 302)
E 5				3.85 (402, 304)
E 6	3.38 W	3.26 (401)	3.63 S	3.69 (401)
E 7				3.30 (400, 404)
E 8			3.03 VW	3.06 (502, 503)
E 9			2.91 S	2.92 (501, 504)
E 10				2.64 (500, 505)
E 11			2.57 W	2.57 (603)
E 12	2.25 W	2.25 (602, 604)	2.52 M	2.52 (602, 604)
E 13			2.40 M	2.40 (601, 605)
E 14			2.19 W	2.20 (600, 606)
				2.19 (703, 708)
E 15			2.13 W	2.14 (702, 705)
E 16			2.04 VW	2.04 (706, 701)
I 1	5.15 S	5.25 (110, 111)	5.56 S	5.56 (110, 111)
I 2	4.44 S	4.44 (211)	4.80 S	4.79 (211)
I 3	4.27 S 4.02 S	4.16 (210, 212)	4.50 S	4.51 (210, 212)
I 4			3.88 S	3.89 (113, 213)
I 5			3.60 W	3.59 (313)
I 6	3.00 W	2.94 (214)	3.25 VW	3.26 (214, 412)
I 7	2.92 W 2.69 W	2.85 (114, 314)	3.13 M	3.16 (114, 314)
I 8	2.65 W	2.63 (414)	2.92 M	2.89 (414)
I 9			2.73 MW	2.74 (215, 315)
I 10			2.64 W	2.63 (115, 415)
I 11			2.34 W	2.44 (515)
II 0	2.92 M	2.92 (020)	3.06 M	3.06 (020)
II 1	2.72 M	2.67 (221)	2.85 S	2.84 (221)
II 2			2.77 W	2.78 (220, 222)
II 3			2.63 MW	2.63 (321, 322)
II 4			2.54 W	2.54 (320, 323)
II 5			2.38 MW	2.40 (422)
				2.37 (421, 423)
II 6			2.26 W	2.26 (420, 424)
II 7			2.17 VW	2.18 (522, 523)
III 0			2.02 VW	2.02 (130)
III 1			1.96 VW	1.95 (230, 232)
				1.97 (231)
III 2			1.89 M	1.83 (330)
				1.88 (331, 332)
				1.85 (333)
III 3			1.79 W	1.76 (433)

VS = very strong; S = strong; M = medium; W = weak; VW = very weak, reflections.
E = equatorial; I = first layer line; II = second layer line; III = third layer line. The arabic numerals denote the sequence of the reflections measured from the meridian.

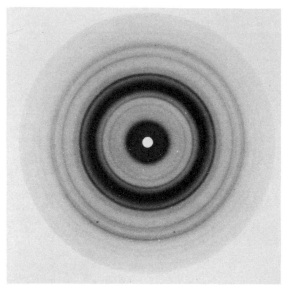

Fig. 9.14 **X-ray diagram of a pellet of holoxylan from** *Caulerpa prolifera* **maintained at 98 per cent R.H. during exposure to the beam.**

13.7 Å wide and 5.85 Å high should be 6.8. The correct number may be either 6 or 7. Number 7 is not acceptable crystallographically, nor would it allow for some water in the lattice. The number may therefore be taken as 6; if each residue holds one molecule of water the density would be 1.57. The crystallographic problem is to station these six residues in the lattice.

Quite simple considerations show the kind of disposition to be looked for, since there are several guide lines:

(1) The xylan chain must be curved on account of the 1, 3-link. In skeletal models rotation around the C^1-O-C^3 link is fairly free but steric hindrance in space-filling models reduces the possible conformations to two or three, all curved. No possible configuration gives a straight chain and the least curved configuration is presented diagrammatically in Fig. 3.8; this shows no steric hindrance.

(2) The curved chains must be fitted together to form a linear microfibril.

(3) The birefringence of the microfibril so constructed must be negative.

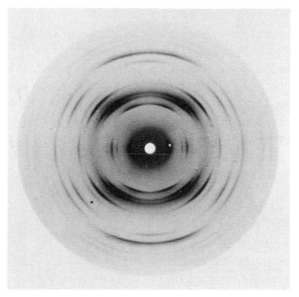

Fig. 9.15 **X-ray diagram of a stack of aligned inner parts of walls of** *Penicillus*
dumetosus **(holoxylan) maintained at 98 per cent R.H. Cell axis $\updownarrow$.**

(4) The length of the xylobiose residue in this chain is about
9 Å. To fit this even in the wet unit cell 6.12 Å high the
length of the residue must be tilted to the fibre axis.

The upshot is — and this is already suggested by the appearance
of the diagram — that the xylan chain is coiled helically, with the
helix axis parallel to the fibre axis. Since the unit cell contains six
xylose residues this disposition could in principle take the form
either of three helices, each with two residues per unit cell, or two
helices with three residues. Frei and Preston (1968) ruled out the
former, for reasons which seemed adequate at the time, and presented a model with two intertwined helical chains. More recently,
Atkins, Parker and Preston (1969) have reconsidered the problem
from the stand point of helical diffraction, and have concluded
that β-1, 3-linked xylan forms a triple, not a double, helix and
Atkins and Parker (1969) have made the necessary calculations of
structure factors. The resulting structure is presented here only
with brief reference to supporting evidence. The original paper

must be consulted for the details concerning this, the second polysaccharide structure shown to be helical (the first was starch), with the most closely defined parameters of all the polysaccharides known.

Three xylan chains are intertwined, each individual helix having exactly six xylose residues per turn in a pitch of 18.36 Å. The axial rise per residue is 3.06 Å as in the alternative model of Frei and Preston (1968). A photograph of the skeletal model is presented in Fig. 9.16a and a space-filling model of one helix in Fig. 9.16b with the helices in a right-hand sense. A novel feature of the structure is that the −OHs in the 2-position point roughly toward the axis so that through operation of the three fold axis a triad of −OH groups is generated, forming a triad of hydrogen bonds, as shown in Fig. 9.17, at intervals of 3.06 Å along the chain, in agreement with infra-red absorption data (Parker, 1969).

Atkins and Parker (1969) constructed both left- and right-handed threestrand helical models and from these, and from the averaged glucopyranose ring coordinates given by Ramachandran, Ramakrishnan and Sasisekharen (1963) (correcting for a sign error in this publication), determined the co-ordinates of each atom in one xylose unit. Computation of the Fourier transform of each helix showed that neither helix could be so constructed as to make absent the 120 reflection. They found, however, that the insertion of one water molecule per xylose unit, in a position at the outside of the structure giving very convincing hydrogen bonding, reduced the calculated intensity to zero for right-hand helices but not for left-hand. This is why the helices are presented in Fig. 9.16 as right-handed. Neighbouring xylose residues are tilted with respect to each other about the C−O−C′ bonds. Using the nomenclature of Fig. 3.7 and taking $\phi = \psi = 0$ when OC−O−CO lie in a plane, the values $\phi = 340°$, $\psi = 65°$ are obtained.

The helical structure of the skeletal polysaccharide perhaps explains some of the unusual features of the walls of these seaweeds. The adherence between xylan microfibrils is much closer than between cellulose microfibrils and this might easily be a consequence of the helical structure. There are presumably 'loose' xylan chains at the surface of the microfibrils and these will be coiled even if not in the same coil as that found in the crystal. These coils could well allow greater entanglement between microfibrils and between these and the matrix which consists in part of a glucan

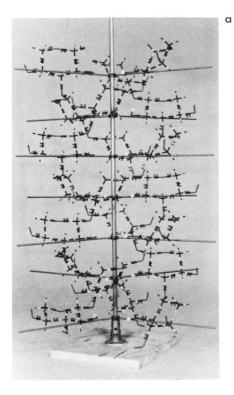

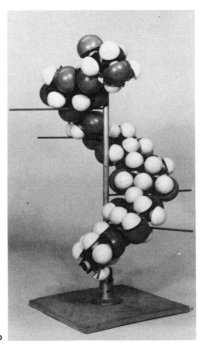

Fig. 9.16 The structure of β-1, 3-linked xylan.
(a) the skeletal model. The central white spheres represent —OH groups. Note that three of these occur together in planes 3.06 Å apart along the fibre axis.
(b) a space-filling model of one chain; black represents carbon; grey, oxygen; white, hydrogen.

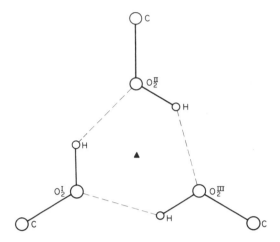

Fig. 9.17 The triad of hydrogen bonds viewed along the axis of the xylan helices. The −C(2)OH of each helix only is drawn. (By courtesy of Dr. K.D. Parker.)

which is 1, 3-linked and therefore probably also coiled. Equally, it is possible that such entanglements might lead to intermicrofibrillar bridges and therefore explain the presence of the short cross bodies of Fig. 9.10. These bodies would then be well developed only where the microfibrils lie parallel to each other and even then would be torn and no longer visible as the lamella became subject to strain during growth. This is precisely as observed. Similarly the coiling explains the negative birefringence of the microfibrils. It is not at present possible to present any quantitative relationships between the birefringence and the structure but it is self-evident that the relatively flat helices are likely to be negatively birefringent.

Bryopsis stands rather apart from the other genera both in the constitution and the physical properties of the wall, as already remarked upon. Removal of the glucan from the thin outer layer of the wall by alkali treatment changes the sign of birefringence from negative to positive. This means that the glucan chains must lie approximately parallel to the xylan microfibrils and that, in the untreated wall, both must lie transversely and the positive birefringence of the glucan (which must be 1, 4-linked) overcompensates the negative birefringence of the xylan. The true birefringence of the skeletal polysaccharide is here revealed only when an 'encrusting substance' is removed. Some time ago, Green (1960) using solely the method of light microscopy on unextracted walls of *Bryopsis*, and assuming the skeletal substance to be cellulose and therefore positively birefringent, delineated the run of the micro-fibrils in a whole plant. By sheer coincidence the overriding bire-fringence of the incrusting glucan led him to the correct answer. In the electron microscope, however, the microfibrils are seen invari-ably to lie almost at random; the polarizing microscope with its high sensitivity to the collective effect of molecular chains lying in preferred orientations, emphasizes what is evidently only a slight tendency toward transverse orientation.

9.3 Xylan-mannan algae

Both polysaccharides discussed in this chapter occur in one and the same plant in two red algae, *Porphyra* and *Bangia*. The Rhodo-phyta are separated into two classes, the Floridiophyceae and the Bangiophyceae (Cronquist, 1960). The algae we are concerned with belong to the latter class, regarded as the more primitive of the two classes from which the other class has been derived.

The plant body of *Porphyra* consists of an extensive mem-branous sheet one cell thick with a cuticle on each side, prominent on the east coast of Britain as deep red sheets draped over the rocks. The wall polysaccharides of this plant have recently attracted a good deal of attention beginning with the isolation by Jones (1950) of a mannan in the alkali extract, a slightly branched polymer of β-D-mannopyranose with 1, 4 links (very much like the mannan of the mannan weeds). Nunn and von Holdt (1957) appear to have been the first to establish the nature of the water-soluble mucilage which makes up over half the weight of the dried alga, and Cronshaw *et al.* (1958) and Turvey and Rees (1958) the first to attempt a definition of the skeletal polysaccharide. Cronshaw *et al.* described the wall as microfibrillar though cellulose is absent. They found, however, that treatment with 24 per cent KOH removed a good deal of xylan and left a residue which was granular and consisted essentially of mannan. They therefore concluded that in this plant cellulose is replaced by mannan and that the alkali-sensi-tive mannan microfibrils are associated with xylan in much the same way that cellulose is with hemicelluloses. Turvey and Rees on

271

the other hand, and later, in a more detailed paper, Peat, Turvey and Rees (1961) came to the conclusion that mannan and xylan, together with some glucan, formed the skeletal polysaccharide.

The situation was cleared up by Frei and Preston (1961 and 1964 b) from whose papers the following statements are drawn. They worked in the main on *Porphyra umbilicalis* but examined also the other genus in the Bangiophyceae in the form of *Bangia fusco-purpurea* in sufficient detail to show that these two plants are very similar from the present point of view.

Hydrolysates of the whole thallus after extraction in 85 per cent ethanol to remove soluble sugars contains mannose and xylose, in relative amounts which vary with the season, abundant galactose and some glucose and 6−O−methyl galactose. Hot water extracts some 50 per cent of the dry weight, mostly as polymers of galactose and 6−O−methyl galactose. The residue after further chlorite treatment, amounting to 19 per cent of the initial dry weight, is enriched in xylose but this is largely extracted by 24 per cent KOH leaving a residue (6 per cent) which is almost entirely mannan. When the cells and the cuticle are separated − as they can be either chemically (2 per cent HCl) or mechanically − and examined separately, it becomes apparent that though each contains both xylan and mannan, xylan predominates in the cell wall proper and mannan predominates in the cuticle, Correspondingly, the X-ray diagram of the cell walls is identical with that of the xylan algae, and can be attributed with confidence to β-1, 3-linked xylan; and the diagram of the cuticle is equally undoubtedly that of β-1, 4-linked mannan. Here, then, are primitive plants which make use, as skeletal polysaccharides, of both of those in the mannan and the xylan algae.

In the electron microscope the walls appear microfibrillar and the cuticle granular, again in harmony with the findings for the other algae. This is best exemplified in material from the basal region of the plant where cells of the thallus grown down as long thin extensions, forming the attachment rhizoids (Fig. 9.18). Here a replica of the surface of a rhizoidal wall may be seen, lying on a replica of the cuticle. with the microfibrils lying parallel to cell length. These rhizoids grow down through the plant, through the intercellular substance (which is largely mannan) and a transverse section (Fig. 9.19) shows that as they do so, they secrete on the outside of the lamellated xylan wall, a sheath which is mainly

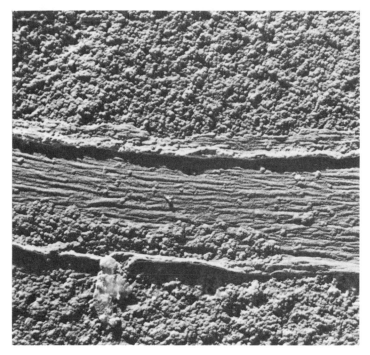

Fig. 9.18 **Electron micrograph of carbon replica of a split face at the base of a plant of *Porphyra*. Magnification 21 700 X, shadowed Pd/Au; showing a rhizoid (xylan) grown through extracellular material (mannan).**

mannan. The microfibrils at the (often globular) tips of growing rhizoids, and in the walls of the basal cell and all other cells of the thallus, lie at random.

9.4 Taxonomic considerations

The green algae considered in this chapter clearly fall into two well-defined groups as far as their walls are concerned, each wall separated from other algae in which the skeletal polysaccharide is cellulose. They are all members of the old order Siphonales (an ordinal name now formally dropped since it is not based on a legitimater family name) which contained a very heterogeneous assembly of plants with, as we now know, a wide diversity of cell-wall composition and organization. Increased knowledge, and greater attention to cytochemical details, have led to the creation of two new

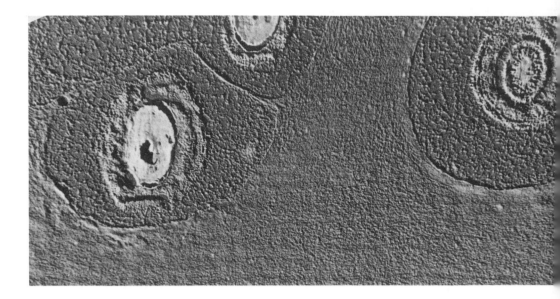

Fig. 9.19 Electron micrograph of a transverse section of the base of a plant of *Porphyra*. Rhizoids (with xylan walls) lie in the intercellular substance (mannan) but separation from it by a secreted layer, also mannan.

orders, the Siphocladales and the Dasycladales, leaving behind the Siphonales *sensu stricto*. While the organisation of the walls is only one of a number of taxonomically definitive criteria, some considerable weight must surely be given to it as the end product of chains of biochemical reactions of importance. In the present context this means that the Caulerpaceae, the Udotaceae and the Dichotomosiphonaceae (all xylan) must be separated from the Bryopsidaceae (xylan and 'cellulose') and, *a fortiori*, from the Dasycladaceae the Codiaceae and the Derbesiaceae (all mannan). The situation seems best to be met by the grouping of Feldman (1946 and 1954) who, for quite other reasons, has created the order Derbesiales, has put into a new order Caulerpales all the true xylan families and has placed the Bryopidaceae with the Codiaceae in a new order, the Codiales. Putting aside the special circumstances of the Derbesiales (p. 228), the Codiales is the only order which, from the point of view of the wall, is heterogeneous and it could be suggested that this could be broken down to two orders, the Codiales *sensu stricto* and the Bryopsidales. It is difficult at any rate to understand why Christensen (1962) should have reunited all these families into one order, the Caulerpales, which in effect reconstitutes the old, unsatisfactory Siphonales.

CHAPTER 10

Flowering plants; secondary walls

Although the morphology of land plants is different from that of
plants growing in an aqueous environment, arising in part from the
need to accommodate to drier conditions, and in spite of a very
different method of development, the fine structure of the walls of
land plants is in general very similar to that already described for
the algae. As far as is known, however, no land plant is dependent
upon either β-1, 3-linked xylan or upon β-1, 4-linked mannan for
its skeletal polysaccharide; this niche is reserved exclusively for
cellulose. Attention will be confined in this chapter to secondary
walls — and chiefly the secondary walls of elongated cells — although brief mention will be made of the structure of the primary
wall as it still exists completely ensheathing the cell. We begin with
the best understood walls, perhaps because they are the most
heavily exploited commercially, the walls of xylem elements and
in the first place with the tracheids of conifers. Attention is first
directed toward *normal* wood as distinguished from the *reaction*
wood on the upper and lower faces of leaning trunks or horizontal
branches.

10.1 Wood tracheids and fibres

Tracheids may be regarded as the primitive cell type in the xylem
from which other cell types have been developed. It has already
been recorded (p. 41) that the walls of these cells consist of
cellulose (*c.* 50 per cent), lignin (*c.* 30 per cent), hemicelluloses
(*c.* 10 per cent, with xylan predominating in conifers) and pectic
compounds, the composition varying to some degree between

species. Tracheids can readily be recognised in macerated material as long cells, (the length/breadth rates varying up to 100 or more) with thick walls, prominently pitted, and ends tapering to a blunt chisel edge which, in the intact wood, lies radially. The cells are usually several tens of μm wide. In section the secondary wall is prominently divided into three layers (Fig. 10.1) first named by I.W. Bailey S1, S2 and S3. The letter S is shorthand for 'secondary' and implies that all three layers are secondary, a view strongly held by Bailey on the grounds of ontogeny. Some later workers have introduced another terminology, using 'transition lamella' for S1 and 'tertiary lamella' for S3. This usage has been contested by Wardrop and Dadswell (1957) and by Nečesany (1957) and there can be no doubt but that Bailey was correct. The terms S1, S2 and S3 have now become standard. These may readily be distinguished in optical sections, particularly if the wood has been stained in safranin or congo red, but are most prominent when thin transverse sections are viewed in a polarising microscope between crossed polaroids. In the 45° position the S1 and S3 layers appear bright and therefore (positively) birefringent while the S2 layer appears dark, or at least not so bright as the other two layers. It will be shown later that this heterogeneity stems largely from differences in microfibril orientation. The wall is also heterogeneous chemically and this will also be considered again (p. 290). For the moment the heterogeneity may be remembered as a precaution against oversimplification. The S2 layer varies considerably in thickness. In the wood laid down by the tree late in the season — the late or summerwood — it is considerably thicker than either the S1 or the S3 layers and often thicker than both taken together; in early or spring wood it is very much thinner and may then be thinner than either S1 or S3. The layer S3 is usually thinner than S1 and in some conifer species, such as *Picea sitchensis* (Sitka spruce) may be missing. The orientation of the microfibrils in these three layers is highly significant in matters concerned with the differentiation of the cell wall and with its physical properties and considerable attention has been given even in recent years to methods whereby the orientations may be measured. This is because xylem is so variable in this as in other features that, although the structure of tracheids is now well known in the general sense, each investigator needs to determine over again the microfibril orientations in any piece of wood of which he is studying properties

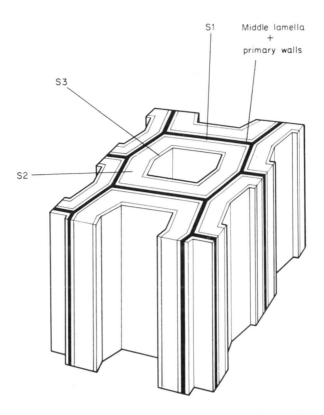

Fig. 10.1 Diagrammatic representation of a segment of a conifer tracheid surrounded by 6 others as in the xylem. The plane walls lie tangentially in the stem.

which are to be correlated with this variable. There are broadly three ways in which the relevant determinations may be made and attention is now given to each of these in turn.

10.1.1 Determination of microfibril angle

(a) *Optical microscopy*
This provides the methods of choice when the spread of orientations over the whole specimen needs to be known and when S1, S2 and S3 must be recognised independently.

Surface markings
In some species the wall carries optically visible striations. As with

the algae, these reflect accurately the direction of the microfibrils; they normally run in fairly steep helices round the cell and mark the microfibrillar helix in the S2 layer (Preston, 1947). They are more easily seen in dry than in wet sections and sometimes are more clearly visualised under a polarising microscope. Equally, the major axis of slit pits, where these occur, lies close to the microfibril direction, though a more reliable measure is the direction of the cracks which sometimes run from each end of the slit (Preston, 1947). The striation direction in one layer, coupled with the m.e.p. of the whole wall, serves to define the microfibril orientation in the other layers (see below).

Orientation markers
With care, elongated crystals can be persuaded to grow in the cell walls and these adopt the orientation of the microfibril. Crystals of iodine and of silver have been used in this way (Bailey and Vestal, 1937; Meylan, 1966). For instance, longitudinal sections, 20 μm thick, treated in a mixture of 10 per cent nitric acid with 10 per cent chromic acid in 95 per cent alcohol, may be stained in a 6 per cent solution of aqueous iodine, may be air dried and then fixed in 40 per cent nitric acid. Iodine crystals may then be observed lying in parallel arrays from which the orientations in S1, S2 and S3 may be detected. In theory the iodine crystals lie in 'elongated porosities of the cellulose matrix' but, regrettably, no such porosities occur in intact wood of sufficient size to accommodate crystals visible in the optical microscope. Only when the crystal lattice forces are sufficient to displace the wood substance (and not merely the cellulose itself) can such crystals grow and the swelling induced in this way, together with what must be drastic effects of the pretreatment necessary to introduce the iodine in the first place, cannot lead to much confidence in this method as a precision tool.

The m.e.p. of the whole wall in face view
It has already been shown (p. 85) that the m.e.p. of the whole wall runs helically around the cell so that for determinations of the helical angle one wall must be removed either from the specimen or at least from the optical path. The angle between the m.e.p. and the direction of cell length then gives the required angle. This is the weighted mean of the angles in S1, S2 and S3 layers and

does not strictly therefore give a meaningful value. In practice, however, the angle is close to the angle in the S2 layer, with an error not normally greater than ± 2°. The tracheids must be separated by maceration of the wood, e.g. by alternate chlorination and heating in 3 per cent Na_2SO_3 to remove lignin. This undoubtedly changes the orientation but only slightly. Subsequent procedures may be as follows:

(a) Following the method of Preston (1934), the macerated cells are fixed to a glass slide by albumin fixative. A sharp razor sliced over the surface of the slide then cuts many tracheids longitudinally in two and the single walls left on the slide are mounted in some standard mountant, e.g. Canada balsam, and the m.e.p. determined. Even with wood selected from one part of one annual ring at a single location in a single tree, the m.e.p. is highly variable for reasons which will be given later. It is therefore necessary to measure at least 50 tracheids and express the results as mean and probable error. The helices are usually, however, all of the same sign, mostly left-handed (S) helices. This method cannot, of course, be used if it is necessary to retain the cells whole. It was the method first used to demonstrate the helical arrangement of the crystallites in tracheids and fibres and to determine the relationship between helical angle and cell length (p. 317).

(b) An ingenious alternative has recently been devised by Page (1969) which allows definitive observations on whole cells or on whole tissues. In this method a beam of plane polarised light is projected downwards through the upper wall of a tracheid lying on a glass slide (Fig. 10.2). A mirror inserted in the lumen reflects the light back through the *same* wall and through a crossed analyser. The cell is extinguished when the weighted mean of the microfibril directions lies in the plane of vibration or at right angles to it and a compensating plate can be used to determined which extinction direction is the m.e.p. As in (a) above, this lies close to the microfibril direction of S2. The mirror consists of a blob of mercury inserted into the cell by submersion in mercury and the application of pressure in the order of 1000 psi. At this pressure the lumen is fully penetrated (Stone, Scallan and Aberson 1966). On release of pressure the mercury column breaks and mercury is withdrawn but an appreciable amount of mercury remains. With whole wood it is necessary that the face exposed shall have resulted from a split

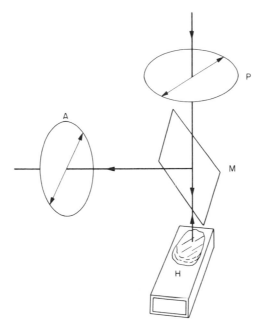

Fig. 10.2 **The method of Page for the determination of the m.e.p. of single walls. P, polarizer; A, analyser; M, half-silvered mirror; H, mercury droplet inside tracheid.**

at the middle lamella; it is therefore necessary to subject the specimen to mild delignification before splitting.

(c) A method suggested by D.R. Morey in 1933, and used at that time by him, seems to have been lost sight of. In this, a technically simple method, isolated tracheids or longitudinal wood sections (again split as under (b)) are impregnated with a dye showing polarised fluorescence and observed in a conventional polarising microscope. The dye molecules line up with the microfibrils, the plane of vibration of the fluorescence lies parallel to m.e.p. of the top wall and this can then be measured.

The birefringence of transverse and longitudinal sections
Following the methods of Chapter 4, the birefringence of a wall in transverse section may be used to calculate the microfibril angle θ if the intrinsic birefringence of cellulose is assumed and the cellulose content of the wall is known. Alternatively, the birefringences of two sections mutually at right angles give the required angle if

281

nothing is known about the crystalline material. This method is tedious except in special circumstances and is better confined to determinations on the individual lamellae in a complex wall and further discussion is postponed until these are discussed (pp. 296, 325).

The birefringence of oblique sections
Again as noted in Chapter 4, the path difference of sections of equal thickness out at increasing angles to the transverse will pass through a maximum when the angle is equal to $(90 - \theta)$. This is again a tedious method better confined to the determination of θ in lamellae inside a complex wall which cannot be examined in other ways. Further discussion is again postponed (p. 296).

(b) *X-ray diffraction*
The above microscopic methods are unavoidable when it is required to know with certainty either the spread of the helical angle between the tracheids of any given wood sample, i.e. the mean, probable error and distribution of the m.e.p., or when the individual wall lamellae need to be identified individually in this respect. They are all destructive, tedious and time-consuming and a search has been made through the years for a method whereby the X-ray diagram of whole wood may be induced to yield at least the mean value of θ, a method which could be used on a specimen say 0.1 cm thick across the grain (Sisson and Clark, 1933; Preston, 1946, 1952; Cave, 1966; Cowdrey and Preston, 1966; Meylan, 1966). With a camera slit 0.05 cm diameter the mean so derived would cover some 200 tracheids. In principle any arc on the diagram can be used but in practice it is preferable to use the equatorial arcs and in particular the strong 002 arcs. The principle of the method can be understood as follows, using the reciprocal lattice concept of Fig. 5.8. The equivalent treatment using the pole figure will be found in Preston (1946, 1952).

Consider first a bundle of parallel fibres in which the microfibrils lie strictly parallel to cell length. Let this, and the reciprocal lattice, be stationed at 0, Fig. 10.3, the fibre bundle lying along FF'. Let P be the reciprocal lattice point corresponding to the 002 planes of any one crystallite (microfibril). In general this will not touch the reflection sphere and there will be no reflection. To generate the diagram of the whole bundle the individual fibril must be rotated about the axis FF', whereby P spreads into a circle PP'.

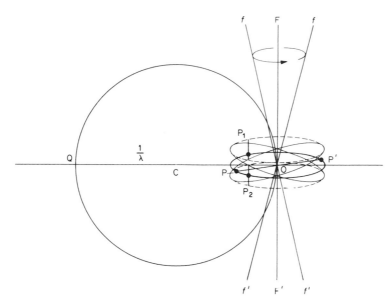

Fig. 10.3 **For explanation, see text.**

This cuts the reflection circle at two points P_1, P_2 on the horizontal circle $P\,P_2\,O\,P_1$, corresponding to two reflections on the equator of the diagram, equally spaced from, and on opposite sides of, the centre. With perfect conditions these are mathematical points of infinite intensity. In practice, the orientation of the microfibrils at least is not perfect and this can be allowed for by allowing the axis FF′ to wobble during rotation to trace out the *solid* cone fff′f′. This causes the points P_1 and P_2 to spread over the reflection circle as in Fig. 10.3 and creates two arcs, rather than points, on the X-ray diagram which are most intense at the equator.

We are now in a position to consider a bundle of fibres in each of which the microfibrils run in a helix making an angle α with the fibre axis and we assume for simplicity that α is invariate among the fibres. Consider any one longitudinal element of any one fibre in which the microfibrils lie in the direction SS′, Fig. 10.4, and let P be the reciprocal lattice point of any one microfibril of this set. To create the diagram of this set we have to make the assumption implicit also in the above argument that there is no strict uni-planar orientation in the wall. We have already seen in the last two chapters that, in those cellulosic walls upon which the relevant

283

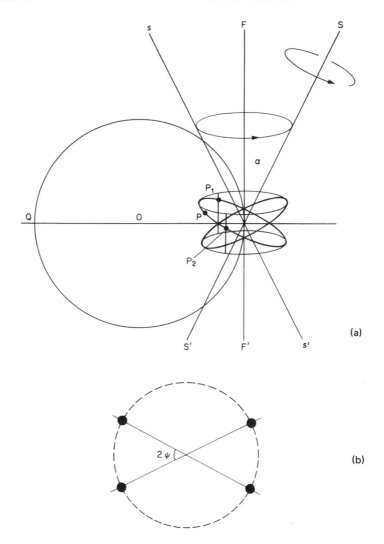

(a)

(b)

Fig. 10.4 For explanation, see text.

observations may be made, there is always some uniplanar orientation in which the 101 planes tend to lie parallel to wall surface and therefore the 002 planes obliquely. In the absence of evidence to the contrary, it is not safe to assume that there is no uniplanar orientation here. Cellulose is, however, commonly uniplanar with an 'error' of ± 70° which means that rotation about SS' is allowable up to about 280° and this is sufficient. P spreads along the circle at right angles to SS' cutting the reflection sphere at P_1 and

P_2 which now lie above the horizontal circle Q 0 so that the reflections lie above the equator on the photographic plate. The diagram of the fibre bundle is now generated by rotation about FF' so that SS' traces the hollow cone S s S' s' and the arcs P_1 and P_2 spread along the reflection sphere as shown. Since the cone is hollow, the arcs are not this time most intense at the centre. Instead they are most intense at the ends so that the 002 planes are represented on the diagram by four points (Fig. 10.4b) again of infinite intensity. Again, a slight wobble of the axis SS' spreads the points into arcs, the most intense points of which lie over the points of Fig. 10.4b. Under these conditions it can be shown that

$$\cos \theta \cos \psi = \sin \alpha$$

where θ is the Bragg angle for the 002 planes. α is therefore determinable.

In practice, α varies from cell to cell in the bundle so that the four points of Fig. 10.4b are very much spread along the circle shown. Provided that the mean α is fairly large, the four arcs remain separate, the diagram is a four point diagram and the above equation may be applied. An example is shown in Fig. 10.5a, representing the wood of *Juniperus virginiana* in comparison with Fig. 10.5b, the diagram of a narrow helix of well-oriented regenerated cellulose threads for which α is 19.8°. Note that in Fig. 10.5a there is apparently an 002 arc on the meridian. This is spurious, being due to an overlap of the tails of the equatorial arcs (Preston, 1946).

Unfortunately, these conditions do not apply either to wood in general or to the fibres of any other plant tissues. There is normally a spread of values of α among the tracheids of even the smallest piece of wood, and the arcs are spread to the extent that they overlap along the equator giving a single arc on each side of the diagram, with an intensity distribution around the circle in Fig. 10.4b as represented diagrammatically in Fig. 10.6 and Fig. 10.7. Sisson and Clark (1933) met this situation by showing that the angular separation between the points on the arc at which the intensity is 40 per cent of the maximum intensity (Fig. 10.6) is closely equal to 2α as measured by other methods. Cowdrey and Preston (1966) used the same convention and checked that the angle so derived was closely similar to the mean derived by method (a) p. 280.

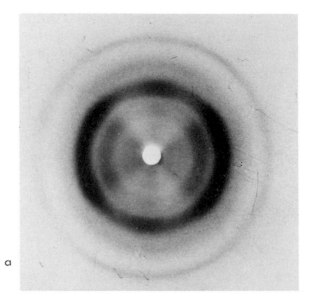

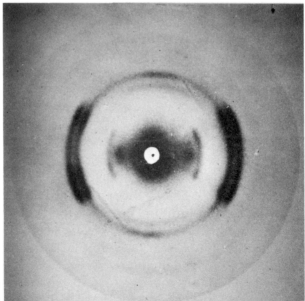

Fig. 10.5 (a) X-ray diagram of xylem of *Juniperus virginiana*, grain ↕, beam normal to grain.
(b) X-ray diagram of a narrow helix of regenerated cellulose threads, helical angle $19.8°$.

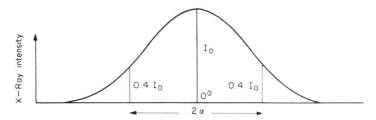

Fig. 10.6 Distribution of intensity along an arc of the 002 circle of Fig. 10.4b for conifer wood when the average angle α is small. $0°$ marks the intersection of the circle with the equator of the diagram.

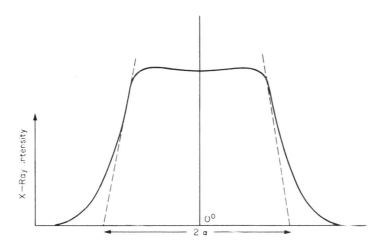

Fig. 10.7 As for 10.6, but α larger though not large enough to separate the two arcs.

There is no theoretical justification for the choice of the 40 per cent points. Moreover, the intensity distribution often has a flattish top, rather than a peak (Fig. 10.7). More recently Meylan (1966) has suggested another parameter, chosen in the first place on practical grounds but with some theoretical basis (Cave 1966). He has used the angular distance between the intercepts of the tangents at the point of inflection on each side of the arc with the zero axis of intensity (Fig. 10.7) as a value of 2α and has checked the reliability of this angle by the iodine impregnation method quoted above. The assumptions made in deriving the theoretical basis for this method, and the theory itself, seem acceptable for

287

tracheids considered to be square in cross-section, but the corresponding theory for cells of circular section leaves something to be desired. Since, however, tracheids — particularly late wood tracheids — are *approximately* rectangular in section, the method would seem to stand as far as conifer wood is concerned. Lofty, El-Osta, Kellogg and Foschi (1974) have recently introduced an elegant new X-ray method. Please see p. 327.

(c) *Electron microscopy*

In the electron microscope the walls of all cells of wood are clearly microfibrillar, the mutual arrangement varying from an almost strictly parallel order to a rather widely dispersed array (Fig. 10.11). When the ordering is good the microfibrillar angle may be determined readily. This carries with it the advantage that any changes in the angle between even the finest lamellae may be detected. As a routine tool, however, it carries the overriding disadvantages that it is laborious and very time consuming and that only exceedingly small volumes of a specimen may be explored at any one time. The X-ray method allows a determination over $25 \times 10^7 \, \mu m^3$ in one operation in, say, 10 hours. Polarization microscopy will yield the mean helical angle and the probable error in, say, four days (allowing time for maceration and the examination of 100 tracheids). A complete survey of the S2 layer alone in 100 of the tracheids contained in the X-ray beam of the size quoted (there are in fact 200) would mean of the order of 10^6 observations at a magnification of 50 000 X; allowing one hour per observation (a gross underestimate) this would occupy more than a human lifetime!

10.1.2 Wall layering

All observation methods agree in demonstrating that the S1, S2 and S3 layers visible in the wall differ both chemically and in their physical organisation. We now consider these in turn.

(a) *Chemical differentiation*

In the first place, lignin is demonstrably not deposited uniformly through the wall. It was demonstrated long ago by removing the cellulose and other polysaccharides from the wall by treatment of sections in strong sulphuric acid (Ritter 1925, 1928 and 1929; Harlow 1932) that the residual Klason lignin, which spreads over the rest of the wall as a loose mesh or even as a powder, is heavily concentrated around the middle lamella region. Bailey (1936), indeed, was able to show by dissecting out the region (middle lamella plus primary wall) that the lignin content here was about 70 per cent as against the 30-40 per cent of the whole wall, though his material probably contained also the S1 layer. Similarly, when wood is attacked by brown rot fungi (*Coniophora* and *Polyporus*) which decompose the polysaccharides, the material remaining shows the lignin concentrated in the middle lamella region (Asunmaa, 1954, *Picea excelsa*; Meier, 1955). Apart from this, the distribution of lignin across the wall varies from species to species (Bailey and Kerr 1935), ranging from radial to concentric and even in discrete lamella (Traynard and Ayroud, 1952; Traynard *et al.*, 1954).

More recent work is roughly in agreement. Using ultraviolet absorption spectroscopy Lange (1954) and Lange and Kjaer (1957) have shown that more than 70 per cent of the wall lignin in spruce lies external to the layer S2 and Ruch and Hentgartner (1960) have added that the lignin which does occur in the secondary wall is uniformly distributed both in spruce wood and in jute fibres. This is agreed by Wardrop (1963) using electron microscope methods and by Côté, Day and Timell (1968). In *Pinus radiata* tracheids and *Eucalyptus* fibres the lignin content is reported to be highest in the middle lamella, especially at cell corners, and higher between radial walls than between tangential walls (Wardrop, Dadswell and Davics, 1961). Most workers agree, however, that there is a major change in lignin content between S1 and S2. Exceptionally, Berlyn and Mark (1965) consider that less than 40 per cent of the lignin lies in middle lamella and primary wall and that the bulk of the lignin is in the secondary wall. Evidently the last word has not been said on the distribution of lignin.

Within the secondary wall itself, Bailey and Kerr (1935) and Kerr and Bailey (1934) showed long ago that the cellulose component and the lignin component in the walls of tracheids and fibres form two interpenetrating matrices in which the lignin may be more concentrated in either concentric lamellae or radial lines, dependent upon species. Concentric lamellation has recently been demonstrated most beautifully by Nečessaný, Jurášek, Sopko and Bobák (1965) and by Bobák and Nečessaný (1967). By electron microscopic examination of the inner face of the wall of fibres of *Populus niger* over a period of 24 h during wall development, they were able to conclude that cellulose is deposited between 12 noon and 6 p.m. and lignin after midnight (0 – 6 a.m.). In the morning hours (6 – 10 a.m.) neither cellulose nor lignin is deposited. Evidence that the 0 – 6 a.m. amorphous layer was lignin depends upon its removal by chlorination. Since this treatment removes other substances in addition to lignin, this alternation cannot yet therefore be regarded as fully established.

The lignin in the secondary wall is certainly deposited around, rather than within, the microfibrils since delignification of a wide variety of lignified cells, including wood tracheids and fibres, causes no change in the X-ray diagram of the cellulose (Astbury, Preston and Norman, 1935; Preston and Allsop, 1939) except for a slight change in microfibril direction in heavily lignified fibres such as coir (Preston and Allsopp, 1939). Accordingly, when cellulose is removed from a lignified wall the direction of orientation of the microfibrils can be detected in the lignin residue (e.g. in normal wood of *Larix laricina* (Côté, Day and Timell, 1968)).

It seems to be a special feature of lignification that any wall layer is laid down and completed long before lignification can be detected so that, for instance, the S3 layer in tracheids is deposited before the layer S2 becomes lignified (Wardrop and Bland, 1959; Wardrop, 1963). Indeed, according to Kremers and Reeder (1963), more than half the cellulose and most of the glucuronoxylan are laid down before much lignin is 'deposited' and most of the lignin appears after all the polysaccharides are present. Lignification therefore occurs at sites remote from the cytoplasm.

The polysaccharides in the wall are not deposited uniformly through the wall either. In the earliest publications defining S1, S2 and S3 Kerr and Bailey (1934) and Bailey and Kerr (1935) using analytical techniques showed that most of the pectin of the wall is present in the middle lamella and primary wall and this was confirmed by Assunmaa and Lange (1953) using spectroscopic techniques. On the other hand, the cellulose content is known to increase from the middle lamella through the wall to the S3 layer (Asunmaa and Lange, 1953b; Meier, 1961). By far the most detailed analysis has, however, been given by Meier (1961), using *Betula verrucosa*, *Picea abies* and *Pinus sylvestris*. Meier proceeded by cutting thin radial longitudinal sections of the young xylem and cambium and dissecting out tracheids or fibres in different stages of development. In this way he obtained a series of cells with middle lamella and primary wall only (M + P), M + P + S1, M + P + S1 + S2, and M + P + S1 + S2 + S3. Hydrolysis of each gave the constituent sugars by chromatography, whence the amounts of polysaccharide could be estimated for each layer by difference, on certain assumptions:

(1) galactose and arabinose are present as pure polymers of each sugar separately
(2) all mannose comes from a glucomannan with a glucose/mannan ratio of 1/2
(3) glucuronoxylan may be obtained from the xylose value by multiplying by 1.52.

This was on the understanding that the 4-O-methyl-D-glucuronoxylan contains 12 per cent 4-O-methyl-D-glucuronic acid and 17 per cent O-acetyl (Bouvery, Lindberg and Garegg 1960). He also made allowance for the fact that glucuronoxylan on acid hydrolysis also yields uronic acid derivatives.

The volume ratios of the (lignin-free) layers was (M + P)/S_1/S_2/S_3 = 3 : 15 : 76 : 6, equivalent, incidentally, to a 3200 per cent increase in wall volume in the mature wall over the cambial walls. The percentage content of the wall fractions could therefore by calculated as presented in Table 10.1. It appears that arabinan is almost completely confined to the middle lamella and primary wall, and galactan to M + P + S1. This parallels the distribution of pectin. On the other hand, glucomannan either increases from (M + P) to S3 (softwoods) or remains approximately constant at a low level (hardwoods), while glucuronoxylan has a higher concentration in S1, S2 and S3 than in (M + P) (hardwood), with a less marked difference, at a lower level, in softwoods. Cellulose content is throughout greater in the secondary layers than in the primary wall. It is therefore apparent that the spectrum of polysaccharides deposited changes during cell differentiation.

(b) *Physical differentiation*
The helical arrangement of the microfibrils in S1, S2 and S3 is reflected in the helical run of the m.e.p. of the wall as a whole and it was in these terms that the structure of tracheids and fibres was first examined. Table 10.2 presents a few examples of the inclination of the m.e.p. to cell length when *single* walls are observed in face view, the sign of the helix being usually left-handed. It has

TABLE 10.1

Percentages of polysaccharides in the different layers of the fibre wall

Polysaccharide	Layers M + P* (%)	Layer S_1 (%)	Layer S_2 (outer part) (%)	Layers S_2 (inner part) + S_3 (%)
Birch				
Galactan	16.9	1.2	0.7	0.0
Cellulose	41.4	49.8	48.0	60.0
Glucomannan	3.1	2.8	2.1	5.1
Arabinan	13.4	1.9	1.5	0.0
Glucuronoxylan	25.2	44.1	47.7	35.1
Pine				
Galactan	16.4	8.0	0.0	0.0
Cellulose	33.4	55.2	64.3	63.6
Glucomannan	7.9	18.1	24.4	23.7
Arabinan	29.3	1.1	0.8	0.0
Arabinoglucuronoxylan	13.0	17.6	10.7	12.7
Spruce				
Galactan	20.1	5.2	1.6	3.2
Cellulose	35.2	61.5	66.5	47.5
Glucomannan	7.7	16.9	24.6	27.2
Arabinan	29.4	0.6	0.0	2.4
Arabinoglucuronoxylan	7.3	15.7	7.4	19.4

* Contains also a high percentage of pectic acid.

already been shown that the X-ray diagram may be interpreted in the same way, namely that in tracheids and fibres the microfibrils lie helically round the cell at an angle to cell length which is closely the same whether measured by the m.e.p. or the X-ray diagram. Moreover, there is no evidence here that there is more than one microfibrillar orientation. On the one hand, when whole tracheids are observed under a polarizing microscope so that the light beam passes through *two* walls, the m.e.p. lies either longitudinally or transversely and the extinction is not perfect; the specimen is therefore clearly a crossed crystal plate combination and can be interpreted as two plates crossing at an angle twice the helical angle measured by the m.e.p. of single walls; whereas *single* walls show no sign that they may also contain crossed crystal plates. On the other hand, the X-ray diagram does not contain arcs other

TABLE 10.2

The angle θ between the cellulose chains and cell length as determined by a variety of methods

Cell type	Method of Determination	A.R.*	θ (°) Range	θ (°) Average	Authority
			Cedrus sp.		
Tracheids	m.e.p.	2	36.0 − 67.0	56.0 ± 1.0	
		4	36.6 − 70.0	49.3 ± 1.1	
		6	37.0 − 62.2	50.4 ± 1.0	
			Larix leptolepis		
		2	36.0 − 73.0	53.3 ± 1.2	Preston (1934)
		4	35.0 − 69.4	50.8 ± 1.2	
		6	34.0 − 58.4	45.5 ± 0.7	
			Abies sp.		
		2	42.8 − 68.8	53.3 ± 1.0	
		4	34.8 − 64.8	47.5 ± 1.0	
		6	35.8 − 62.4	49.8 ± 0.9	
		8	30.0 − 52.8	44.7 ± 0.9	
		11	23.0 − 47.6	34.7 ± 0.8	
			Pinus longifolia		
		2	39.0 − 74.0	56.3 ± 1.3	
		4	34.5 − 61.0	43.7 ± 1.1	Misra (1939)
		6	22.0 − 57.0	35.5 ± 1.2	
			Pinus sylvestris		
	X-rays	2	———	36.2	
		4	———	30.1	Preston (1948)
		8	———	19.3	
		11	———	14.0	
			Cannabis sativa (hemp)		
Fibres	m.e.p.		0.0 − 5.0	2.3 ± 0.3	Kundu and
	striations		0.0 − 6.0	2.0 ± 0.3	Preston (1940)
			Corchorus capsularis (jute)		
	m.e.p.		0.0 − 23.0	7.9	Preston (1941)
			Sisal		
	m.e.p.,		8.0 − 32.3	20.4 ± 7.2	Preston and
	X-rays			18°	Middlebrook (1949)
			Bamboo		
			Dendrocalamus longispathus		
	m.e.p.		2.0 8.0	5.0 ± 1.2	
			Dendrocalamus strictus		Preston and
			3.0 − 7.0	6.4 ± 1.7	Singh (1950)
			Bambusa arundinacea		
			2.0 − 8.0	5.4 ± 1.3	
			Melocanna bambusoide		
			7.0 − 14.0	10.4 ± 1.9	

*A.R. = annual ring.

than those representing a single rather steep helix; the claim by Bailey and Berkeley (1942) that meridional arcs at 3.9 Å and 5.4 − 6.1 Å as seen in Fig. 10.5 a demonstrate the presence also of flat helices fails since these are clearly overlaps of the tails of the arcs corresponding to the steeper helix (Preston, 1946 and 1952).

Nevertheless, Bailey and Kerr (1935) had already drawn attention to a phenomenon which led them to a different view. When thin sections of tracheids or fibres are examined under a polarizing microscope between crossed polaroids, the layers S1, S2 and S3 show different optical properties; S1 and S3 are bright in the 45° position and S2 is dark. On the other hand, S2 is bright when viewed in longitudinal section. These authors therefore concluded that what are now known to be microfibrils are inclined in a slow helix in S1 and S3 and in a steep helix in S2, and they produced strong corroborative evidence for this view. Thus, if isolated cells or whole sections are swollen in alkali then striations can often be observed running more or less transversely in S1 and longitudinally in S2; iodine crystals grown in the wall take up two orientations, one flat and one steep; and attack by soft rot fungi produces diamond-shaped cavities which take up the two orientations (Bailey and Vestal, 1937). There were clearly alternative explanations possible for all these phenomena but subsequent work, by polarization microscopy and X-ray diffraction as well as electron microscopy, has proved this particular explanation to be in substance correct.

The evidence from polarization microscopy is of several kinds.

(1) When tracheids are macerated and split open following method (a), p. 280, a fine lamella may sometimes be seen protruding to one side as shown diagrammatically in Fig. 10.8. This is clearly a lamella which passes under the single wall and may be taken to represent such a lamella though the part visible is probably folded back from the side wall which has been in part cut away. The single wall is then clearly a crossed crystal plate combination. If the m.e.p.s of the whole wall, Δ_t, and of the fine lamella (Δ_1) are measured, together with the angles between the corresponding m.e.p.s (θ_t and θ_1), then following the methods of p. 104, the values Δ_2 and θ_2 for the second component of the whole wall may be determined. This has been done for a large number of tracheids of *Pinus radiata* (Preston, 1947). The angle θ_1 and θ_2 can be

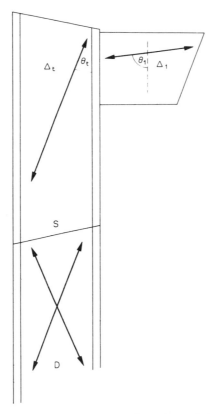

Fig. 10.8 Diagrammatic representation of a split tracheid, D, double wall; S, single wall.

checked against the striation directions visible on the wall. To give an example, in one tracheid θ_1 was 79° and Δ_1 was 9.9° ± 0.8°. For the whole wall, Δ_t was 51.3° ± 2.8° and θ_t was 28.5°. The striations visible on the whole wall were inclined at 32.5° to cell length so that the m.e.p. and striation direction did not coincide. To determine Δ_2 and θ_2, a plane triangle may be used instead of a spherical triangle since Δ_1 is small. A distance Δ_1 is therefore laid out and, at an angle of $2\theta_{1t}$ (81° in this case) a length Δ_t is marked, using the internal angle since Δ_t in the subsequent triangle is the path difference required to close the triangle and this is at 90° to the Δ_t required, Δ_2 turns out to be 50.0° (or, by calculation, 50.0° ± 2.0°). From θ_{12}, the inclination of the second plate may be calculated; the value is 34.5° ± 2.0°, close enough to the striation angle of 32.5° to be acceptable. At the time it was thought that the fine outer lamella was the primary wall; it is now fairly clear that it was S1 so that the values $\Delta_1 = 9.9°$ and $\theta_1 = 79°$ refer to S1. Equally the values $\Delta_2 = 50.0°$ and $\theta_2 = 34.5°$ refer to the combination S2 and S3 and the discrepancy between 34.5° and 32.5° (the striation direction), if this is significant, could mean that the microfibrils of the S3 layer also lie in a flat helix.

(2) Oblique sections, p. 282, have also been used to the same effect with xylem (Wardrop and Preston, 1947 and 1951). The sections were cut at various tilts from the transverse about the radial direction and tangential walls only were examined. With tracheids of *Picea* average values for θ turned out to be about 52° for S1 and 18° for S1; with fibres of *Nathofagus cunninghamii* equivalent values were 90° and about 5° respectively.

(3) The use of two sections at right angles, p. 281, has been applied to the complex fibres of bamboo (Preston and Singh, 1950 and 1952). Unlike the tracheids and fibres studied so far, these monocotyledonous fibres develop from the same ground meristem as do the parenchyma cells, so that here we have an interesting contrast. In transverse section the wall has at least three bright layers with intervening dark layers (the optical behaviour being reversed, of course, in longitudinal section). We may name these for conformity S1 (outer, bright), S2 (dark), S3 (bright), S4 (dark), S5 (bright) in a wall with three bright layers.

Ideally the birefringence of all these layers should be determined for one and the same cell both in transverse and longitudinal section. This cannot be done in practice. The regime is, therefore, to determine the path differences (and thence the birefringences) of the layers in thin transverse section of a large number of cells and to determine the averages. Identical material is then macerated and the refractive indices of the outermost layer (S1) determined in longitudinal view by the immersion method described on p. 86 . The value of the larger of these two refractive indices $n_\gamma^{\parallel}$ depends upon the length of the fibre (as will be discussed later, p. 317). It is therefore necessary to construct a graph relating index to length and to read off the value of $n_\gamma^{\parallel}$ for the average length. For S1, therefore $n_\gamma^{\parallel}$, n_γ^{L} and n_α are known and Equation 4.16 and Equation 4.17 on p. 98 may be used to calculate n_γ and θ for S1. Using this value for n_γ, the value of θ may be calculated for all the other layers (Table 10.3). S1 clearly

TABLE 10.3

Wall structure in bamboo fibres. Angles in degrees ± standard errors.

Species	Angle between fibre length and m.e.p. ($^\circ$)				
	S1	S2	S3	S4	S5
Dendrocalamus longispathus	34.0 ± 1.5	5.00 ± 0.24	19.0 ± 2.0	5.00 ± 0.24	10.0 ± 2.0
D. strictus	34.5 ± 4.0	6.44 ± 0.34	20.5 ± 5.5	6.44 ± 0.34	12.0 ± 2.0

resembles the S1 layer of wood tracheids and fibres and S2 the corresponding S2 layer; in these fibres, however, the sequence then repeats itself with the microfibrils of the layers bright in transverse section lying in progressively steeper helices.

(4) A most elegant method was devised by Meier (1957) which demonstrates the essential difference between S1 and S2 in conifer tracheids though without yielding values for the helical angles. Thin radial longitudinal sections of the new sapwood and cambium initials with only the primary wall (P), then differentiated tracheids with P + S1, then cells with P + S1 + S2, as referred to already in discussing the chemistry of these layers. Such a section, placed between crossed polaroids under a polarising microscope with tracheid length at 45°, shows the phenomena demonstrated in Fig. 10.9. Cells with (S1 + P) walls are bright and negatively birefringent. Cells with (P + S1 + S2) are at first darker, but as the thickness of S2 increases (from right to left in Fig. 10.9) the cells become brighter and positively birefringent. This is clearly in harmony with a slow helix in the S1 layer and a steep helix in S2, the point of minimum brightness corresponding to the point at which the negative birefringence of S1 is most nearly compensated by the positive birefringence of the developing S2.

Although, as we have seen, the X-ray diagram of mature cells does not normally yield detectable arcs referring to S1, these could be found in a system like that used by Meier. Such a system occurs with the leaf fibres of sisal and has been examined (Preston and Middlebrook, 1949 a). Sisal (*Agave sisalana*) is monocotyledonous and the leaf continues to grow by a basal meristem throughout the life of the plant. As a consequence, fibres pass through the same sequence from the leaf base upwards as tracheids do from the cambial zone inwards. It is therefore merely a question of taking

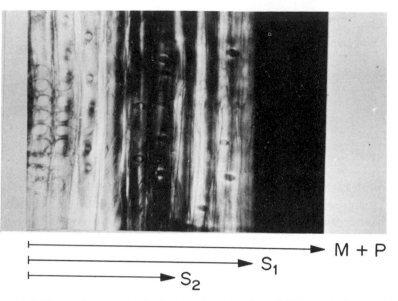

Fig. 10.9 **Photomicrograph of a longitudinal section of differentiating conifer xylem between crossed polaroids; the cambium is on the right, the mature sapwood on the left. (By courtesy of Dr. H. Meier.)**

material from the correct level of the leaf to find fibres with the S1 layer only. These can then be selected in amply sufficient quantity to yield good X-ray diagrams; they are four-point diagrams from which the helical angle may safely be calculated (p. 284). Fibres with, in addition, a fully developed S2 layer yield a diagram with dispersed equatorial arcs only, swamping the diagram of S1, so that the helical angle may be measured only as the angular distance between points on the arc at 40 per cent of the central intensity (p. 285). Some results are given in Table 10.4 which gives also supporting evidence including determinations of the m.e.p. in face view of the wall. Sisal fibres are clearly built on the same plan as are wood tracheids and fibres.

The model of tracheid and fibre structure reached at that time is depicted in Fig. 10.10. The first confirmation by electron microscopy of this very general type of structure was given by Hodge and Wardrop (1950) and by Wardrop (1950) using wood finely ground in a blender. This was confirmatory, however, only in the sense that the run of the microfibrils in random isolated wall lamellae could be understood in terms of the model. This was

TABLE 10.4

Microfibril orientation in sisal fibres

Leaf	Length of leaf (cm)	Distance from base of leaf (cm)	Wall type	Fibre length mm	Helical angle (°)			% cellulose
					X-ray data	m.e.p.	pits	
1	153	0 − 3	S1	1.06 ± 0.34	49			65.4
		54 − 58	S1 + S2	2.58 ± 0.78	18			78.3
2	79.9	0 − 3	S1	2.78 ± 0.82	40	40.2 ± 8.4	39.5	77.3
		10 − 13	S1 + S2	2.78 ± 0.89	18	20.4 ± 7.2		83.2
		76 − 79	S1 + S2	2.50 ± 0.94	20			83.2

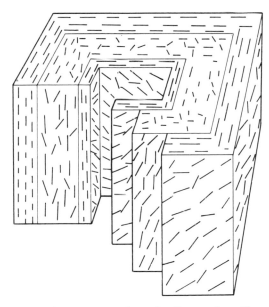

Fig. 10.10 **Diagrammatic representation of a dissected conifer tracheid showing structure of S1, S2 and S3 layers as determined by polarization microscopy and X-ray diffraction analysis in about 1950. Short lines show run of cellulose crystallites.**

rapidly followed by the more decisive pronouncement made possible by the use of ultrathin sections (Wardrop, 1954 and 1957) and of surface replication (Harada, Miyasaki and Wakashima, 1958) (Fig. 10.11). Indeed, Wardrop and Dadswell (1957) have shown by this method that over several hundred species of conifer the

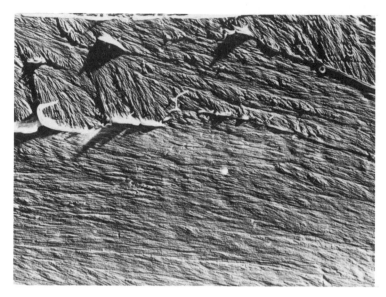

Fig. 10.11 Electron micrograph of cellulose acetate replica of a split tracheid wall of *Picea sitchensis*; magnification 9 000 X, shadowed Pd/Au. This shows the S1 layer with the S2 layer revealed on upper left and right.

wall layering of the type S1, S2, S3 with flat, steep, flat helices is common. Electron microscopy soon showed, however, that the walls are even more complicated than polarization microscopy had suggested. Bailey and Kerr (1935) had already demonstrated by light microscopy a fine lamellation in S2 and Wardrop and Dadswell (1957) and Frei, Preston and Ripley (1957) established that the microfibrils of this layer, all lying in steep helices, were inclined at an angle varying slightly from lamella to lamella. With regard to S1, Bosshard (1952) was early in suggesting that there are here two crossed sets of microfibrils, confirmed by Wardrop (1954), Meier (1955), Emerton and Goldsmith (1956), Frei and Preston (1957) and Wardrop and Harada (1965). In *Pinus radiata*, at least, these occur as two flattish helices of opposite sign (Wardrop, 1957; Frei *et al.*, 1957). A similar situation has been claimed also for the S3 layer (Wardrop, 1964). Harada *et al.* (1958) have observed lamellae between S1 and S2 and between S2 and S3 with an intermediate orientation, though Wardrop and Harada (1965) have shown that these are not invariably present.

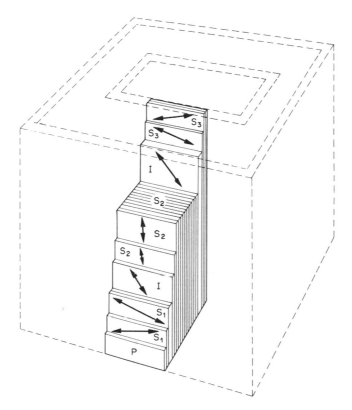

Fig. 10.12 **Diagrammatic representation of the tracheid wall revealed by electron microscopy. Double arrows show the general run of the micro-fibrils in the various layers. The intermediate layers I are not always present, and S3 is absent in some species.**

The structure of tracheids and fibres is therefore in general revealed by electron microscopy to be as in Fig. 10.12. The two crossed helices of layers S1 and S3 cannot be of equal value because under the polarizing microscope the m.e.p. is tilted in a slow helix; one of the microfibrillar helices must therefore preponderate over the other. In some instances (as in the xylem of *Fraxinus*, *Alnus* and *Picea*) S3 is either absent or only poorly developed, and a similar situation arises in the compression wood of conifers (i.e. the wood developed on the lower side of an horizontal branch or leaning stem) (Wardrop and Dadswell, 1950). Côté, Day and Timell (1968) have fairly recently reviewed the special features of compression wood. The tracheids are normally shorter than are

301

those of the neighbouring wood of the same age and, typically, have a rounded outline, prominent intercellular spaces, and fissures in the S2 layer which run helically, parallel to the microfibrils. Côté et al. support for *Larix laricina* those workers who had previously concluded that only a small proportion of the lignin in compression wood lies in the middle lamella region and that there is between the S1 and S2 layers a thin layer which is especially highly lignified. Within tension wood (on the upper side in angiosperms), the innermost wall layer, next to the lumen, is thick, gelatinous, relatively little lignified and with microfibrils lying almost longitudinally; calling this the G layer, the sequence through the wall from outside to inside the cell may be S1, S2, G; S1 S2 S3 G; or S1 S2 S3 S4 G (Wardrop and Dadswell, 1955). The gelatinous layer may be lamellated (Casperson, 1962, *Aesculus*; Scurfield and Wardrop, 1962, *Acacia*) radially banded (Jutte, 1956) or homogeneous (Sachsse, 1965, *Populus*; Robards, 1967, *Salix*). It often contains an oxidising enzyme which may be a laccase and may be associated with the lack of lignification (Wardrop and Bland, 1959). The chemistry of the G layer has recently been examined, against the background of earlier studies, by Scurfield (1972). These striking modifications in wall structure clearly derived from an orientation of the developing cell in the earth's gravitational field, different from that in the normal wood situation, present an intriguing problem which will need to be taken up later.

The model of Fig. 10.12 illustrates the striking similarity between the structure of fibre and tracheid walls in higher plants and the walls of algae of the *Valonia* type. Here again is the slow helix (S1 and S3), the steep helix (S2) and the third orientation not always present (I). The structure here passes through only one cycle (or two in the case of fibres like bamboo) but this may be consequential upon the brief duration of wall thickening in higher plant cell walls compared with that in the algae. The similarity is surely too close to be trivial.

The walls of tracheids and fibres are, however, not always and everywhere uniform as so far figured, and attention now needs to be paid to three features, one of which locally disturbs this uniformity. These are the bordered pits which affect all wall layers, the 'warts' which are a feature of the inner surface of the S3 layer in some species, and helical thickenings.

10. 1. 3 Bordered pits

Pits occur in these secondary walls usually over areas in the primary wall which were pierced by plasmodesmata and are known as primary pit fields (p. 388). Over these areas the primary wall may or may not be thinner than elsewhere. At the onset of secondary wall thickening, wall deposition fails to occur over the primary pit field so that on either side of the septum separating two cells (middle lamella plus two primary walls) canals appear in the wall. These are the pits. In tracheary cells successive secondary lamellae overarch on the original open area so that the canal becomes narrower toward the cell lumen, forming a dome-shaped cavity in each secondary cell wall (Fig. 10.13a). These pits are called *bordered pits*; like all other pits they invariably occur in pairs as shown in Fig. 10.13, one at the same locus in each abutting cell and are said to *correspond*. Pitting of this kind was well recognized during the last century and Dippel and Sanio identified problems in their construction and development, many of which are still unsolved.

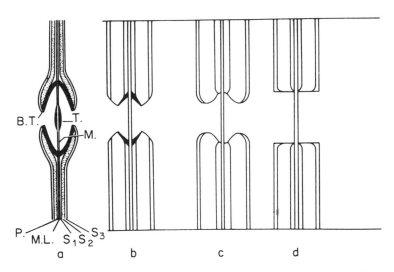

Fig. 10.13 Diagrammatic representation of the disposition of cell wall layers in the region of the pit, seen in longitudinal section (a) gymnosperm tracheid (b) angiosperm fibre-tracheid (c) angiosperm vessel (d) parenchyma cell; M, pit membrane; T, torus; B.T. initial pit border. (By courtesy of Professor A.B. Wardrop.

In conifers bordered pits occur only on the radial longitudinal sides of the tracheids, except in the last one or two tracheid layers cut off from the cambium at the end of each growing season, when a few may occur on tangential faces also. Even with tracheids, pits can be very numerous; in *Pseudotsuga* for instance, there may be 90 pits per tracheid in early wood though only about 10 in late wood (Phillips 1933). They are commonly clustered into files of a few or many leaving stretches of wall undisturbed between them, In vessels they may be so closely clustered that the whole wall structure is disturbed (Preston, 1952). The shape of the cavity is readily seen in longitudinal sections of tracheids, particularly when the lumen of the cells, and the pits, have been filled with an opaque substance as by the deposition of silver following upon the impregnation of the wood by silver nitrate solutions followed by hydrazine hydrate solutions (Bailey and Preston, 1969) (Fig. 10.14). In surface view the outline of the primary pit field is still visible as the outer edge of the border, forming a circle with a diameter approaching the width of the cell and therefore perhaps $20\,\mu$m, while the pore bordering on the lumen may be only $5\,\mu$m in diameter or so. In conifers, the pore is sometimes also circular but in some species is elliptical, the direction of the major axis lying close to the microfibrillar direction in S2. In fibres of angiosperms the pore becomes closed in to a slit (Fig. 10.15) which now accurately represents the run of the microfibrils in S2, the slits on adjacent walls (or opposite walls) crossing at an angle twice the helical angle (Fig. 10.15).

Under a polarizing microscope the border shows a phenomenon best understood with the circular pits of *Pinus sylvestris* (Fig. 10.16)(Frontispiece). The m.e.p. runs round the border in a circle, so that the border shows a maltese cross separating sectors with alternations of addition and subtraction colours (p. 84). The arms of the maltese cross lie parallel to the vibration directions in the polarizer ana analyser (crossed) so that they remain fixed in attitude as the section is rotated. It will be clear that the cellulose microfibrils of the border must therefore circle round the border. The circle is, however, neither perfect nor complete. This is evidenced by the fact that the path difference of the border, measured along a circular path midway between the pore and the outer edge of the border, is not constant (Fig. 10.17 a) (Preston, 1939). It reaches a minimum at each end of a diameter of the circular path which is inclined to

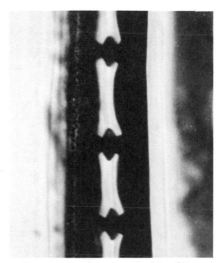

Fig. 10.14 Photomicrograph of two tracheids of *Pseudotsuga taxifolia* impregnatcd with silver as in the text. Note the cavities of the bordered pits between them.

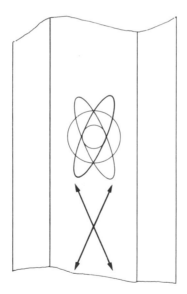

Fig. 10.15 Diagrammatic representation of crossed bordered pits in fibres, seen in face view. The vertical lines represent the side walls. The larger circle represents the outer edge of the pit, on the primary wall side. Passing inward through the secondary wall, the wall overarches to a narrow orifice (small circle) which then spreads to an ellipse (and eventually to a narrow slit, not shown). The double ended arrows show the microfibrillar direction of the two contiguous walls.

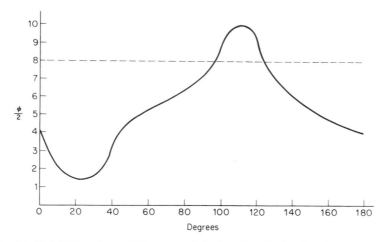

Fig. 10.17 (a) **The phase difference of the border of a bordered pit measured at various azimuths around the (circular) border. The diameter from 0° to 180° is parallel to the cell axis.**

tracheid length at an angle very close to that of the m.e.p. in the neighbouring wall and therefore to the microfibril direction in S2. The explanation is clearly that, as the microfibrils of the wall approach the pit border, they diverge at a point and to an extent which depends upon the depth in the wall at which the microfibrils lie and the degree of development of the border at the time of their deposition (Fig. 10.17b). Circulation of this type has been amply verified in the electron microscope (Fig. 10.19), and the above interpretation of the optical properties of the border has been supported by Harada (1973).

It might be expected that the layering in the pit border could be complex since there would be expected to be a cell wall/cytoplasm interface on both the inside and the outside of the border, each depositing wall layers. This is not, however, precisely the case. The plasmalemma does not follow the contour of the border (Wardrop and Foster, 1964; Wardrop, 1965) (Fig. 10.18) and this apparently means that there is no wall synthesis from inside the pit. In any event, the pit in conifers is initiated as a ring of cellulose on the primary wall (Sachs, 1882; Bailey and Vestal, 1937; Frey-Wyssling, 1955, Wardrop, 1954a), formed before the S1 layer is developed and structurally separate from it (Wardrop, 1965). This has been confirmed by electron microscope studies (Harada *et al.*

306

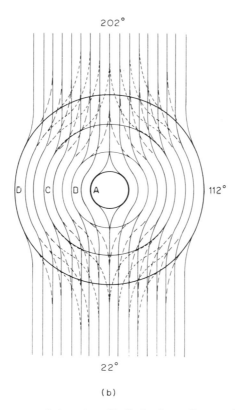

Fig. 10.17 (b) **The run of the microfibrils in the wall around the pit border, in explanation of (a).** Solid lines, the microfibrillar direction in layers at the inside of S2, nearer the lumen; dotted lines, run of microfibrils in layer on the outside of S2, nearer the primary wall, with the size of the pit pore at that time marked by the outer faint circle; broken lines, the run of microfibrils in an intermediate layer when the pit pore was as shown by the inner faint circle.

1958; Liese and Hartmann-Fahnenbrock, 1953) and via the optical properties of the border (Wardrop and Davies, 1961). According to Wardrop and Davies (1961) and Wardrop (1965) the layers S1, S2 and S3 run into the border of the pits, as shown in Fig. 10.13 together with similar interpretations of pits in other cell types to be considered later. Jutte and Spit (1963), however, have given a different interpretation. In this, the 'initial pit border' of Wardrop is indistinguishable from S1 and this part of S1 becomes completely enveloped by S2 and S3 layers as the border develops. More recently, Harada and Côté (1967), working with five species of

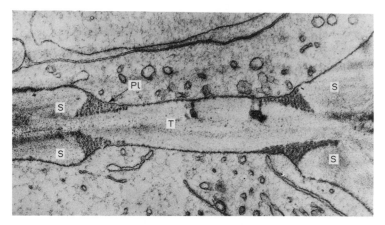

Fig. 10.18 Electron micrograph of a median section through a bordered pit in *Pinus radiata*. T, developing torus; S, secondary wall; Pl, plasmalemma. Magnification *c*. 4 000 X. (By courtesy of Professor A.B. Wardrop.)

Pinus and with *Pseudotsuga menziesii*, *Tsuga heterophylla* and *Abies balsamea* and combining results from polarization and electron microscopy have presented an intermediate view. It need hardly be emphasized that when three experienced groups examine virtually the same evidence and come up with three different answers, the evidence itself cannot be as definite as could be wished. The fine details of border structure still remain to be resolved.

The closing membranes of the pit consists initially of walls of the two contiguous cells, in intimate contact with the cytoplasm and, in the electron microscope, with a typical primary wall organisation (Frey-Wyssling, Mühlethaler and Bosshard, 1956; Wardrop, 1958). During development, the membrane thickens at its centre, in a region, it may be noted, where it is in contact with the plasmalemma, through the deposition of microfibrils with a tendency toward a circular disposition to form the torus (Fig. 10.19). It was at one time thought possible that the formation of the torus might be associated with pit aspiration (i.e. the 'closing' of the pit by a lateral displacement of the membrane onto one border, see below) but this now seems unlikely. The evidence against it has been summarized by Tsoumis (1965). In the surrounding annulus of the membrane – which has come to be called the *margo* – the microfibrils of the primary walls normally become rearranged to

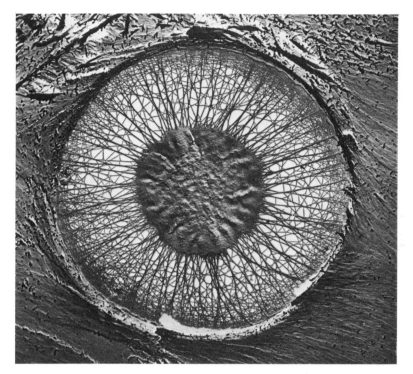

Fig. 10.19 Electron micrograph of a bordered pit in *Abies grandis*. Magnification 5 000 X. Material solvent/exchange dried using ethanol and acetone. (By courtesy of Dr. J.A. Petty.)

form radial strands suspending the torus centrally across the chamber and with interstices between them which appear to provide direct communication from cell to cell (Fig. 10.19). The mechanism of this development is not at all understood. Bauch, Liese and Schultz (1972) have recently reviewed the variability of bordered pits in this regard over a range of 120 species of gymnosperms.

The whole structure is very much as envisaged by Bailey (1913) who was also the first of several (Frenzel, 1929; Liese, 1956) to determine the size of the interstices in the margo by way of the size of the particle suspension which can be made to pass through them. According to Liese particles up to 200 nm can pass in this way from tracheid to tracheid in a number of genera of conifers. It seems therefore somewhat odd that a number of workers have reported the presence of a diaphanous lamella closing off the interstices [Jutte and Spit, 1963 (*Pinus* and *Picea*); Bailey, 1966

309

(*Pseudotsuga*); Murmanis and Sachs, 1969 (several genera); Jutte and Levy, 1971 (*Pinus strobus*)]. In the sapwood, cellulose is reported to dissolve the margo (which is unlignified (Bamber, 1961)) and to remove the diaphanous layer (Bauch, Liese and Berndt, 1970; Jutte and Levy, 1971) which therefore seems to be a real structure. Nevertheless even in green sapwood particles in suspension do pass across bordered pits under pressure (Liese and Bauch, 1964) so that the paradox is not resolveable by the concept that the continuity of the diaphanous membrane may be destroyed as the wood is dried. If it is a reality — and clearly it could represent the remnants of the plasmalemma — then either it does not cover all the interstices or it is disrupted by the pressure gradients required experimentally to move solutions through wood although these have sometimes been not more than a few tens of cms of water. Whether this membrane constitutes a real barrier to movement in standing trees is not clear; its presence would, however, explain some findings otherwise obscure (p. 369).

10.1.4 Pit aspiration

In sections of fresh sapwood which has never been dried the pit membrane lies centrally ass hown in Fig. 10.13. In dried wood, however, the membrane in early wood tracheids is drawn to one side into contact with the border and the pit is said to be *aspirated*. This begins at about fibre saturation point i.e. at the water content (*c.* 30 per cent) at which the lumen is just empty. Late wood pits are not in the main aspirated under these conditions presumably on account of the greater movement required with the thick walls in this region. Aspiration reduces very considerably the permeability of timber and a good deal of work has gone into seeking its cause and devising preventive measures.

Until quite recently the view had prevailed that aspiration is caused by lateral movement of the membrane under the influence of surface tension forces, involved in the evaporation of the last drop of water contained between the membrane and one side of the border. The basis for this is that aspiration can be prevented by replacing the water by an organic liquid of low surface tension. Quantification was reached by Liese and Bauch (1967) using ethanol-water and acetone-water mixtures; they found that a surface tension of less than 2.6 dynes mm^{-1} apparently prevented aspiration. Hart and Thomas (1967) have discussed the surface pressures which might develop during aspiration and Bailey and Preston (1970) have shown that the mechanical forces developed during the necessary stretching of the microfibrils of the margo are of the same order of magnitude as the forces available in surface tension. The relation between aspiration and surface tension is not, however, a simple one. Comstock and Côté (1968) have shown aspiration to occur at less than 2 dynes mm^{-1}, when this surface tension was reached by adding surfactants to water, and not to occur at 4.4 dynes mm^{-1} when an organic liquid was used. Clearly only those pits are regarded as aspirated which remain aspirated when replaced in liquid for examination; and Comstock and Côté suggest that water, unlike organic liquids, allows the membrane to be bonded to the border so that one aspirated a pit remains aspirated. This view has been supported by Thomas and Kringstad (1971). In a most recent and elegant analysis of the mechanics of pit aspiration, Petty (1972) has strengthened the agreement considerably by showing that a surface tension less than 0.12 dyne mm^{-1} would move the membrane to the border (most liquids have a surface tension > 1.5 dyne mm^{-1}). Evidently the pit must aspirate in organic liquids but then regain its central position when the whole liquid has evaporated. The only condition under which the pit should then remain aspirated would be when the membrane is held against the border for a long period so that the margo fibrils could 'creep' (see p. 401). Petty therefore supports Comstock and Davies in concluding that the critical factor is a bonding between the membranes and the wall and he considers the possibility that the bonding involved is hydrogen bonding.

10.1.5 Helical thickenings

In some genera the last wall layer deposited (upon the S3 layer if this is present) is not continuous but takes the form of thickened bars which run helically around the tracheid in a direction which is not normally the same as that of the m.e.p. of the underlying layer. These thickenings are usually more prominent near the centre of a cell and tend to disappear toward the two ends. Optical studies show that the thickenings must contain aggregates of parallel microfibrils and this has been verified in the electron microscope (Hodge and Wardrop, 1950; Liese and Hartmann-Fahnenbrock,

1953; Wardrop, 1964). It would seem reasonable to regard them as homologous with the layers of the secondary wall and to represent an attempt to continue secondary wall formation at a time when the life of the cell is growing to its close.

10.1.6 The warty layer

This represents another form of sculpturing on S3, first described by Kobayashi and Utsumi (1951) and by Liese (1951), from electron microscope observations, as small protuberances 0.05 — 0.5 μm in diameter. The larger warts can accordingly be seen in the optical microscope (Liese, 1957). According to Wardrop, Liese and Davies (1959), Cronshaw, Davies and Wardrop (1960) and Wardrop (1962 and 1964), these small protuberances have both a wall component in the form of a very localised thickening and a cytoplasmic component. The latter is said to consist of the denatured remnants of cell organelles together with the remains of both the plasmalemma and the tonoplast. The warts are of no significance from a structural point of view though they may have some taxonomic value.

10.2 Other mature cell types

10.2.1 Collenchyma cells

Formally, collenchyma cell walls are primary and are therefore in this sense taken here out of order. The walls are unusual for primary walls in that they thicken markedly while the cell is still elongating and develop a structure very similar to that of secondary walls. For this reason, and for want of a better term, Majumdar and Preston (1941) have classed them as thickening primary walls, a term which has been accepted widely.

Collenchyma occurs in the outer region of the cortex of many plants and develops early as a tissue giving mechanical stability while not inhibiting longitudinal growth of the shoot. The cells are usually similar in shape to the fibrous cells considered already except that in some species, though not all, they end in a transverse end wall. Collenchyma walls proper are not, moreover, lignified, though occasionally a lignified secondary layer develops (see below). The characteristic feature by which collenchyma cells are always recognised is that the wall is not uniformly thick. During differentiation, the initially thin primary wall begins to thicken in the neighbourhood of intercellular spaces between three or more neighbouring cells. This causes the cells in transverse section to appear thickened at the corners. The thickening may spread to the whole of the walls lying tangentially in the stem, but never to the radial walls, or may develop so enormously at the corners surrounding an intercellular space as to give the mistaken impression that this space is the lumen. In longitudinal view the thickenings run down the cell as bars. Presumably as the cell develops new wall lamella are deposited over the whole surface of the wall but these are thin except at the bars.

The pectin content of the walls is high and this is perhaps why the walls shrink in thickness so markedly on dehydration. Collenchyma cells of *Petasites vulgaris* after dehydration in alcohol are somewhat difficult to distinguish since the wall thickenings have shrunk to about the thickness of the rest of the walls. On replacement in water the bars swell again by about 150 per cent to the thickness observed in fresh material (Preston and Duckworth, 1946). Shrinkage and swelling in length is, however, negligible, of the order of 0.5 per cent, so that the swelling of these walls indicates that the structural units in the wall lie in the main longitudinally, as pointed out long ago by Haberlandt.

The wall structure of the three main types of collenchyma — corner collenchyma, plate collenchyma and tubular collenchyma — was examined in some detail more than thirty years ago by polarization microscopy and by X-ray diffraction analysis and turned out at the time to be in complete harmony with the swelling properties (Anderson, 1927 (*Solanum lycopersicum*); Majumdar and Preston, 1941 (*Heracleum*); Preston and Duckworth, 1946 (*Petasites*)). The m.e.p. of the wall lies steeply, the helical angle being $0.6° \pm 0.8°$ in *Heracleum* and $2.0° \pm 0.3°$ in *Petasites*. In this, collenchyma walls resemble the S2 layer in walls of very long tracheids and sclerenchyma cells. The presence of this steep helix is confirmed by the X-ray diagram. In transverse section the wall has again three layers, a thick central layer bordered by two narrow layers. The thick layer is only weakly birefringent in transverse section but strongly so, when viewed longitudinally and unquestionably the steep helices may be located in this layer. The outer layer, however, unlike the corresponding layer in tracheids,

313

is almost isotropic in all sections; it does not stain in I_2/H_2SO_4, in ruthenium red or in sudan III. The innermost layer is bright in transverse section between crossed polaroids and it was thought even in these early days that this might imply a slow helix in this layer comparable with the S3 of tracheids.

In transverse section the central layer is finely lamellated with lamellae alternately bright and dark when viewed between crossed polaroids. These early workers concluded, however, that this variation in brightness was not due, in the main at least, to an alternation in what is now known to be the microfibrillar direction. If the wall is swollen and stained in ruthenium red (shown recently by Sterling, 1970) to react with molecules containing two negative groups 0.42 nm apart) then alternate lamellae take up the stain much more deeply than the rest. Alternately, when walls of *Solanum* or *Heracleum* cells are stained with I_2/H_2SO_4 for cellulose, the lamellation is again enhanced. There seems little doubt here, therefore, that, as first pointed out by Anderson, the lamellae are alternately cellulose-rich, pectin-poor and cellulose-poor, pectin-rich. With the tubular collenchyma of *Petasites*, however, staining does not reveal the lamellation so that here cellulose must be distributed more nearly uniformly through the wall.

The structure so far derived was in essence confirmed by Spurr (1957) for the collenchyma of celery (*Apium graveolens*) petioles in an investigation of the effect of boron on wall structure. Examination in the electron microscope has, however, somewhat changed the situation. The first observations (Beer and Setterfield, 1958) were confined to transverse sections and in general were confirmatory, especially of the microfibrillar orientation in the innermost layer. Subsequently, however, Roland (1966) (*Sambucus nigra*); Wardrop (1969) (*Eryngium*) and Cox (1971) (*Apium*) have shown that the central layer is more complex than had been thought since there appear here some lamellae in which the microfibrils lie in a helix much flatter than in the remainder. In *Eryngium* the central layer was found to contain 6 − 10 lamellae (compared with 5 − 7 found by Majumdar and Preston (1941) and about 20 by Beer and Setterfield (1958)) which were alternately thick and thin. In general the microfibrils of the thick lamellae run longitudinally and those of the thin lamellae transversely. The thickness of the thicker layers decreases towards the lumen so that lamellae with transversely oriented microfibrils tend to crowd around the

314

lumen. This no doubt accounts for the inner layer observed by the earlier workers to be bright between crossed polaroids in transverse section, a layer which Wardrop also detected. In Wardrop's material the 'transverse' microfibrils become the further removed from the transverse plane the further the lamella carrying them is from the lumen, a phenomenon also indicated in the model presented by Roland. This could be a consequence of the longitudinal growth of these cells. A peculiar feature of *Eryngium* collenchyma is that at a late stage of development a lignified layer is deposited on the collenchyma wall as a complete, uniform thickening. In this the microfibrils lie helically at an angle of $40° − 45°$ to the cell axis; it presumably represents a secondary layer.

The structure of the wall is not therefore in essence different from that in tracheids and fibres and in *Valonia*-type algae, perhaps tending more nearly toward the latter. A striking difference, of course, is the high water content of the wall as evidence by the pronounced shrinkage on drying (down to one per cent of the fresh thickness in *Eryngium*). This no doubt explains the high extensibility of collenchyma cell walls, allowing the longitudinal microfibrils to slip past each other through a 'pectin' gel which must be dilute.

10.2.2 *Phloem fibres* and/or 'pericyclic' fibres do not differ essentially in structure from wood tracheids and fibres and this applies to jute, ramie, flax and hemp fibres. Similarly the walls of latex vessels are built upon a very similar plan. It is not proposed to linger over these. In the monocotyledons, fibres of *Pandanus* (Bailey and Kerr, 1935) and, more recently, of sorghum (*Andropogon sorghum*) and sugar cane (van Oordt-Hulshof, 1957) show features similar to those of the bamboo fibres already described and therefore provide no exception either.

10.2.3 *Vessel elements* of the xylem of angiosperms, however, merit a little attention. As far as the writer is aware, knowledge of structure in these cells dates back some 35 years (Bailey and Vestal, 1937; Preston, 1939 and 1952) and no electron microscope evidence is available, though a brief statement has been made by Wardrop (1964). These broad cells (up to 0.5 mm diameter) have many contacts with other cells; as a consequence the wall is so heavily pitted that the consequential divergence around the pits makes the structure very complex. As with fibres, the mouths of slit pits (on areas adjacent to elongated cells) denote the preferred direction of the microfibrils. With this as a criterion, the microfibrillar orientation is subject to sudden, abrupt changes. In general, however, in those areas free of pits the m.e.p. in face view lies transversely or nearly so. In transverse section the wall often appears homogeneous under the polarising microscope so that the wall structure may be homogeneous in these cases though small differences in microfibril orientation would be difficult to detect. In less specialised dictoyledons such as *Sassafras officinale* and in some vessels in some more specialised species such as *Fraxinus americana* and *Castanea dentata*, a layering may be observed which recalls strongly the layering in tracheids and bamboo fibre and has presumably the same origin. Wardrop (1964) has shown that some species with apparently homogeneous walls in the optical sense nevertheless show three distinct layers in the electron microscope, with small differences in orientation. The origin of the layering in vessels of *Fraxinus* is confirmed by the appearance on the walls of slit pits with mouths which twist through the wall – the so-called *spiral pits.* It is an interesting circumstance that such vessel elements are confined to individual vessels; if one element is heterogeneous then so are all the elements in the same vessel.

10.2.4 *Parenchyma cells* often fail to develop a secondary wall and will be considered in a later chapter dealing with the structure of primary walls. Of those in which the wall is secondarily thickened only a few have been given attention and even then in far less detail than obtaining with tracheids and fibres. In parenchyma cells from *Avena* coleoptiles, *Allium* roots and *Apium* petioles the microfibrils lie almost longitudinally (Setterfield and Bayley, 1958) and this may be a feature in many parenchyma cells. It recalls strongly the condition found in algae other than of the *Valonia*-type. In ray parenchyma of xylem the existence of a helical organisation in the secondary wall was suggested long ago by Tuszon (1903) with a helix relatively flat with respect to the major cell axis. This has been confirmed by polarization microscopy (Ritter and Mitchell, 1939) and by X-ray diffraction analysis of isolated cells (Gross, Clarke and Ritter, 1939). In optical section the walls are optically heterogeneous much as are those of tracheids

(Wardrop and Dadswell, 1952) and electron microscope studies by Harada and Wardrop (1960) have confirmed that the heterogeneity receives the same explanations, as an inner and outer relatively flat helix and a central relatively steep helix. There is, however, some diversification from this general plan in some species. In *Persoonia* the wall shows the same complexity as found in bamboo fibres, with many lamellae with alternately a slow and a fast microfibrillar helix (Wardrop and Dadswell, 1952). The cell-wall organization of vertical wood parenchyma cells has been shown to be similar to that in ray parenchyma.

10.3 Structural variation in homologous cells

A rather precise model has therefore been reached by which the structure of all elongated cells may be related in a simple pattern. The structure itself is not, however, precise. It is implied already in the above descriptions of cell types that in general the helical angles of the various wall layers, and the m.e.p. of the wall as a whole, varies among the cell types. In those which are broad and whose lengths may be measured in micrometres, the helices are flat; in tracheids which are narrower and whose lengths may be measured in millimetres, the helices are moderately steep; in fibres which are still narrower and whose lengths may be measured in centimetres, the helices are very steep. There is clearly here a possible correlation between helical angle and cell dimensions and attention was early directed to the chance that this might extend to variations among cells of the same type (Preston, 1934).

In this first investigation, advantage was taken of the circumstance, known since the time of Sanio and firmly established during the early decades of the century by I.W. Bailey, that in the trunks of conifer trees tracheid length in general increases from inside to outside the annual rings. This arises because in any one growing season the tracheids are produced by longitudinal division of a fusiform initial in the cambium (Fig. 10.20) which is itself steadily increasing in length. The daughter cell B which is to produce a tracheid enlarges laterally by 200 − 300 per cent (and elongates by a few per cent) and then differentiates by the deposition of the secondary wall. At the same time, the other daughter cell, A, swells radially sufficiently to regain its original dimension and then divides again. Each fusiform initial thereby produces a radial file of tracheids, running across the annual rings, each one of which records the approximate length of the fusiform initial at the time of that particular tracheid's initiation. It is in principle therefore a simple matter to examine the helical angles in a single radial file of tracheids in relation to tracheid length.

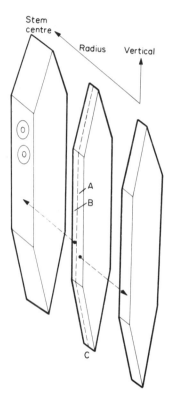

Fig. 10.20 Diagrammatic representation of the differentiation of a conifer tracheid. All the cells are very much foreshortened. The cambial initial, C, divides longitudinally into two cells, A and B. A increases radially in dimension to recover the thickness of the mother cell. B increases very much more in radial dimension to produce a differentiating tracheid.

In practice, observations at that time could not be quite so straightforward. Instead, the average length at any one location in the trunk was measured and the average helical angle for the same location determined on a separate matched sample. This was necessary if only for the reason that the helical angle was measured from the m.e.p. of the wall i.e. by method (a) (p. 280) which is destructive. The method of differnetiation ensures that only the tangential longitudinal walls are new walls at each division and on this account it was thought desirable to examine radial and tangential walls separately. For this reason observations were made only on early wood in each annual ring where the walls can be distinguished since only radial walls carry bordered pits.

Accordingly, the length of not less than 50 tracheids was measured in each annual ring of a narrow section cut from a disc from the trunks of a number of trees, and the m.e.p.'s measured of a similar number of tracheids from closely neighbouring parts of the same sector. The specimens included a 27-year old branch of *Cedrus libani*, the stems of two 7-year old saplings of *Larix leptolepis*, and a 12-year old trunk of *Abies nobilis*. Typical results for length measurements, and for the m.e.p.'s on radial walls are presented in Fig. 10.21. The average helical angle decreases across the wood from one annual ring to the next (and therefore the helix steepens) though somewhat irregularly. Equally, average tracheid length (s) increases, again irregularly. In general as $\overline{L}$ increases, θ decreases and the relation between $\overline{L}$ and θ may be expressed as $\overline{L} = K \overline{\cot} \theta$, where K is a constant [c. 1500μm for *Cedrus*, 880μm and 1050μm for the two *Larix* samples, and 2200μm for *Abies* when $\overline{L}$ is expressed in μm (Preston, 1934)]. At the same time, the helical angle for tangential walls was always less (and the helix steeper) than for radial walls and the decrease across the annual rings was less.

$\overline{L} = K \overline{\cot} \theta$ is the equation for a helix which is allowed to lengthen under constant girth and on this basis the equation could at that time be understood on the assumption that (a) the tracheid wall is homogeneous in helical angle (b) the primary walls of a fusiform initial at any time contains a helix of the same pitch as that on the secondary walls of the tracheid just cut off from it so that the secondary wall was produced by a pseudocrystallization process, since the fusiform initials are constantly elongating. Now that both assumptions are known to be untenable the rationale of this explanation disappears. There is no doubt about the phenomenon, however. This was confirmed for conifer tracheids by Phillips (1941) and, using X-ray diffraction by Preston (1946); a confirmation which has since then been repeated many times, for

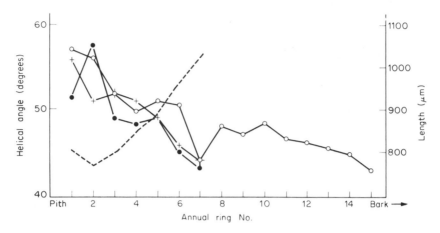

Fig. 10.21 The variation of the orientation of the m.e.p. of radial walls in conifer tracheids across the annual rings, measured as the angle between m.e.p. and cell length.

○ *Cedrus* + *Larix* A ● *Larix* B.

The broken line gives the length variation for *Larix* B.

this and for other cell types, as will be shown.

There are clearly two lines along which an explanation of the phenomenon might be sought. Firstly, the steepening of the helix, in tracheids produced by a cambium becoming progressively older, might represent an 'ageing' process and the length/angle relationship might be fortuitous. Secondly, the cytoplasm at any point in a differentiating tracheid might in some way possess information about the dimensions of the whole cell and the mechanism to lay down cellulose accordingly. In this case, the relationship should hold for the tracheids of different length within a single small piece of conifer wood and not be confined to tracheids in a radial file across the wood. These possibilities are not, of course, mutually exclusive. They have been tested in the following ways (Preston, 1948).

One mm thick longitudinal sections from the late wood in various annual rings at various levels from the ground, in a 15-year old sapling of *Pinus sylvestris*, were examined for helical angle by the X-ray method (p. 285). The results are presented in Fig. 10.22, referring specifically now to the S2 layer. It is clear that: (1) the helix steepens from inside to outside the trunk at any level and (2)

320

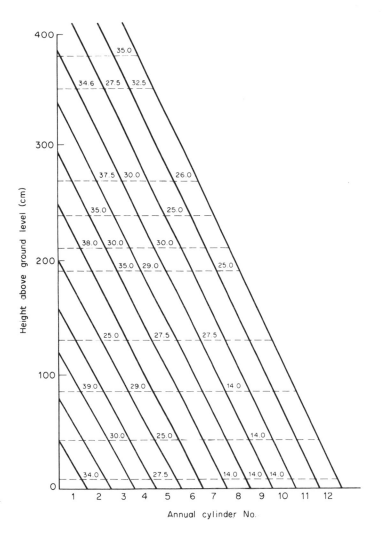

Fig. 10.22 **The average helical fibrillar angle in the S2 layer of conifer tracheids at various heights in a tree trunk and in various annual rings at each height.**

the helix also steepens on passing from top to bottom of the tree in any annual cylinder laid down in one year (i.e. parallel to the lines in Fig. 10.22). There is, however, no general trend if the same annual ring is observed at each level (figures in vertical columns, Fig. 10.22). There is here, therefore, no evidence for an ageing effect not associated with tracheid length increase and all changes in the helical angle could be expressed in a length/angle relationship.

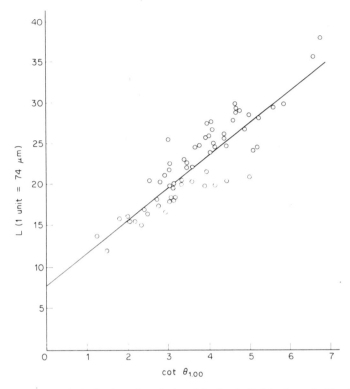

Fig. 10.23 Length/helical angle relationship for individual tracheids chosen at random from a small block of *Picea* wood.

The second possibility was examined for *Picea sitchensis* and *Abies*. It was necessary to assign helical angle θ for the S2 layer of a tracheid individually and to determine the length of the same tracheid. Accordingly, θ was measured either from striations (when visible) or from the tails at the ends of the mouths of slip pits; and the two genera used were chosen because this is then possible. The results presented for *Picea* in Fig. 10.23, containing data for tracheids chosen at random across the annual rings from 2 to 10, are fully conclusive. For individual tracheids within and among the annual rings, irrespective of their position in a radial file, the regression relation is of the form $L = a + b \cot \theta$. Incidentally the relation of L with other possible parameters (θ, $\cot^2\theta \, \cos \theta$, $\cos^2\theta$) is not linear. In Fig. 10.23 θ is expressed as $\theta_{1.00}$. This is because in any one tracheid the value for θ changes as the tips are approached and the girth (g) of the cell decreases in such a way that $g \propto \sin \theta$. Accordingly for each tracheid it is necessary to correct for breadth to a standard breadth of one unit; this is $\theta_{1.00}$. The corresponding quantitative expressions for the two species are:

$$Picea \quad L = 568 + (293 \pm 20) \cot \theta_{1.00}\mu m$$
$$Abies \quad L = 500 + (670 \pm 58) \cot \theta_{1.00}\mu m$$

It is important to note in these and later expressions that these are *statistical* relations. It does not follow — and it is not true — that *any* shorter tracheid shows a flatter S2 helix than *any* longer tracheid.

More recently, similar relations have been found for the S2 layer in other conifer species;

Pseudotsuga taxifolia (three trees grown at different rates) (X-ray method; Wardrop and Preston, 1950).

$$L = 1645 + 515 \cot \theta \; \mu m \quad \text{(specimen A)}$$
$$L = 774 + 597 \cot \theta \; \mu m \quad \text{(specimen B)}$$
$$L = 567 + 713 \cot \theta \; \mu m \quad \text{(specimen C)}$$

Pinus radiata (two samples, different growth rates) (X-ray method; Wardrop and Dadswell, 1950).

$$L = 585 + 1484 \cot \theta \; \mu m$$
$$L = 1655 + 735 \cot \theta \; \mu m$$

Indeed, wherever the relationship has been looked for, it has been found (Wardrop, 1951; Echols, 1955; Dadswell and Wardrop, 1959; Bisset, Dadswell and Wardrop, 1951; Cowdrey and Preston, 1966; Cave, 1968 and 1969; Meylan and Probine, 1969). Echols (1955) found that 91 per cent of the change in helical angle is accounted for by change in length. Exceptionally, Smith (1959) has shown in *Pinus caribaea* a linear relationship between L and θ rather than $\cot \theta$.

The relationship $L = a + b \cot \theta$ has been found to apply not only to S2, but also to S1 (Preston and Wardrop, 1949) for:

Pseudotsuga taxifolia

$$L = 2340 + 860 \cot \theta \; \mu m \; \text{(S1)}$$

compared with $L = 1600 + 520 \cot \theta \; \mu m$ for S2 of same cells, and

Pinus radiata

$$L = 1480 + 3280 \cot \theta \; \mu\text{m (S1)}$$

compared with $L = 570 + 1090 \cot \theta \; \mu\text{m}$ for S2 of same cells.

In this investigation, θ was determined for S1 from the birefringence in transverse section (method (iv), p. 281) and for S2 by the X-ray method.

In all these investigations there is a considerable spread of the experimental points about the regression lines given by the above equations, a circumstance not surprising in a biological investigation. This means, however, that factors other than cell length may also be correlated with helical angle if only secondarily. It is known, for instance, that the chemical composition of the tracheid wall changes across the annual rings from inside to outside. An increase in cellulose content was first noticed by Wardrop (1948) in *Pseudotsuga taxifolia* and more recently, Larson (1966) has shown that over the 58 annual rings of a 58-year old tree of *Pinus resinosa*, glucose and mannose increased rapidly in hydrolysates from inside to outside while galactose, xylose, arabinose and lignin decreased. The change in cellulose content has been confirmed by Dadswell, Watson and Nicholls (1959) for 15 trees of *P. radiata*. It cannot be denied that such changes might be associated with minor variations in the helical angle and perhaps this should be looked at. A further possibility that rates of growth mediated by environmental factors might be a modifying factor has been examined by Wardrop and Preston (1950) in an investigation from which the three equations above (specimens A, B and C) for *Pseudotsuga* were drawn.

These authors showed that for all three specimens, tracheid length L and time in years from the centre of the stem, T (the annual ring number) are related by an equation of the form

$$L = k_0 + k_1 \log T$$

Hence dL/dT, the tracheid growth rate, may be calculated for any value of T. Similarly, from

$$L = a + b \cot \theta$$

$d\theta/dL$ can be calculated again for any value of θ and therefore of T. It was found that, for any value of T,

$$dL/dT_A < \text{either } dL/dT_B \text{ or } dL/dT_C$$

and $\qquad d\theta/dL_A >$ either $d\theta/dL_B$ or $d\theta/dL_C$

This means that in tracheids which have elongated more rapidly the microfibrillar helix tends to remain flatter than in tracheids which have elongated more slowly. Growth rate is therefore a modifying factor though not, of course, necessarily a direct one.

Not that there is necessarily a causal relation between L and θ. Acceptance of a causal relationship would be perhaps easier if the equation $L = a + b \cot \theta$ could be shown to imply that something about the cell thereby became constant. Neither a nor b is constant within a species, between trees of the same species, or, necessarily, throughout a single tree. Formally a is the tracheid length at which the helix would be completely flat ($\theta = 90°$) and is a length normally shorter than that of any fusiform initial in the cambium from which the specimen yielding the value was derived. If anything, it must represent a cell of the procambium. In this sense, perhaps the equation is better written:

$$(L - a) = b \cot \theta$$

This equation does not imply, as Wardrop and Dadswell (1953) seem to believe, that the number of turns of the helix in a cell of length L should remain constant. It should mean, however, that the number of turns over a length $(L - a)$ should remain constant. Perhaps the cytoplasm is sensing not the full length L but a 'reduced' length $(L - a)$.

Finally, relations of the type under discussion have been found applicable to fibrous cells other than those in xylem. By dissecting out fibres in length classes from the bamboos (*Dendrocalamus longispathus*, *D. strictus*, *Bambusa arundinacea* and *Melocanna bambusoide*) Preston and Singh (1950 and 1952) have shown that the helical angle (whether measured as the m.e.p. or by X-rays) is related to fibre length by the relation

$$L = 750 + 200 \cot \theta \ \mu\text{m} .$$

For the outermost lamella of these fibres, using transverse and longitudinal sections p. 281 for determination of θ they demonstrated a relation of the form

$$L = 5980 \, (\cot \theta - 0.95) \, \mu\text{m}$$

with the interesting corollary that no bamboo fibre should have an outermost helix flatter than $\theta = 45°$. In the work on sisal fibres already mentioned (p. 299) it was also found that longer fibres have steeper helices both in S2 and S1; no mathematical relations were, however, derived. Stern and Stout (1954), however, have found a relation of the same type to hold for a number of vegetable textile fibres. The same qualitative behaviour has been reported for what is presumably the secondary wall in parenchyma cells of *Avena* coleoptiles, (Preston, 1938; Veen, 1970).

A connection between L and θ therefore seems to be widespread among plant cells. Though the precise formulation of this relationship has no more meaning at the moment than have, for instance, the mathematical representations of growth curves, and though these may not be thereby implied a causal relation, it is nevertheless clear that the secondary wall is deposited under close delicate control. This is a matter for scrutiny when we come to examine the biosynthesis of the wall. At a less elevated level, the differing orientation of microfibrils between cells of different length is relevant to studies of the mechanical properties of the secondary wall as related both to the role of secondary walls in the mechanical stability of plants and to exploitation, in for instance, wood technology and paper making. The mechanical properties of these walls are therefore taken up next, both for their own intrinsic interest and as a lead into a study of similar properties of the primary wall necessary for an understanding of wall extension and cell growth.

Lotfy, El-Osta, Kellogg and Foschi (1974) have recently introduced an elegant method whereby the 040 meridional arc may be used to determine the microfibrillar angle, applicable for any value of the mean angle in a specimen which can be introduced into an X-ray beam. This involves an iterative procedure using a computer programme (which is available) and is somewhat time-consuming; it may well, however, become the method of choice among the X-ray methods.

CHAPTER 11

Viscoelastic properties of secondary cell walls

The physical properties of cell walls which arise as a consequence of their structure and which are of importance in understanding the functions of cell walls and the commercial usages of cell walls, as well as giving a lead into the properties of the growing cell wall are, in turn, the tensile properties, the swelling properties, and the associated permeability properties. These are collected here, for want of a better term, as viscoelastic properties.

11.1 Tensile properties

When a solid is subjected to a *stress* (force per unit area) the dimensions normally change almost instantaneously by an amount which may be measured as the *strain* (the ratio of the change in size to the initial size). The ratio stress/strain is called the *modulus* and the inverse the *compliance*. The modulus is qualified by an adjective defining the distribution of stress and therefore strain. If the stress is a straight pull on a wire, for instance, the modulus is a tensile modulus; if a solid is uniformly compressed, the modulus is the bulk modulus; if the stresses are such as to cause shear (in which one plane in the solid tends to move over a neighbouring plane) the modulus is the shear modulus or the rigidity. At sufficiently small distortions of these kinds the original dimensions are restored when the stress is removed, and reimposition of the stress produced the same strain. The deformation is then called *elastic* and the modulus (Young's modulus in the case of axial tension in a wire) is a property of the material only. With sufficiently large strains, however, the process ceases to be reversible; on removal

327

of the stress the body still retains some strain, called the plastic strain. On re-application of the stress, the resultant strain is not the same as that found initially. The modulus is then a function not only of the material but also of its mechanical history. With plant cell walls the elastic limit is reached after only very small strains and this leads to difficulties as we shall see. Moreover, even the value of Young's modulus is time-dependent, in the sense that the precise value depends upon the rate of application of stress, and history-dependent in the sense that repeated stretching even within the elastic limit may so change the structure that the modulus changes. It was shown long ago for cotton hairs, for instance, that repeated stretching causes the modulus to increase (Brown, Mann and Pierce, 1930). Moreover, all the mechanical properties of cell walls depend upon the water content (Table 11.1).

For isotropic bodies strained within the elastic limit, three moduli serve to define the body; one tensile modulus, one bulk modulus and one shear modulus. With anisotropic bodies, however, many more moduli are required. For monoclinic symmetry, for instance, three tensile moduli are required for three axes mutually at right angles; in the cellulose crystallite, in a direction parallel to the chains (defined below as the x_3 direction) and in two directions perpendicular to the chains and mutually at right angles (x_1 and x_2). Fortunately, although cellulose is monoclinic, the cell wall, owing to the angular distribution of microfibrils about microfibril length (i.e. lack of perfect uniplanar orientation), can be considered orthorhombic and the number of moduli thereby reduced. Complete analysis is still a formidable problem.

As an example of a modulus, consider a rod of circular section (Fig. 11.1) and of length L loaded axially. At first the relation between load and extension is linear (OA Fig. 11.2) and reversible. Under these conditions, the ratio of stress to strain is,

$$\frac{W/\pi a^2}{\Delta L/L} = E$$

E is called Young's modulus, and the bulk of the discussion below will be concerned with this modulus. If the rod is made of cell wall material, strain beyond A produces a curvilinear relation with stress (AC) which is a quasi flow. Removal of the load causes the stress/strain relation now to move over the broken curve of Fig. 11.2, the elastic (recoverable) strain at C being ML and the

Fig. 11.1 **For explanation, see text.**

plastic strain OL. Beyond C comes a region CD of rapid extension with considerable internal slip which in cellulosic bodies leads to an increased parallelism of microfibrils (Murphy, 1963) sufficient to introduce increased stiffness and the curve transiently turns over (DE). Finally, beyond E, the rod extends rapidly and fractures, giving two more parameters, the extension at break and the breaking stress. In a real situation, that is to say in dealing with whole cells and tissues, the distribution of stress is not so simple as this and other moduli, especially the bulk modulus and the shear modulus, must be measured or calculated in order to understand the mechanical behaviour. These are defined in much the same way. The treatment then, however, becomes unavoidably complicated and highly mathematical. These complications will be introduced below as the argument proceeds but the intention will be to give an insight into the physical process at work rather than to demonstrate the precise logic of the mathematics; for that the reader will need to consult the orginal literature quoted.

329

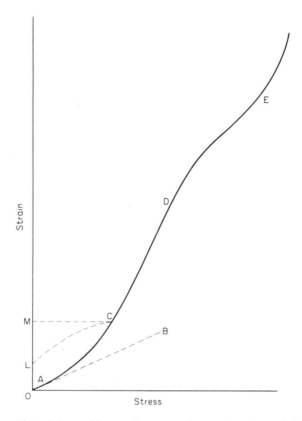

Fig. 11.2 **Load/extension curve for cell wall material.**

Apart from observations on primary walls which will be discussed in the next chapter, little work has been done on cells other than tracheids and fibres. This is in part for technical reasons, in part because only with these cells is the structure of the walls in higher plants known in sufficient detail, and in part on account of the high economic interest in the tissues of which these cells form a part. The strength of secondary cell walls of parenchyma is of importance in, for instance, the extraction of oil from oil seeds and in the quality of potatoes. Tissues of this kind are, however, intractable from the point of view of technique. Observations have so far mostly been confined to applications of a 'penetrometer' or of the deflection of beams of tissue loaded at one end, with results which are difficult to interpret in any meaningful way. Some exceptions will, however, be discussed below.

11.1.1 Tensile properties of individual cells

It is not to be expected that the mechanical properties of a tissue such as wood should be the simple sum of the properties of the cells of which the wood is composed. Account has to be taken in addition of the presence of the cementing layer — the middle lamella — and of interactions between cells in view of their mutual arrangement. Moreover, in the isolation of single cells from a tissue wall components like lignin are removed so that the properties of the isolated cell cannot be the same as of the cells *in situ*. Nevertheless treatment of single cells forms a guide to the relationship between mechanical properties and cell wall structure such as must be involved in tissue deformation and is therefore important.

The fundamental question to be asked concerns the nature of the adjustments at the molecular level involved in the absorption of strain energy during deformation. What proportion of the energy, for instance, is absorbed by the cellulose component and what proportion by the matrix; is the cellulose component so strong and the matrix so weak that the matrix shears, allowing microfibrils either individually or in groups to slip past each other; or does the wall deform as a whole with no intermicrofibril slip? If the microfibrils, or aggregates of them, do slip then the helical organisation makes it possible that a whole tracheid or fibre might behave under axial tension as an array of helical springs in parallel. In that case a possible model would be a helical spring of which the properties of the winding would be in part the properties of cellulose and in part the properties of the cellulose-matrix interaction. Considerable attention has been paid to this possibility both for primary walls (p. 411) and for secondary walls.

The importance of helical structure for mechanical properties was early recognized by, for instance, Brown *et al.* (1930). They were faced with the complicated situation in cotton hairs in which the microfibrillar helix ($\theta = c.\ 30°$) changes in sign periodically along the fibre with a very much disturbed segment between the helical segments. They noted, however, that the load/extension curve for stretched hairs was concave to the load axis, unlike the situation with viscose rayon which is not helically structured, and attributed this to the helix. This relationship between helical structure and mechanical properties has since then been amply confirmed. Meredith (1946), working on the same material, showed a close correlation between E and θ, although the small range of helical angles available in cotton is not sufficient by itself to prove such a relationship. The wider range of angles available with sisal fibres, however, makes the situation perfectly clear (Spark, Darnborough and Preston 1958) (Table 11.1). Young's modulus for $\theta = 10°$ is some 30 times that for $\theta = 50°$. With these (delignified) fibres, moreover, the helical angle θ decreases (i.e. the helix becomes steeper) as stretching proceeds, as to be expected for a helical spring model (Balashov, Preston, Ripley and Spark 1957). The change in θ for a strain of 5 per cent reveals that, assuming that the microfibrils slip past each other, the slip is only about 16 Å.

More recently Hearle (1963) has made one of the first attempts formally to check predictions from a helical spring model against the experimentally observed values of E reported by Spark *et al.* (1958) for sisal fibres, Meredith (1946) for cotton hairs, and Stout and Jenkins (1955) for a number of other plant fibres. In common with later workers, he assumes that the matrix is isotropic (which is certainly not at least wholly correct, see p. 189) and is very much weaker than the cellulose component. His model is based upon a solid fibre (with no lumen) built of microfibrils lying helically at an angle θ which is uniform throughout the fibre, the microfibrils being imbedded in, and at the molecular level continuous with, an isotropic, weak matrix. He assumes extension to occur by (1) an increase in the length of the fibrils; calculation shows that this may predominate at low values of θ ($\theta < 10°$ say) (2) extension like a helical spring with bending and twisting of the microfibrils; it can be seen intuitively, and proved by calculation, that this should predominate at large values of θ (3) a reduction in volume of the matrix by lateral contraction of the helices as they elongate longitudinally. Since in practice fibres are not solid, and since they also twist as they are stretched (and this is not allowed for), the effects of volume reduction will be smaller than assumed and, in effect, the bulk modulus will be less than the assumed value. However, volume reduction through compression of the fibre lumen will be resisted by other smaller forces involving reduction in volume of the fibre wall and compression of void spaces in the wall, so that the form of the equation involved may well be the same. The effective modulus, for stretch parallel to the fibre axis is then

$$E_\theta = \frac{EF_1(\theta)\,[\,K(1 - 2\cot^2\theta)^2\,]}{EF_1(\theta) + K(1 - 2\cot^2\theta)^2}$$

TABLE 11.1

Specimen	Young's modulus			Breaking strength		Breaking strain	Reference
	$\parallel$ microfibrils x 10^{10} N m^{-2}	$\perp$ microfibrils x 10^{10} N m^{-2}	$\parallel$ cell axis x 10^{10} N m^{-2}	$\parallel$ microfibrils x 10^{10} N m^{-2}	$\parallel$ cell axis x 10^{10} N m^{-2}	%	
Cellulose crystallite	5.65						Trelvar (1960)
Theoretical		0.2 − 0.4					Srinivasan (1941)
	12.46						Mark (1964)
From lattice distortion	13.4						Sakurada *et al.* (1964)
Calculated from intrachain bond energies				0.72			Mark (1943)
				0.75			Mark (1964)
Calculated from H bond energies for slip between chains (depends on overlap assumed)				0.116			Mark (1943)
				> 2.0			Mark (1964)
Lignified wall, xylem airdry	> 5.0						Cave (1969)
	1.0	0.02					Cowdrey and Preston (1966)
Sisal; single fibres (calc.)	2.1						Hearle (1963)
Cotton hairs. Air-dry ($\theta = 30°$)			0.85			5.4	Brown *et al.* (1930)
Wet ($\theta = 30°$)			0.29			9.4	
Sisal fibres (delg.) Room humidity $\theta = 50°$			0.3		0.008	14.5	Spark *et al.*
$\theta = 10°$			9.8		0.05	2.0	(1958)
Pinus radiata late wood $\theta = 25°$					0.013		Wardrop (1951)
$\theta = 10°$					0.033		

where E is the Young's modulus of the microfibril and K a bulk modulus equal to three times the bulk modulus of the matrix if the cellulose component is regarded as incompressible. $F_1(\theta)$ is a function of the angle θ defined earlier in a theory of twisted yarns (Hearle, 1958) and it is not clear how well this matches the situation in plant fibres.

Fig. 11.3 gives the corresponding curve fitted to the two sisal fibres measured by Spark *et al.* (1958) for values of $E = 2.1 \times 10^{10}$ N m^{-2} and $K = 0.2 \times 10^{10}$ N m^{-2} (the value for E is very much lower than any other recorded value (Table 11.1)). Values for other fibres are included in Fig. 11.3 and show a wide spread from the theoretical curve. It may be relevant that all the measurements in Fig. 11.3 were made on dried fibres. Since the moduli of fibres decrease significantly as water is absorbed, both for fibres with steep and fibres with flat helices, then in dried fibres the matrix must be resisting extension. The value $E = 2.1 \times 10^{10}$ N m^{-2} therefore contains an element due to the matrix and would be even smaller if this element were removed. Nevertheless in view of the lack of rigor in the theoretical derivation — admitted by Hearle — the fit must be considered good. It is not expected that a curve fitted to one fibre should fit other fibre types investigated by other workers with different strain rates, at different relative humidities and with different chemical compositions which certainly affect the moduli. The behaviour of single fibres therefore is not inconsistent with a helical spring model.

11.1.2 Tensile properties of whole tissues

We now turn to the xylem as the tissue investigated most intensely. Both Cowdrey and Preston (1966) and Cave (1968 and 1969) have measured Young's modulus variations across the growth rings of conifer trees and have derived structural models, upon which these variations may be explained, which differ from each other and from that developed by Hearle for single cells. Both agree, however, that the modulus varies with the helical angle θ and in much the same way.

Cowdrey and Preston (1966) examined the early wood of two discs of *Picea sitchenis*, one (B) with 31 annual rings and the other (P) with 34. Since in any piece of wood the angle θ varies widely, they chose to examine thin longitudinal strips of wood, 180 — 150 μm wide tangentially, 150 — 200 μm radially with an

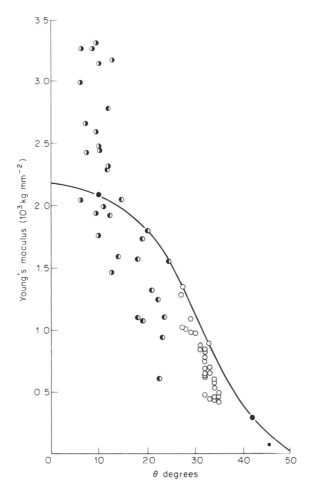

Fig. 11.3 The relation between the Young's modulus and fibril angle for a number of fibres (after Hearle, (1963)). ○, cotton; ●, sisal; • coir; ◑, bast fibres; ◐, leaf fibres.

experimental length of 2 mm, in order to minimise this variation. A similar limitation was imposed by Mark (1965 and 1967) in a study of another mechanical property (p. 347). This carries with it the acceptance of some uncertainty in interpretation due to the presence of cut tracheids on the flanks of the specimen but this was expected to be small. θ was measured either through the m.e.p. of matched specimens or by the X-ray method and therefore refers to S2. Stress/strain curves were recorded on the apparatus of Spark

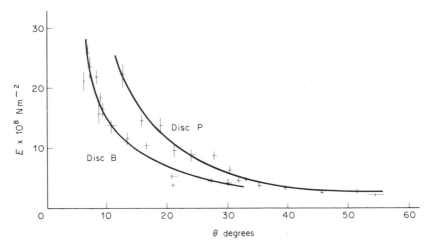

Fig. 11.4 The experimental relation between Young's modulus of *Picea* wood strips from two discs and the fibril angle of the S2 layer (after Cowdrey and Preston, (1966)). The arms of the crosses give the standard error.

et al. (1958) and strains recorded up to 0.5 per cent; plastic extension intervened at strains in excess of 0.01 per cent and therefore Young's modulus was always calculated for strains less than this (initial Young's modulus). Relations between the modulus and θ are given in Fig. 11.4; each tends to a minimum of about 2.0 x 10^8 N m^{-2} for large values of θ.

Two models oc cell structure were considered in interpreting these results. In each of them the tracheids were taken to be long smooth tubes with no overlapping. The layer S3 is absent in spruce and the S1 is thin, with slow microfibrillar helices unlikely to bear much of the stress. Accordingly only the S2 layer was included in the models, and for this θ was taken as constant through the wall and invariate from cell to cell. Rays and the middle lamella were ignored as inducing only secondary effects. Similarly, although Cowdrey and Preston referred to the twisting of individual cells under axial stress, prohibited in whole wood samples, they did not include corresponding energy factors since these, too, were considered to be of only secondary importance.

The first model was similar to that of Hearle, except that the compression factor was ignored since the walls are relatively thin and the lumen wide. The wall was taken as a series of helical fibrils

336

with fibril diameter a, fibril Young's modulus q and torsional rigidity n, free to bend and twist as the cell comes under strain. The resultant expression for the compliance (J) related to longitudinal strain is

$$\frac{J \ln (r_0/r_i)}{r_0^2} = v(q \tan^2 \theta + 2n)/a^2 qn$$

where r_0 is the outside radius and r_i the inside radius of the S2 layer, and v is a space factor to allow for the fact that not all the transverse section of a cell is covered by microfibrils. Provided all other factors are constant or nearly so, the relationship between J and $\tan^2 \theta$ should be linear. The results presented in Fig. 11.5 for disc P only (because this had the greater spread of $\theta - 12.5°$ to $54.4°$) show that for $\theta < c.\ 35°$ the relationship is indeed linear. Moreover, the above equation shows that for this line,

$$\frac{\text{slope}}{(\text{intercept})_{\theta=0}} = \frac{q}{2n}$$

which, from Fig. 11.5 is equal to 22.2. Because $q/2n > 1$, the fibrillar helices will tend to coil up under small extensions. A value of the same order (84.4) was found for sisal (Preston, 1955) using a similar model. The value of the intercept itself gives a value of q, assuming that v may be taken as 0.5. Taking $a = 5$ nm (the radius of a microfibril), $q = 2.0 \times 10^{17} \mathrm{N\,m^{-2}}$, six orders of magnitude too great (Table 11.1). Taking the calculation in reverse, if q is taken as $10^{11} \mathrm{N\,m^{-2}}$ then $a = 5\,\mu\mathrm{m}$, equivalent to about one quarter of the cross-sectioned area of a tracheid. This could be taken to imply that with a helical spring model large bundles of microfibrils form the helical winding and only these, not the microfibrils of which they are constituted, can slip past each other. It is difficult to see how otherwise the behaviour can be rationalized on a helical spring. A more serious objection is, however, the lack of agreement at $\theta > 35°$. It will be realized that the material is such that observations are few for higher values of θ. Nevertheless, the relationship between J and $\tan^2 \theta$ is clearly not linear over the whole range of θ so that, although wood behaves as if each tracheid comprised four helical springs for the smaller values of θ, the helical model fails as a general expression of the stress/strain relationships.

Attention was therefore turned by Cowdrey and Preston (1966) to a second possible model, the 'anisotropic homogeneous wall

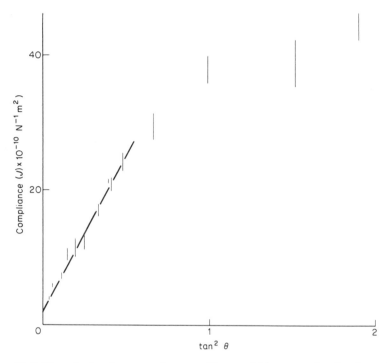

Fig. 11.5 The relation between the compliance of *Picea* wood strips from disc P, and $\tan^2 \theta$. Vertical lines represent experimental values (giving the spread between individual specimens as the standard error).

model'. In this the wall is taken as a single phase system with anisotropic elastic properties with orthorhombic symmetry. The axes of symmetry are defined by the angle θ. The basis of the model can be understood as follows.

Suppose a cell is slit down its length and opened out into a strip (Fig. 11.6), with the x_3' direction parallel to cell length and x_3 direction parallel to microfibril length. A stress σ_{33} is applied to the top of this strip while the bottom is held stationary. Stresses and strains will develop in the x_3 and x_2 directions (and in the x_1 direction, perpendicular to the page but this is ignored since the wall is thin in this direction).

Let C_{22} be the ratio of strain in the x_2 direction to stress in the
    x_2 direction,
    C_{33} be the ratio of strain in the x_3 direction to stress in the
    x_3 direction,

338

C_{23} be the ratio of strain in the x_2 direction to stress in the x_3 direction,

C_{66} relate shearing strain to shearing stress in the x_2x_3 (wall) plane.

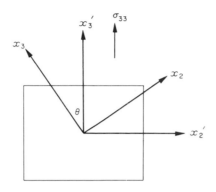

Fig. 11.6 Diagrammatic representation of a segment of a tracheid slit longitudinally and opened out to a plane. The cell axis lies parallel to the longer edge of the page.

Then it can be shown (Cowdrey and Preston, 1966) that the ratio of strain to stress in the x_3' direction, the compliance, is

$$J = \frac{e_{33}}{\sigma_{33}} = (C_{22} + C_{33} - C_{23} - C_{66}) \sin^4 \theta + (2C_{23} + C_{66} - 2C_{33})$$

$$\sin^2 \theta + C_{33}$$

$$= a_0 + a_1 \sin^2 \theta + a_2 \sin^4 \theta$$

A curve of this kind fitted to the experimental results for disc P is given in Fig. 11.7 and is almost linear with $\sin^2 \theta$ since the term in $\sin^4 \theta$ is relatively small. The fit can only be considered as good. From a_0, a_1 and a_2 for the fitted curve the values of Young's modulus parallel and perpendicular to the microfibrils may be cal- lated. These are:

$$E_{\parallel} = 1.0 \pm 0.2 \times 10^{10} \, \text{N m}^{-2} \text{ (Disc B) or}$$
$$1.2 \pm 0.4 \pm 10^{10} \, \text{N m}^{-2} \text{ (Disc P)}$$

$$E_{\perp} = 1.4 \pm 9.3 \times 10^{10} \, \text{N m}^{-2} \text{ (Disc B) or}$$
$$2.9 \pm 0.7 \times 10^{9} \, \text{N m}^{-2} \text{ (Disc P).}$$

These are now in principle the moduli of the wall substance, not of

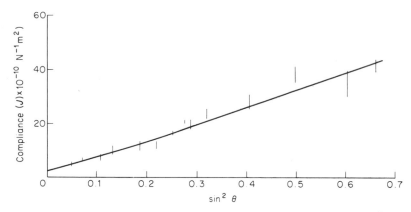

Fig. 11.7 As in Fig. 11.5 except that the relation is in terms of $\sin^2\theta$.

cellulose alone. They are nevertheless well within the acceptable range (Table 11.1). There can be no doubt but that the samples behave as if the walls were reacting according to the model. In passing, it may be noted that the model also suggests that the relative shear in the wall induced by longitudinal extension reaches a maximum at about $\theta = 10°$ and zero as $\theta \rightarrow 90°$. It also says that the microfibrillar helix should tend to uncoil at small values of θ.

In a more recent re-examination of the problem Cave (1968 and 1969) has adopted a model similar to the second model of Cowdrey and Preston (1966) but has added a device whereby allowance can be made for the inhibition in whole wood of the tendency of tracheids to twist on extension. Moreover, rather than evaluate his derivations by curve fitting, he assumes the value of moduli given by Mark (1967) and calculates the values of Young's moduli to compare with his observed values. Stress/strain curves are obtained on an Instron testing machine and θ is derived by the X-ray method of Meylan (1966). Specimens measure 2 mm x 2 mm x 60 mm (grain direction) and are chosen from the early wood of two discs of a *P. radiata* trunk. The cross-sectional area of the specimen is therefore some 200 times greater than that used by Cowdrey and Preston (1966) and therefore about 200 times more tracheids are loaded.

Within such a sample, the angle θ will vary from tracheid to tracheid through several 10s of degrees. In a sense, Cave attempts to cover this by allowing for a Gaussian distribution of θ not

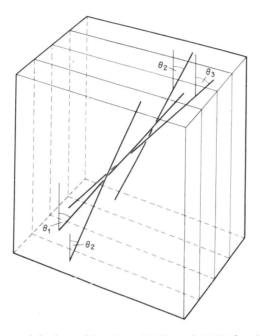

Fig. 11.8 The model adopted by Cave (1968 and 1969) for the wall of a wood tracheid (cell axis parallel to longer edge of page). The microfibrils are supposed to lie at different angles (θ_1, θ_2 etc.) between *lamella* to allow for a spread of θ between *tracheids*.

among the cells but within a single wall. His model is therefore as represented in Fig. 11.8 (modified from Cave, 1968). A slab of wall is considered to consist of several elementary slabs, bonded intimately to each other, within each of which θ is constant but different from that of any other slab. He allows for the inhibition of twisting by introducing a second helix of opposite sign into the wall (which would tend to twist in the opposite direction) since this 'closely resembles the bonded wall pairs of adjacent cells'. The fibril distribution then changes from $f(\theta)$ to $f(\theta) + f(-\theta)$. The computed values for the longitudinal Young's modulus are given in Fig. 11.9 in comparison with his experimental results. The fit is clearly good at least for values of $\theta > c.\ 15°$.

In comparing his model with the second model of Cowdrey and Preston, Cave makes the valid point that each of them leads to quartic functions in $\sin \theta$ and $\cos \theta$ (though in point of fact he nowhere explicitly gives his relation of E to θ) so that the fitting

341

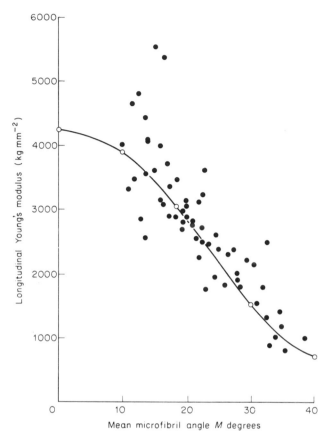

Fig. 11.9 The relation between Young's modulus and θ according to Cave (1968) (by courtesy of Dr. I.D. Cave).

procedure might hide deficiencies in the model of Cowdrey and Preston. On the other hand it must be said that, ingenious though the mathematical trick certainly is of introducing a crossed fibril structure into the model, it does not comply with the realities of the situation. The principal divergencies will be discussed later (p. 36) in dealing with the swelling of walls as interpreted by Meylan and Barber (1964). We note here only that, in dealing with the real tracheid, twisting is inhibited only at the outside of the wall where the cell is in contact with its neighbours; whereas lamellae on the inside of the wall may retain the tendency to twist, causing sheer stresses. The twist is in any case not enormous. On the helical spring model, the angular movement, $\Delta\phi$ radians, of the bottom of a helical spring L cm long, held stationary at the upper end and extended by a length of ΔL cm, is given by:

$$\Delta\phi/\Delta L = [\cos\theta \sin\theta (1 - 2n/q)]/a [\cos^2\theta + (2n/q) \sin^2\theta]$$

where q is the Young's modulus and n the torsional rigidity of the winding, a being the radius of the helix. For $a = 20\,\mu\text{m}$, $\theta = 45°$, $2n/q = 0.5$, $\Delta\phi$ is such that the lower end of the helix moves through only a fraction of a micron when a specimen of the dimensions used by Cowdrey and Preston is extended 0.01 per cent; decrease of θ or increase of $2n/q$ will decrease the movement still further. Whether it is better to ignore these small distortions or to remove them by interposing lamellae which do not exist is a moot point. The chances are, however, that the observed spread of E for any single value of θ is so large that critical experimental comparison of the two models may never be possible. Each of these models, incidentally, assumes that the matrix is isotropic, and this is a weakness. As we have seen (p. 189), some molecular species at least in the matrix are lined up with the microfibrils. Perhaps in a sense this is not all that important. What *is* important is that all workers agree that there is a relation between E and θ such that E decreases as θ increases and that the relation is quartic in trigonometrical functions of θ. This relation is undoubtedly of importance in the industrial use of timber and in designing breeding regimes where timbers may be grown with required tensile properties. It is probably also important in an understanding of the growth of isolated thick-walled cells such as *Cladophora* and *Chaetomorpha*, to which these theories have not yet been applied in any extensive way, and for which the twisting of the helix becomes important (p. 215). Otherwise, the mechanical properties of these dead cells appear to have no particular significance for themselves. It may be argued that it is a blessing that a tree should, as it grows older with a crown offering increasing wind resistances, place the strongest wood at the periphery of the wood cylinder; but even then it is questionable whether the greater stability is conferred by the compressive strength of the xylem on the side away from the wind.

The true physiological significance of this work on dead cells is the understanding it may give of the mechanical features of living cells such as leaf and root parenchyma, so closely associated with water uptake, and the pointers it gives to an understanding of the

343

function of primary walls. Living parenchyma cells present an intractable problem as far as the wall itself is concerned, even when the cells are mature and growth has ceased. These complexities may be exemplified by the work of Falk, Hertz and Virgin (1958) and Nilsson, Hertz and Falk (1958). These authors addressed themselves to the problem of the relationship between cell turgor pressure and the mechanical properties of a tissue. Broadly speaking, the first paper presents the experimental results and the second the underlying theory, using strips or discs of the potato tuber as experimental material. In determining Young's modulus *of the tissue* they used two methods, (1) straight forward linear extension of a strip of tissue under load, (2) a dynamic measure using the resonance frequency method first introduced by Virgin (1955). The latter method had already been used by Middlebrook and Preston (1952) in determining the moduli of the *wall* of *Phycomyces* sporangiophores. These authors used a flattened sporangiophore in the shape of a ribbon, imposed a vibration of variable frequency at one end by means of a beat frequency oscillator and observed the frequency at which the strip resonated in the fundamental mode. Young's modulus may then be calculated from (Lochner, 1949),

$$\nu = (m^2 k/4\pi l^2)\sqrt{q/\rho}$$

where m = 1.875 for the first mode,
   l = length of ribbon
   ρ = wall density
   q = Young's modulus parallel to length of ribbon
   k = radius of gyration of cross section of ribbon
      = $d/2\sqrt{3}$ for a ribbon, d being thickness of ribbon

They found q to vary widely among sporangiophores around a mean of 20 x 10^{10} N m^{-2} (15 observations). This figure refers to chitin, not cellulose, and is measured parallel to the molecular chains.

Falk *et al.* (loc. cit.) used discs or strips of potato tissue initially in the fully turgid condition in water (containing however 10^{-5} M $CaCl_2$ and K_2SO_4). Osmotic uptake of water will have caused a swelling with a stretching of the wall which puts the tissue at point C, say, (Fig. 11.2) though the shape of the curve cannot be certain. On placing in say, 0.4M mannitol, the load on the wall is reduced, the cell shrinks, passing down the dotted line to L. On replacing in water, the disc returns to C and repetition of this

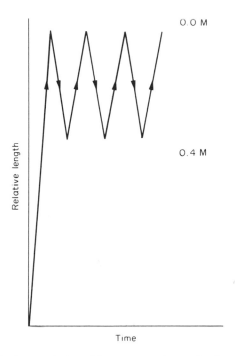

Fig. 11.10 **Length changes induced in potato tuber strips by alternating immersion in water and 0.4 M mannital (after Falk** *et al.***, 1958).**

sequence causes the disc to alternate between C and L, though it cannot be certain that the upward path is the same as the downward path. The corresponding oscillation in disc diameter is given in Fig. 11.10. The extension is reversible so that this treatment introduces a device, used on many subsequent occasions by other workers, whereby plastic deformation and elastic deformation may be separated. Subsequent resonance frequencies (using strips of tissue) were measured on such 'conditioned' material in which the wall is not physically in the same condition as that obtaining with the fresh untreated cell. Under these conditions a close relationship was found between v and turgor pressure, and therefore between Young's modulus of the tissue and turgor pressure. On the basis of a model of the tissue as a cubical array of spherical cells with elastic walls (each cell in contact with 6 others) Nilsson *et al.* (loc. cit.) were able to establish for longitudinal stress the relation:

345

$$E = 3 \left[1 + \frac{7 - 5\mu}{20(1 + \mu)} \right] p + \frac{3(7 - 5\mu)E_c d}{10(1 - \mu^2)r_0}$$

where E is the Young's modulus of the tissue
E_c is the Young's modulus of the wall
μ is Poissons's ratio for the wall
d is wall thickness
r_0 is cell diameter
p is turgor pressure

Each term in this equation changes when polyhedral cells are considered, when freedom of movement between cells varies and so forth. The authors conclude on the basis of measured and assumed values of various constants that the relation is likely to be roughly

$$E = 3p + 1.5 \times 10^6 \, \mathrm{N\,m^{-2}}$$

Both terms, especially the second, are uncertain. The first term derives from cell deformation on swelling and does not explicitly contain the Young's modulus of the cell wall. The second term involves E_c though this is not closely defined, and the authors themselves appreciate that deficiencies in the model are such that cell wall moduli cannot be derived with any certainty. Moreover, the experimentation shows that the Young's modulus of the tissue depends upon the tissue cross-section so that this is only a notional Young's modulus; in the real tissue, irregular disposition of cells will involve complex shear stresses even on uniaxially applied stress and these may vary with tissue dimensions.

11.2 Mechanical failure

The behaviour of wood under pressure below the breaking pressure has not, as far as the writer is aware, been treated in the same detailed way as has the tensile behaviour, perhaps because of the complex events developing when wood is compressed. These events are now reviewed briefly if only because they are responsible for cell wall deformations familiar to anatomists which must be recognized in some instances as artefacts. This involves some discussion of the fracture of wood and we turn first to fracture under tension.

11.2.1 Failure under tension

It is well recognized in materials science that failure under tension is initiated as a crack at some point of stress concentration. Whether this crack develops, and how rapidly it spreads if it does develop, depends upon the stress distribution at the end of the crack and immediately beyond it. The propagation of a crack, and therefore failure, is therefore a highly complex matter particularly in an anisotropic material such as wood. Development of the relevant theory has not so far been attempted but observations have been made which are relevant for the theme of this chapter.

The findings up to 1963 have been reviewed by Wardrop and Addo-Ashong (1965). Failure under tension involves the fracture of the cell wall itself, for which in wood breaking stresses range from 5 to 15 x 10^7 N m^{-2}, compared with the value for the long ramie fibres, with a correspondingly very steep microfibrillar helix, of 110 x 10^7 N m^{-2}. The values for wood are small presumably on account of the slow microfibrillar helices involved, though the figures quoted are for individual cells so that some damage due to isolation may be involved. The breaking load, however, is known to increase with cell length (Garland, 1939; Wardrop, 1951) and this is a good indication of a relationship with θ. With the steeper S2 helices of longer fibres a major component of the tensile stress lies parallel to the strong microfibrils and the shear direction makes a large angle to them. The opposite is true for the S1 layer and it may therefore be significant that failure begins in the S1 layer or at the S1/S2 boundary. An indication that (M + P + S1) strongly influences the strength of wood (and therefore of other tissues perhaps) is given by the effect of water content. Dry wood is mechanically stronger than wet wood in spite of the fact that tracheids and fibres isolated from it are stronger wet than dry (Hermans, 1949). The fibre effect is clearly due to a lubrication of the structures of the wall, allowing a more equable distribution of load in wet cells; consequently the wood effect must involve the middle lamella region, perhaps by the formation of additional internal bonds on drying.

The localization of failure initiation in or outside of the S1 layer has more recently received the strongest support from the elegant theoretical and experimental approach of Mark (1965 and 1967). Like Cowdrey and Preston (1966) Mark used thin (*c.*

22 μm) sections of wood from which to punch out specimens with a central neck free of ray cells in order to localise the fracture plane under tension. These specimens were stretched wet. Using the known chemical composition, and known elastic moduli, and measuring θ by the tilted section method (p. 282), he was able to calculated from experiments the stresses in the various layers of the wall and compare these with stresses based upon theory. Only the shear stress in S1 lay close to the theoretical stress on fracture; no other calculated stresses attained even 1/3 of the theoretical strength. Hence the cells must fail initially in the S1 layer (where, be it noted, the microfibrillar helices are slow). Following such a failure, redistribution of stress causes fracture due to another mechanism not yet investigated. Mark was further able to show that the strength of the middle lamella is such that stresses reach the failure level first in S1.

Nevertheless, the mean breaking strength of whole wood varies with the microfibrillar angle in S2 as shown by Cave (1969). He has recorded an increase of breaking strength from 10 Kp mm^2 at $\theta = 40°$ to about 50 Kp mm^2 at $\theta = 10°$. In between these values for θ there is a considerable spread of values for strength and Cave does not attempt to define the relation in terms of a model or even by a regression line. In terms of the above findings of Mark it could be that the connection is an indirect one since, as the microfibril angle in S2 becomes steeper, so does the angle in S1, the layer in which fracture appears to be initiated. In most recent times, the fracture faces of wood broken under tension have been examined under the electron microscope by Saiki, Furokawa and Harada (1972). They have observed that the fibrils in S2 are broken, that the layer S1 is sheared along microfibril length and that the middle lamella is flat and smooth. Although there is some shear between cells, shear failure in the most part occurs within S1. This is clear visual confirmation of the conclusions reached by Mark.

11.2.2 Compression failure

When wood of a conifer is compressed in a direction normal to the grain the walls buckle (if the direction of compression lies tangentially or radially and therefore parallel to tracheid walls) or concertina (if the direction of compression lies obliquely) (Frey-Wyssling and Stussi, 1948; Moscaleva, 1957). Failure in the walls occurs either as compression or tension failure; but the whole wood then becomes compacted and fails as a whole (under much higher compression stress) perhaps due to wall shearing stresses. Compression failure parallel to the grain is structurally much more complex.

The first microscopically recorded event on such compression was seen by Robinson (1920) as failures in the wall called 'slip planes', and these were for a long time thought of as the initial stage of damage. Until 1968, only two workers, Kisser and Frenzel (1950), had recorded an earlier phase in the form of slight, local wall thickenings resulting from small deformations of the wall fibrils. These could not be differentiated in the optical microscope either by staining or polarization methods; they have now, however, been confirmed by Dinwoodie (1968) both in the optical and electron microscope. The 'slip planes' which then develop (Fig. 11.11) result from a gross buckling of the microfibrils of the wall as seen in longitudinal section, which can clearly be recognised in the polarising microscope on account of the different alignment of the microfibrils in the buckling zone. They have nothing in common with the slip planes of the crystallographer. Perhaps for this reason they are given other names by some workers (*compression failure* (Wardrop and Dadswell, 1957), *microstauchlinien* (Frey-Wyssling, 1953 b), *dislocation zone* (Page, 1966), *misaligned zone* (Hartler, 1969). When the wall under a polarizing microscope is extinguished, the slip plane shows up as a bright line running obliquely across the wall at an angle to cell length variously recorded as 57° (Rayne, 1945), 62° (Frey-Wyssling, 1953b), 69.5° (Kisser and Steininger, 1952) and 61° (Dinwoodie, 1968). In an isotropic body such a failure would run through the body at 45° to the direction of stress and the greater angle clearly stems from the anisotropy in the elastic properties of the wall to which reference has already been made. Slip planes may be observed in any longitudinal section of wood, whether prestressed or not, and arise as artefacts due to sectioning, which can with care be recognised or in part avoided (Dinwoodie, 1966; Keith and Côté, 1968). A good deal of the earlier work must therefore be regarded with caution.

Slip planes due to compression begin to develop under loads about one half of the breaking load (Kisser and Steininger, 1952) and are clearly due to local concentrations of stress. They may be regarded as an attempt to spread the stress more uniformly. A

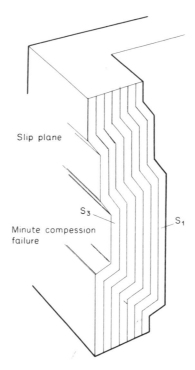

Slip plane

S_3

Minute compession
failure

S_1

Fig. 11.11 **Diagrammatic representation of the deformations in wall layering seen in section of slip planes and minute compression failures.**

failure of this kind, however, brings higher stresses to neighbouring cells as the load is increased and slip planes develop either transversely or longitudinally, or both, away from the initial point of failure. The transverse development across many walls leads to what Dinwoodie (1968) calls a compression crease. He has followed the development both of slip planes and compression creases with increasing load and finds both present at loads only 25 per cent of the failing load. In perhaps the most intensive investigation so far, Dinwoodie (1972) has recorded the angle of slip planes in four softwoods and two hardwoods, examining both early and late wood at a variety of moisture contents and temperatures. He has demonstrated that differences in the microfibrillar angle, θ, of S2 and in the ratio (longitudinal stiffness)/(transverse stiffness) of the wood account for 71.6 per cent of the variability in the slip plane angle while the microfibrillar angle alone accounts for only 48.4 per cent. There is therefore a correlation between θ and the angle of shear failure in the wall, but a change in θ is not always reflected in a change in the shear angle. Perhaps for this reason the shear angle is the same on tangential and radial walls of conifers in spite of a difference in θ, and is the same in normal and compression wood where differences in the ratio of the moduli are not sufficient to affect differences in θ so that some third factor is indicated. At any rate the speculation of Garland (1939) that the angle of shear and the value of θ are connected appears at last to have been confirmed and the way is now perhaps open for a (highly complex) mathematical approach.

Following extensive development of slip planes, the fibres and tracheids themselves begin to buckle at increasing loads as described by Wardrop and Addo-Ashong (1965). This leads to the development of 'buckling planes' whose traces on longitudinal radial faces of wood run at $90°$ to the grain but on tangential faces at $45° - 60°$ (usually nearer $60°$). On the radial face a small 'step' develops as the wood above the line is moved tangentially with respect to the wood below the line, the tracheids buckling mostly in the tangential plane. Again, deviation from an angle of $45°$ on the tangential face is due to anisotropy, this time anisotropy of anatomy though not to the distribution of the rays. The anatomy of failing specimens has been examined by Bienfait (1926) and Frey-Wyssling (1953). Since buckling planes lie at $45° - 60°$ the buckled tracheids lie at $45° - 30°$ to the grain and the buckling involves shear in the cell walls. According to Wardrop and Addo-Ashong (1965), the major zone of failure is between layers S1 and S2, at a point of weakness already noted above.

The only features of compression failure which appears to have received attention from the theoretical side is this last feature, the development of buckling planes (Grossman and Wold 1971). The argument is based on a model of conifer wood as an array of parallel tubes of rectangular cross section lying in radial rows, the rows being bonded side by side with the cells overlapping as observed. It is assumed that the intercellular bonding materials (including the S1 layer) are weaker than the walls, as shown by Mark (1967). The walls are regarded as elastic and with no anisotropy (the weakest part of the argument) and the slip planes and compression creases are taken to be important in loosening intercellular bonding. The argument then goes that the radial arrangement of the tracheids favours a tendency for each radial row to buckle as a whole because

351

this corresponds to a minimal surface area of fractured bonding material. It therefore follows that in a tangential longitudinal plane each cell represents a whole row, and the whole row will behave as this cell does. When a length of such a cell becomes unstable under the compression stress it adopts a configuration such that most of the strain energy is released, releasing the load on to its neighbours, while the distance between the ends of the unstable region remains constant. Adjacent tracheids then become unstable but these are liable to buckle over a length displaced along the tracheid direction to allow for closest packing and a minimum of deformation and fracture energy. On buckling, each tracheid, as it were, sits in the lap of the next. Calculations then predict an angle of $67° - 79°$ for the inclination of the buckling plane on the tangential longitudinal face of the wood, compared with the observed $45° - 60°$. This seems satisfactory in view of the crudity of the model. The stresses involved are, however, several times greater than those actually observed. More recently, Scurfield, Silva and Wold (1972) have observed in compression failure, slip planes and fibre buckling much as described by Grossman and Wold (1971).

11.3 Growth stresses

It is well recognized that in herbaceous plants growth stresses develop whereby, in some petioles for instance, stresses induced by the turgor pressure of internal parenchyma cells are transferred to the epidermis. When a longitudinal sector of such a tissue is dissected out it then curves with the epidermis on the concave face. Less well known, apparently, are the much larger compressive stresses which can develop over and above the stresses due to the weight of the plant. In trees these can reach levels at which the onset of compression failure can be observed. It was shown long ago by Jacobs (1938) (see also Jacobs, 1945) that when a plank is sawn across the diameter of a trunk it may split explosively along the line of the pith, and the resulting halves bend away from the pith; the wood toward the centre of the trunk has expanded in length and away from the wood has contracted. This central compression he attributed to a slight longitudinal shrinkage of newly differentiated tracheids just inside the cambium, imposing compressive stresses on the internal wood which progressively increase as the trunk increases in diameter. In theory, very high stresses can be reached in this way. According to Boyd (1950) these are not reached because the wood accommodates itself through 'minute compression failures' (slip planes and compression creases) which are very numerous toward the centre of a tree, leading in some cases to the well-known 'brittle heart'. Dinwoodie (1965) and other workers have more recently become involved with the consequent effects on timber processing. Gillis (1973) has recently extended the theoretical treatment of the development of these stresses given by Kübler (1959) and has shown that the phenomenon is understandable in a quantitative sense if the new wood, for reasons as yet unspecified, develops both a longitudinal tension and a circumferential compression. Stresses of this kind are considered by Boyd (1972) to be related to a swelling of the tra cheids or fibre wall induced by lignification.

11.4 Swelling and shrinkage

As they occur in fresh tissues, cell walls are saturated with water and even in the dead cells of freshly felled timber the water content of the wall is normally about 30 per cent on a dry weight basis. The relative amount of water is, of course, much higher with living tissue and with cells like collenchyma cells with a high pectin content. As water is removed, the cell wall shrinks in volume and on rewetting swells again. The swelling curve is not necessarily, or usually, identical with the shrinkage curves so that there is some hysteresis and, if a wall is held bone dry for a long period, it may not recover its original dimensions on rewetting; in that case presumably bonds such as hydrogen bonds have been satisfied internally and are no longer available to water. It will be recalled that the crystal lattice of cellulose I is not penetrable by water so that swelling and shrinkage *per se* are confined to the matrix. The phenomena involved are important wherever plant cells are subjected to violent alternations of relative humidity and, of course, for all fibrous materials of commercial importance such as cotton, hemp, ramie, and of course, wood.

Consider first the swelling of an isotropic gel. We do not ask how the water comes to be imbibed but accept that it *is* imbibed and causes an increase in volume. If the gel is in the form of a hollow tube (Fig. 11.12) and the outside diameter swells x per cent then the inside diameter also swells x per cent, the wall thickness swells x per cent and the length swells x per cent. This is not commonly what happens to the walls around elongated cells.

353

Normally, the length swells much less than the diameter and the wall swells in thickness more than either the inside or the outside diameter. It has been for many years a question whether such anisotropy of swelling is due more to the anatomy of a tissue or more to the presence of microfibrils and, if the microfibrils play a part, whether or not there is a connection (not necessarily causative) between the anisotropy in any particular cell and the helical angle θ.

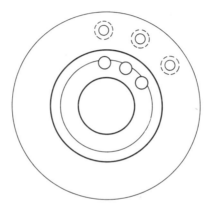

Fig. 11.12 Transverse section of a tube in the unswollen (thick lines) and swollen (thin lines) condition. The small circles represent the transverse section of non-swelling rods embedded in the wall of the cylinder. The broken circles represent the diameter which the 'holes' occupied by the rods in the unswollen wall would reach on wetting if the gel were not bonded to the rods.

If an array of non-swelling microfibrils is embedded in the gel and bonded to it, all parallel to the length of the tube, then these clearly impose a restriction on swelling. In transverse view (Fig. 11.12) on swelling the 'holes' occupied by the microfibrils move outwards and increase in diameter (broken circles). Since the microfibrils are bonded to the gel then the gel comes under tensile stresses unless the bonds break, a matter which will be discussed further below in terms of some modern theories of swelling. In the longitudinal direction, however, the gel will come under compressive stresses. Anisotropy of swelling is therefore likely in principle. If, moreover, the tube is surrounded by a sheath with a low swelling capacity, then on uptake of water the whole gel comes under compression and the relative increase in diameter of the lumen will be less than that of the wall thickness. If the considerations are added (1) that the matrix of the wall may not itself be isotropic and (2) the wall, particularly of the cells so far most investigated — namely wood cell walls — are structurally complex then it may be appreciated that attempts to understand swelling present formidable problems.

The effects of hydrostatic pressure on swelling gels in terms of net water uptake appears not to have been considered by most theorists in this field but it could be of consequence since pressure gradients are certainly set up as a wall swells or shrinks. Attention was called to these effects in wood cells by Barkas (1949) on thermodynamic grounds. He was able to show that the shrinkage of wood between equilibrium with saturated air and air at 25 per cent R.H. (i.e. between about 25 per cent and 15 per cent water content in the wood) is only about 27 per cent of the calculated stress-free value. This restraint is due to the anatomy of wood, namely the present of an outer resistant sheath to every cell (probably S1 + M + P) which leads to the situation that the lumen does not shrink as much as would be expected (in a homogeneous body) or actually expands (Clark, 1930; Koehler, 1931). The effect arises, of course, because hydrostatic pressure increases the vapour pressure of absorbed water. In principle water may be taken up into cell walls by either or both of two mechanisms (a) through hydrogen bonding to —OH groups and the setting up of chains of water molecules and (b) passage into 'capillary space'. As pointed out by Barkas (1949) the implied distinction is meaningless below about 60 per cent R.H. since then the radius of a capillary giving a meniscus curvature such that the capillary water is in vapour pressure equilibrium with the atmosphere is 2×10^{-7} cm; the 'surface of the capillary' is then meaningless. Indeed by immersing wood flour in a series of sugar solutions, whereby molecular absorption of water would cause an increase in sugar concentration, Barkas was able to construct a sorption isotherm for vapour pressures close to saturation and to show that near saturation well more than half of the water was held by adsorption. As the wall dried, removal of the adsorbed water causes shrinkage not only in the matrix itself but also in such void 'capillary spaces' as there are and these may become occluded; correspondingly, on rewetting, swelling will tend to open out the pores again.

355

Since the presence of microfibrils in a wall is likely to induce anisotropy of swelling it might be expected that the angle at which the microfibrils are tilted to cell length in tracheids and fibres should be, in part at least, related to the variation of swelling properties found in these cells and in the tissues which they compose. It was therefore at one time discouraging to find an authority like Frey-Wyssling (1940) denying to cell walls any part in the anisotropic shrinkage of wood. Investigation since that time has brought into prominence both variations in swelling and shrinking associated with variations of wall structure and possible mechanisms by which the association may be validated in quantitative terms. There can now be no doubt about the facts; whether or not the relation is causal, there *is* a relation. Quantification of the relation in physical and mathematical terms depends, however, on the validity of the model of the cells and the tissues upon which the theoretical argument is based. All the authors of theoretical evaluations of the relation admit without cavil that their models are deficient in one way or another, and perhaps the greatest value of the more recent attempts is that they point to the parameters of the wall which it is necessary to know before a model can be thoroughly checked against the behaviour of the real structure.

Although the evidence in the case of single isolated cells is scanty, it seems that these behave in some ways differently from whole wood. This is perhaps not surprising since in isolating the cells almost all the lignin and some of the matrix polysaccharides have been removed. Single cells will be dealt with first as historically the first to be investigated from the present point of view.

11.4.1 Single cells

As far as the writer is aware the only relevant results were published almost precisely thirty years ago (Preston, 1942) in an examination of longitudinal shrinkage in tracheids and fibres from both softwoods and hardwoods. The intention was to attempt an explanation in geometrical terms only of longitudinal shrinkages measured some years earlier. These were made on cells isolated by maceration in 0.5 per cent chromic acid from sections cut obliquely to the longitudinal. The ends of the cells had therefore been cut away leaving a long hollow tube cut obliquely at each end so that θ could be determined. Shrinkages induced by transferring the cells from

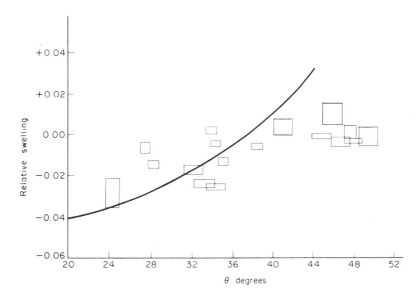

Fig. 11.13 Relation between swelling (or shrinkage) and θ in softwood tracheids and hardwood fibres. The sides of the rectangle give the spread of the experimental errors at each experimental point. The curve represents the equation derives in the text.

a dilute glycerine-water mixture to absolute alcohol are presented in Fig. 11.13 in terms of microfibrillar angle θ. It might be expected intuitively that cells with slower microfibrillar helices (larger θ) would shrink more than cells with steeper helices but this is not apparently the case; shrinkage becomes reduced as θ increases from about 10° to about 40° and thereafter remains approximately zero. The geometrical explanation of this was based upon a model of the cell consisting of the layer S2 only (the slower helices in S1 and S3 had not at that time been established) and it was assumed that the cellulose crystallite formed a continuous thread coiling helically round the cell (microfibrils were not discovered until 6 years later). It was observed that, on shrinking under the conditions used, the cells did not twist and this lack of twisting was an essential explicit part of the geometrical theory (not *implied* by the theory as Kelsey (1963) was later to state). It was therefore taken that the number of turns of the helix did not change as the cell contracted so that the length of cell *per turn* changes in proportion to the whole cell length. In that case, if the length of the

357

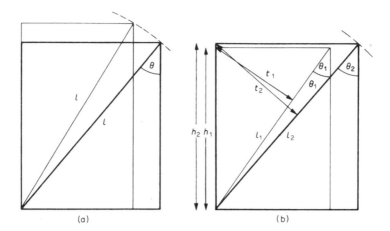

Fig. 11.14 Each rectangle with heavy lines represents a segment of the cell corresponding to one turn of the microfibrillar helix cut open longitudinally and laid out flat. The rectangles with light lines represent the same cell on shrinking laterally if (a) the helical winding (length l) does not shrink and if (b) it shrinks materially.

cellulosic spiral winding does not change as the cell shrinks laterally then the cell must increase in length. Shrinkage in length can occur only if the helical winding itself shrinks (Fig. 11.14). It is immaterial whether this shrinkage is due to a contraction of the winding by desorption in the amorphous zones of microfibrils or to a compression due to the contraction of the matrix whereby it is surrounded. No assumption is made about the matrix either; it is immaterial whether this is isotropic or anisotropic. From Fig. 11.14b a relation can be deduced quite simply between h_2, the length per turn in alcohol, to h_1, the length per turn in glycerine — water in the form

$$\left(\frac{h_2}{h_1}\right)^2 = \frac{(1 + x)^2/(1 + y)^2 - \sin^2 \theta_1}{(1 + x)^2 (1 - \sin^2 \theta_1)}$$

where x is the relative shrinkage in girth of the cell, y the relative shrinkage in length of the helical winding and θ is the helical angle. Taking x as 0.13 (the average measured over 100 cells) y may thereby be determined for any value of θ_1. The corresponding curve fitted for a value of y (0.056) determined for $h_2/h_1 = 1$ at $\theta_1 = 38°$ is included in Fig. 11.13. It clearly gives a reasonable fit for values of θ up to about $40°$ but beyond that predicts an *elongation* of the cell on drying which does not occur. In this range tracheids and fibres should in theory resemble rope or string. These are made of twisted yarn, the yarn itself consisting of fibres twisted at about $40°$ and therefore invariate in length on dessication; the whole rope or string accordingly lengthens on drying and shortens on wetting.

Although with whole wood Koehler (1931), Welch (1933, 1935) and Cockrell (1943) have reported longitudinal expansion on drying and later workers (below) have predicted that this should happen under some circumstances, it does not seem to do so with single cells; the theory has clearly left something out of account. This may be traced to two sources. Firstly, defining t as in Fig. 11.14 it is clear that this should contract precisely as the girth b contracts unless there are restraints. It can be shown quite simply that

$$\frac{(t_1 - t_2)}{t_2} = \left(\frac{1}{K} \frac{1+x}{1+y} - 1 \right)$$

where K is b_2/b_1. In the range from $\theta = c.\ 20°$ to $\theta = c.\ 45°$ this decreases from $0.9x$ to $0.5x$, i.e. the helical organization requires that the shrinkage of the wall in a direction at right angles to the helical winding is less in the plane of the wall than at right angles to it, and the difference becomes greater, the slower the helix. This in turn implies a tension in the plane of the wall which requires on thermodynamic grounds a diffusion of water from elsewhere to a site between the turns of the helix. The effective reduction in t will, therefore, be a compromise which would at a certain angle (a certain value of $(t_2 - t_1)(t_2 x)$) inhibit the change in length. Secondly, although the maceration process has removed some of the outer constraints, the S1 layer is still present and this must impose restriction on girth increase in such a way that, as in whole wood, any cylindrical shell in the wall will not be permitted to contract as much as does the whole wall thickness. It is difficult to allow for this. It will readily be seen, however, that if any such shell is not allowed to contract in girth while the helical winding still contracts, then the length of the shell (along the fibre axis) will decrease whatever the value of θ; restriction of this kind will reduce the elongation which would otherwise occur for $\theta > 40°$. This effect must be much more marked in whole wood.

11.4.2 Whole wood

As with single cells, the contraction of wood as a whole on drying is much less in the direction parallel to the grain than it is in the transverse plane. Vintila (1939) was early in attempting to connect this also with the helical organisation of the constituent cells. He did this by constructing a vector diagram relating the longitudinal contraction to the transverse contraction and calculating the resultant which he assumed to be at right angles to the helical winding. This type of diagram has, as pointed out by Frey-Wyssling (1940), no theoretical basis and, as also in the later work of Turnbull (1940), gives far too small a value for θ. At that time Frey-Wyssling also pointed out that, as for the individual cells discussed above, transverse contraction of a helix implies for some values of θ, a longitudinal expansion, whereas whole wood always contracts in length when taken from the fresh condition to oven dry. This is why he then abandoned the cell wall as a controlling factor and attributed anisotropic contraction to the properties of the middle lamella. On the other hand, from the figures given by Vintila for the early wood of *Larix* and *Abies* the values of h_2/h_1 for whole wood may be calculated as 0.997 and 0.998 respectively, and the mean of 154 determinations by Turnbull on *Pinus patula* yield a figure of the same order. In terms of the theory presented above for individual cells this implies a helical angle of 39°, well within the range to be anticipated.

In the early 1940s it seemed, therefore, that the basis for the longitudinal/transverse anisotropy of wood and other tissues was understood in terms of wall structure if only in a vague sort of way and that the theory merely needed refinement.

Since that time evidence has accumulated that wall organisation and this type of anisotropy are indeed associated. It has, however, become apparent that whole wood shrinks and swells more at high values of θ than at low values, the reverse of the apparent situation with individual cells. This was demonstrated conclusively by Cockrell (1943), making use of the fact that the value for θ is greater in inner annual rings than in outer, and showing that shrinkage in inner annual rings can reach 2.5 per cent ($h_2/h_1 = 0.975$) whereas for rings beyond 25 the value has fallen to *c.* 0.25 per cent ($h_2/h_1 = 0.998$). Koehler (1931) and Welch (1933 and 1935) had already claimed that some wood samples show no decrease in length on drying and even swell on drying to air-dry but then shrink to oven dry. Cockrell (1943 and 1947) supported this observation for small values of θ and interpreted it as due to a coming together of cellulose chains linked at each end so that the linked ends are pushed apart. This might equally be explained along the lines developed above for single cells. Similarly, Smith (1959) and Dadswell and Nichols (1959) have supported the connection between θ and swelling anisotropy though they attempted no explanation since they considered the effect an indirect one. Boutelje (1962), however, attempted an explanation in terms of a distribution of water mainly between concentric lamellae of microfibrils.

A new dimension has more recently been added to this story. This begins perhaps with the review of Kelsey (1963) pointing out the inadequacy of these earlier theories and calling attention to the discrepancy between the behaviour of whole wood and that to be expected by the only quantitative interpretation so far attempted, that of Preston (1942), for individual cells already dealt with. Subsequently, Barber and Meylan (1964) have attempted a highly sophisticated explanation, based upon a model relating swelling properties to mechanical properties which they admit is too simple and which they in this first paper and subsequently properly regard as a first attempt which appears, however, to have promise.

The authors consider a model of the wall in wood as part of a large flat sheet consisting of the S2 layer only, ignoring the tubular form of the cells and cell corners. The microfibril angle is assumed to be constant over the whole specimen and the matrix is assumed to be isotropic in its swelling properties. As the matrix of dry wood swells it therefore puts the non-swelling microfibrils under tension and itself under compression; correspondingly, on dehydration from the fresh condition the microfibrils come under compression. The argument then proceeds as follows using swelling rather than shrinkage.

If the microfibrils are parallel to cell length (Fig. 11.15 a) the necessity to stretch them allows only a small longitudinal expansion (x direction). The swelling in the y direction (and the z direction) will then be relatively greater, greater indeed than it would have been without the longitudinal restriction. The authors base this on the assumption that the total volume swelling of the gel

361

will be the same as it would be in the absence of microfibrils. This cannot be precisely true because the compression of the gel consequent upon the presence of microfibrils will increase the vapour pressure of the contained water and the gel will therefore come to equilibrium with pure water at a lower water content. This may, however, be a minor matter.

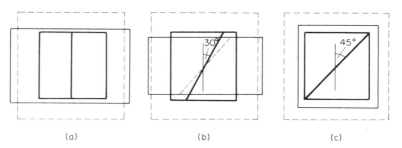

(a)   (b)   (c)

Fig. 11.15 Swelling of wood with each of three microfibril angles. The small square indicates the area of the equivalent dry wall sheet, the broken square the effect of isotropic swelling, the rectangles (medium square in (c)) the shape of the wet wall when the swelling has anisotropy dictated by the microfibril angle. For further explanation, see text. (After Barber and Meylan, (1964)).

When the microfibrils make an angle of 30° to cell length (Fig. 11.15 b), swelling should be skewed. This is prevented in the real wood by reason of the fact that each cell wall is cemented to another through the middle lamella. To allow for this, the sheet of wall is considered to be bonded to a second sheet with the same value of θ but in the opposite sense so that the microfibrils are crossed at an angle 2θ. This has the additional advantage that it allows mathematically for the circumstances that the cells in whole wood cannot twist, and is a device we have already seen to be adopted in dealing with mechanical properties. The microfibrils restrict expansion in the longitudinal direction and to some extent transversely. Since, moreover, the transverse expansion is greater than the longitudinal, the angle θ will increase slightly on swelling (broken line in b). Consequently, even if the longitudinal expansion is zero the microfibrils will come under tension and lengthen; zero longitudinal expansion is possible. If the elastic modulus of the microfibrils is high enough, or the modulus of rigidity of the matrix is low enough, so that the microfibrils can withstand tension

362

without much elongation, the material will contract in length as it expands transversely. This is the case exemplified in Fig. 11.15 b. When, on the other hand, $\theta = 45°$ (Fig. 11.15 c) longitudinal and transverse expansions are equal and positive; no change in θ is possible. Both expansions may be small if the matrix is weak and the swelling may appear largely as an expansion in wall thickness, in the z direction. This establishes a basis for a relation between swelling in wall thickness and θ (for which, however, there is as yet no experimental evidence). Beyond $\theta = 45°$ the x and y components interchange their behaviour.

On the basis of this model, Barber and Meylan (1964) show that the relative expansions e_x, e_y, e_z in the three directions are related to the expansion e_0 of the unrestricted gel by the equations

$$e_x/e_0 = [1 - (E/S) \sin^2\theta \cos^2\theta]/\Delta$$

$$e_y/e_0 = [1 + (E/S) \sin^2\theta \cos^2\theta]/\Delta$$

$$e_z/e_0 = [1 + (E/2S)(1 + \cos^2 2\theta)]/\Delta$$

where E is the elastic modulus of the microfibril, S is the shear modulus of the matrix and

$$\Delta = 1 + E/3B + (2E/3S)(1 - 3\sin^2\theta \cos^2\theta)$$

(B is the bulk elastic modulus of the matrix)

The resultant curves for x and y are very similar to the broken lines in Fig. 11.16, (though these are taken from a later paper by Barber (1968) to be discussed below). Barber and Meylan compare their predictions with the experimental findings of Cockrell (1947) and with some of their own; Fig. 11.16 includes the experimental curves of Harris and Meylan (1965) based upon 750 samples of *P. radiata* (full lines) and of Meylan (1968) (half dotted) together with experimental points obtained in the writer's laboratory by Dr Ranganathan (Preston, 1963). The ordinate of the theoretical curves is adjusted to give a fit with the experimental curve and the comparison is to be made in a qualitative, not a quantitative sense.

There is a good deal of scatter of all the experimental points but otherwise the fit is encouraging. The wood itself, however, shows a rapid increase in the (very small) elongation at smaller angles ($15° - 20°$) than predicted by the theoretical curves of Barber and Meylan ($40° - 50°$), as pointed out by Harris and Meylan (1965). Moreover, the minimum observed longitudinal shrinkage occurs

363

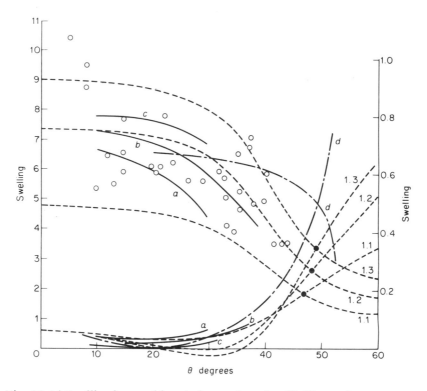

Fig. 11.16 Swelling in wood in relation to the microfibrillar angle; curves starting at upper part of graph represent transverse swelling, at lower part of graph longitudinal swelling. ---- theoretical curves due to Meylan (1965); curve *a* juvenile early wood, *b* juvenile late wood, *c* mature late wood (experimental, Harris and Meylan, 1965); ----- experimental (Meylan, 1968). ○ experimental points by Ranganathan (see Preston, 1963). Solid circles mark the points at which transverse and longitudinal shrinkages should theoretically be equal. The ordinate to the right refers to experimental results; the ordinate to the left refers to the theoretical curves, adjusted to give a match.

at 20° instead of the predicted 30°. These authors attribute this to the fact that wood with different values of θ is selected for experiment from a variety of annual rings so that the known changes in wall composition across the annual rings may involve a change in e_0, (which is taken as constant in the theory). There is, however, another more probable explanation inherent in the model (see below).

The model has more recently been made more sophisticated by

Barber (1968) by converting the plane sheet of wall into a hollow tube, internal diameter r_i and external diameter r_0. The two-ply bonding between adjacent walls is allowed for by considering the wall itself to contain both microfibril directions, and an external sheath is added which restricts the swelling. The broken curves of Fig. 11.16 represent the relationships, deduce for $E/S = 20$ and with an isotropic sheath, for r_0/r_i equal to 1.1, 1.2, 1.3 as marked. The curves are very similar to the earlier curves except that the predicted elongation at $20° - 30°$ is much reduced. Since that time Cave (1972a) has reformulated the theories of Barber and Meylan in terms of his treatment of the shrinkage of a fibre composite (Cave, 1972b). He uses as a model the same multiply sheet used by him in interpreting the mechanical properties of the wall (p. 342). His curves are closely like those presented here.

Assessment of these theories resolves itself into an assessment of the validity of the models used. The authors of the models are commendably prudent and fully realize that this exciting preliminary work can come to fruition only if the models can be modified to meet more nearly the structure of the real wood, and can be made quantitative only if the parameters of the wall which they show to be necessary can be measured. As it is, the question has to be asked whether the qualitative agreement which they have so far achieved constitutes a proof that the connection between θ and shrinkage is dependent upon wall properties in the way they assume. Agreement with experiment, especially qualitative agreement, is no proof of the correctness of an argument if the model upon which the argument is based is false. For instance, one might assume that a circular area of the wall would swell to an ellipse and this ellipse might be regarded as a section of an index ellipsoid. The swelling at any angle θ to the minor axis of this ellipsoid (the helical angle) could then be calculated exactly as for refractive indices as discussed in Chapter 4. The resultant curves (Fig. 11.17) are of the correct shape but are meaningless since the model is unacceptable.

Some of the differences between observation and the predictions from the model of Barber and Meylan possibly arise through conceptually trivial errors in the model. For instance, we have seen that in any piece of real wood there is a wide range of values of θ, whereas in the model one invariate value is assumed, equated with the mean value in the real wood as determined by X-ray methods.

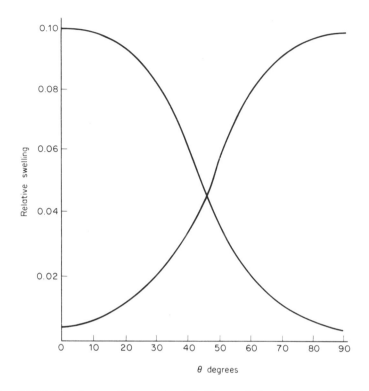

Fig. 11.17 Relative swelling v microfibrillar angle based upon the 'index ellipsoid'. Curve starting at upper right represents transverse swelling, the other longitudinal swelling.

Since flatter helices swell and shrink more longitudinally than do steeper helices, the steeper helices will be put under tension in the real wood and the wood will move more than would be expected from the 'mean helix'. In other words, the mean value for θ appropriate for the swelling and shrinking equations may be greater than the mean derived by X-rays and this may explain why the theoretical curves for longitudinal shrinkage turn upwards at higher values of θ than recorded by experiment. In other ways, neither the behaviour nor the morphology of the model seems acceptable and it is not easy to see how these can be improved while retaining the same model.

For instance, the argument is based on the concept that the swelling matrix can stretch the microfibrils. Taking Young's modulus for a microfibril as c. 10^{12} dynes cm^{-2} and the energy of a hydrogen bond as 3.5×10^{-13} ergs and assuming that three hydrogen bonds per cellobiose residue are available for bonding to the matrix (an overestimate, if anything) it is a simple matter to show that the energy required elastically to extend a microfibril by 0.1 per cent is several orders of magnitude greater than the energy of bonding to the matrix; so that even assuming that the H bonds were correctly oriented they would break and the microfibrils would not extend. No avoidance of this can be achieved by considering the microfibrils to be slightly sinuous (as they are) so that the initial Young's modulus may be very much reduced, since in molecular terms swelling of the matrix can stretch only that part of the microfibril in contact with it, irrespective of a slightly different orientation in neighbouring lengths of the microfibril. In morphological terms, too, the double-ply (or multi-ply) model seems too far from reality. With radial walls, particularly of later wood, a considerable sector of the wall of one cell faces, not a similar flat sheet of the next, but an intercellular space on each side of which the walls lie sensibly in a plane at right angles to it (Fig. 10.1). With tangential walls also the walls of neighbouring cells curve away from each other at the sides so that here they do not even remotely constitute a balanced laminate. In total, not more than about 40 per cent of the wall laminate could be taken as a balanced laminate. Other assumptions implicit in the arguments, such as a negligible volume proportion of microfibril to matrix and an effect of an external sheath only on internal stress without regard to the reshuffling of either microfibrils or matrix, or both, in a lamella which though swelling is constrained to move inwardly, are probably trivial in the sense that refinements of theory can perhaps take care of them. The assumption of isotropy in the matrix seems, however, again a more serious matter.

11.5 Porosity

This examination of the swelling and shrinkage of cell walls leads directly to considerations of the loci in the wall in which water molecules find their places, the mechanism by which they are held, and of the spaces in which they can move. Movement of water within and through cells has long been a matter of interest in plant physiology and concern to wood technologists by virtue of the need to force large quantities of liquid through wood during wood preservation processes. The physiological importance has recently become emphasized by the growing belief that, in its passage through a higher plant, water moves through roots and through leaves, not across plasmalemma and therefore *through* cells, but along the cell walls. Such a belief can be firmly established only when the flow path through the wall is defined, the permeability measured or calculated, and flow equations devised and found to be acceptable.

A water-saturated cell wall is, by definition, in equilibrium with an atmosphere at 100 per cent R.H. and contains free water which is mobile, and it has already been seen that in wood cell walls this mobile water constitutes less than half the total water at saturation. In principle, it may be held in several ways. There may be in the wall voids which are permanent (in the sense that they do not close as the wall is dried) and sufficiently large (say more than a few nm in diameter) that they may be regarded as 'capillary spaces' though not necessarily long or straight or of circular cross section. Again, there may be 'capillary spaces' which progressively close on drying (transient capillary spaces); these might be expected to be smaller in width than permanent voids and a high percentage of the water within them may be held to the wall of the voids, and to each other, by molecular forces with binding energies which decrease from the wall of the void inwards. Finally, the water may be in solution in the 'solid' material of the wall. With fine 'capillaries', less than a few nm in diameter, the distinction between these last two categories becomes meaningless.

When pores which are permanent can be seen in the wall so that they are *known* to be present then, however irregular they may be in size, shape or distribution, parameters defining their availability for water flow may be deduced from the correct flow equations. Proof of the presence of finer pores, which cannot be seen, whether or not they are assumed to be transient, encounters precisely the difficulties met with in considering flow through the plasmalemma. On the assumption that such pores are present, their parameters may be determined from flow equations and by other more sophisticated means. Such determinations can clearly not be accepted as proofs of the existence of pores without the most careful scrutiny.

In view of the conceptually simpler situation of permanent pores, these will be taken first. Both with these and the transient capillaries the evidence refers exclusively to xylem cells but no doubt the principles may be referred to other cell types also though the details need care in translation. The permanent pores are those in the margos of bordered pits.

11.5.1 Permanent pit membrane pores

It has been shown in Chapter 10 that the margo of the bordered pits is penetrated by pores of various shapes and dimensions, and that the passage through wood of particles in suspension of the order of 200 nm in diameter, and the filtering off at the membrane of particles of diameter larger than this (Bailey, 1913), has led to the currently accepted belief that most of a water flow induced across the wall of a conifer tracheid passes through these pores. Since there are alternative paths through the wall, and since some or all of the pit membrane pores may be occluded by a thin membrane, the relevance of these pores for flow can only be established by experiment. Accordingly, the pore radius has been determined for flow of air under pressure through wood (Stamm, 1935), from the pressure required to cause an air-water meniscus to pass through water-saturated wood (Stamm, 1929), and from pressure permeability and electro-osmotic flow (Stamm, 1932). These all give values between 10 and 100 nm. More recently, Stamm has turned to Berthold's method by measuring the pressure required to cause increasing amounts of octyl alcohol to flow through water-saturated wood (Stamm and Wagner, 1961; Yao and Stamm, 1967); with this method, 0.5 in thick specimens yield a higher value of about $0.35\,\mu$m. These values can nowadays be compared with the apertures in the pit membrane which range up to about $1\,\mu$m across.

Sebastian, Côté and Skaar (1965) appear to have been the first to attempt a direct comparison between visually measured and experimentally determined pore radii. They obtained values ranging from 0.7 to $2.5\,\mu$m for sapwood and from 0.8 to $4.8\,\mu$m for heartwood of white spruce, considerably larger than those given by Stamm.

The values were reached, however, by application of the Adzumi theory (Adzumi, 1937) for the flow of gases through tubes which are expressly very much narrower than they are long. The theory is therefore not strictly applicable to pit pores, so that pore

369

diameters measured in this way are not completely acceptable. Petty and Preston (1969) adopted a different approach. They compared the longitudinal flow of dry air and of liquid n-hexane through air-dried transverse sections of several species of conifer wood, the sections being appreciably thicker than the mean tracheid length. The validity of the methods and arguments used were checked on ultrathin 'millipore' filters $25 \pm 5\,\mu m$ thick and with pores $0.45 \pm 0.02\,\mu m$ wide.

In liquid flow, they used the modified Poiseuille equation

$$V = \frac{N\pi a^4 \Delta p}{8\eta(l + na)} \qquad\qquad 11.1$$

shown by Bond to be applicable both to tubes (l = 4 to 13 x a) and orifices (l = 0.102 x a) by use of the correct value for n in the Couette correction. Here V is the volume of liquid, viscosity η, passing through a plate under an applied pressure difference Δp, if the plate has N pores of radius a and length l; n was taken as 1.146. With gaseous flow, the flow is molecular if the mean free path (λ) is appreciably greater than a or l. The volume of gas flowing per unit time at pressure p is then

$$Q_P = \frac{N\pi a^2 \Delta p}{\sqrt{(2\pi \rho_1)}} \cdot \frac{1}{1 + 3l/8a} \qquad\qquad 11.2$$

where ρ_1 is the density of the gas at unit pressure and at the experimental temperature. In the experiments the mean air pressure used was less than 1 cm Hg (at 0.76 cm H_8, λ = 10 μm).

Rearranging the equations,

$$\frac{N\pi a^3}{1 + l/na} = \frac{8\eta V n}{\Delta p} = P \qquad\qquad 11.3$$

$$\frac{N\pi a^3}{1 + 3l/8a} = \frac{Q_P\sqrt{(2\pi \rho_1)}}{\Delta p} = D, \qquad\qquad 11.4$$

from which

$$a = \frac{P(1 + l/na)}{D(1 + 3l/8a)}$$

P/D gives an approximate value for a since the ratio of the bracketed terms varies only between 1 and 2.3 for the whole possible range of l/a. This can be refined by substituting this value and a possible value for l (say 0.15 μm) on the right-hand side, giving a much more accurate value for a. This can then be repeated if necessary.

In fact, if a is of the order of $1\,\mu$m, changing l by a factor of 2 changes the value of a by only 5 per cent. Substitution of the value of a in 11.1 or 11.2 gives a value for N. Average values of a found for sapwood were $0.85\,\mu$m (*Picea sitchensis*), $0.51\,\mu$m (*Thuja plicata*), $2.33\,\mu$m (*Larix leptolepis*), $1.05\,\mu$m (*Pseudotsuga menziesii*), the values for heartwood being somewhat smaller. From the value of N, the number of open pores per tracheid was calculated and was mostly less than unity.

The method allows no calculation of the number of *conducting* tracheids and therefore no estimate of the number of pores *per conducting tracheid*. The calculations assume that all the resistance to flow is at the pore, but Petty and Preston (1969) showed that if this number is much higher than unity, then the resistance of the tracheid lumen would be appreciable both for gas and liquid flow.

Subsequently Petty (1970) was able to confirm this through the gaseous permeability of the sapwood of *P. sitchensis*, determining the number of conducting tracheids by staining with reduced basis fuchsin. As expected, the conducting channels in his (dried) specimens were all in the latewood. The average value of a (now corrected for resistance in the lumen) was $0.14\,\mu$m. Both Petty and Preston (1969) and Petty (1970) examined the effect of a non-uniform distribution of pore size on the mean pore size and concluded that the mean pore sizes given might be overestimated by about 40 per cent. More recently Petty (private communication) has corrected this value for a by applying a Couette correction analogous to that used by Petty and Preston (1969). The corrected value is $0.35\,\mu$m, with the number of pores per *conducting tracheid* at about 40. The radii of the pores calculated in this way are similar to those measurable on the equivalent electronmicrographs of the type illustrated in Fig. 10.19.

It has already been seen that air-drying of wood leads to aspiration of early-wood pits and possibly to other changes affecting the permeability of the pit pores. Petty and Puritch (1970) have examined this by comparing the mean radius in air-dried and solvent-exchange dried wood. In the solvent exchange process, the water in the wood is replaced by methanol and this in turn replaced by hexane which is then evaporated in a stream of N_2; this has the effect both of reducing the aspiration of pits and, as will be shown later, of retaining in the wall 'capillary spaces' occluded by normal air-drying. They give values of $a = 0.26\,\mu$m for an air-dried wood and $0.16\,\mu$m for solvent-exchange-dried wood (both corrected as above). The number of pores is increased in the latter, however, by a factor of more than 100 so that the total permeability is enormously increased.

The most recent findings in this field are by Smith and Banks (1971) and offer a nice example of the influence of initial assumptions upon the interpretation of data. Whereas in the three papers just discussed the flow-controlling part of the pit structure was assumed to lie in the margo, these authors make the more general assumption that resistance to flow is contributed by all parts of the pit including the aperture in the pit border. On this assumption Smith and Banks (1971) found it possible to apply a viscous/slip equation for gas flow, without needing to assume a value for pore length.

The appropriate equation is derived from Poiseuille's equation corrected for slip and takes the form

$$K = \left(\frac{K_v}{\eta}\right)\bar{p} + \delta K_s\bar{v}. \qquad\qquad 11.5$$

K is the specific flow or permeability $\left(= pv\,\dfrac{1l}{A\Delta p}\right)$; η is the viscous flow of the gas; $\bar{p}$ the mean pressure; δ a dimensionless coefficient ($\simeq 1$); $\bar{v}$ the mean molecular thermal velocity of the gas ($= \sqrt{(8RT/\pi M)}$; R the gas constant; T the temperature (K) and M the molecular weight of the gas. K_v is the permeability constant for viscous flow K_s for slip or Knudsen flow (representing the contribution due to slip of the gas molecules over the surface of the pore wall). The equation has been shown to apply to a wide range of porous materials down to pressures which bring the mean free path to an order of size similar to that of pore diameter (commonly down to a few mm absolute or below). It is obvious from equation 11.5 that K should be a linear function of $\bar{p}$, as it is found to be for many porous solids. For wood, Smith and Banks (1971) found the function to be quadratic in $\bar{p}$ and traced this to the circumstance that K is the sum of *two* linear functions of $\bar{p}$, k_1 and k_2, such that $1/K = 1/k_1 + 1/k_2$. They consider k_1 to represent flow along tracheid lumina and k_2 takes the form

$$k_2 = a\bar{p} + \bar{b} \qquad\qquad 11.6$$

the first term representing viscous flow and the second slip flow. This may be compared with the relation proposed by Carman (1956) for non-cylindrical capillaries,

$$k = \frac{em^2}{C_0 q^2} \cdot \frac{1}{\eta}\,\bar{p} + \tfrac{4}{3}\delta\,\frac{em}{C_1 q^2}\,\bar{v} \qquad\qquad 11.7$$

where e is the porosity, C_0 the shape factor for viscous flow in non-cylindrical capillaries, m is the mean hydraulic radius, C_1 the factor for slip flow, q the length of the flow path divided by the length of the sample and $\delta = 3\pi(2 - fq)/16f$, f being the fraction of gas molecules which undergo diffuse reflection at the capillary wall.

From Equations 11.6 and 11.7,

$$\frac{a}{b} = \tfrac{3}{4}\frac{C_1}{C_0 \delta \eta\,\bar{v}}\,m_1 \qquad\qquad 11.8$$

is a function of the pore radius.

Using longitudinal flow of H, He, Ne, Kr and Xe through 2 cm long pieces of bone-dry wood of *Abies grandis*, *Picea Abies*, *Pinus sylvestris* and *Abies alba*, Smith and Banks report values of pore radius as $0.8\,\mu$m, $0.85\,\mu$m, $0.85\,\mu$m and $0.85\,\mu$m respectively. The error here is likely to be within ± 20 per cent, dependent upon the constants assumed in Equation 11.8.

These values are considerably larger than those of Petty and Preston (1969), Petty (1970) and Petty and Puritsch (1971). Smith and Banks take the view that they represent not pores in the margo, but the annulus between the torus and the pit border (which amounts to about $2\,\mu$m as against c. $1.7\,\mu$m). Oddly enough, the idea that the major resistance to flow lay in this annulus had been already suggested by Bailey and Preston (1970). These authors based their conclusions on measurements of the rate of flow of water under pressure gradients applied longitudinally. Experiments of this kind are beset with the difficulty, first pointed out by Kelso, Gertjejansen and Hossfeld (1963), that following the Bernouilli principle water is liable to release gas bubbles at the low pressure regions around restrictions. Under a constant pressure gradient the flow rate then falls off with time owing to an air embolism. Correspondingly, if an increased pressure gradient forces

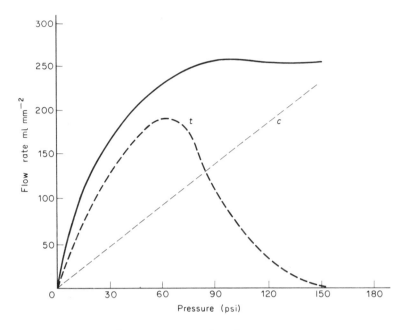

Fig. 11.18 **For explanation, see text.**

an embolism through the constriction the flow is disproportion-
ally increased. Using *initial* flow rates only, flow/pressure relations
were found for 5 mm thick (fresh) transverse slices of *Pseudotsuga
menziesii* sapwood, a relevant example of which is shown in Fig.
11.18. Flow rate increases with applied pressure (solid line) up to
about 90 psi but thereafter levels off or decreases slightly. The
only possible explanation seemed to be the aspiration of some
pits at the higher pressures. By considering the forces required to
displace the pit torus laterally, they were able to show that flow
through a bordered pit annulus is proportional to Pw^3 where P is
the pressure gradient and w the width of the annulus between the
border and torus. This means that with increasing applied pressure
the rate of flow will increase until the torus is displaced half-way
towards the border and will then decrease, as shown in curve t
Fig. 11.18. Curves of this kind were indeed found for the flow of
petroleum distillate (which does not penetrate the wall) through
dry wood. This seems to show that under the pressure gradients
used the major resistance to flow lies in the annulus and not in
the pores in the margo.

374

The situation concerning permanent pores which are certainly present is therefore even now not completely understood. Even with these it seems evident that theories tend to support the initial assumptions upon which they were based. This difficulty may well be accentuated when dealing with transient cell wall capillaries.

11.5.2 Cell wall capillaries

Returning to Fig. 11.18, the difference between the flow/pressure relations of wet and dry wood can only mean that paths are available to water in wet wood which are not available to non-polar liquids in dry wood. These can only be cell wall 'capillaries' which may be occluded by drying (see below) or be unavailable to non-polar liquids. Assuming that flow through these follows the straight line c in Fig. 11.18, then summation of $t + c$ could well give the experimental curve for water flow. This constitutes an argument for the presence of capillaries in wet walls which could contribute to flow even if only to a relatively minor extent.

The structure of the walls described in previous chapters and particularly the *précis* of structure presented in Fig. 7.13 should make it clear that the so-called capillaries can at best be only tortuous fissures between the microfibrils and among the molecules of the matrix. It has been shown on many occasions that metals and other substances can be deposited from solution in plant cell walls in aggregates visible in the electron microscope or even under the light microscope. An outstanding example is provided by the work of Frey-Wyssling and Mitrakos (1959) on sections of ramie fibres within which gold, silver or copper solutions were reduced, e.g. by hydrazine hydrate. The crystallites so deposited measured less than 10 nm across and tended to lie in lamellae parallel to the wall surface. Similar aggregates were found in wood cell walls by Bailey and Preston (1969) to occur in all three layers S1, S2 and S3 and an example is shown here (Fig. 11.19) from the work of Chou (1971). It was, indeed, deposits of this kind which led Frey-Wyssling to the multi-capillary wall structure of Fig. 7.4. Although such deposits could be nucleated on matrix polysaccharides, molecules of which might pass through them, and even though lattice forces in the growing deposit may distort any capillary space in which they grow, they do provide some evidence for the existence of void spaces which are continuous through the wall. They

Fig. 11.19 Electron micrograph of longitudinal section of a tracheid wall in a freeze-dried block of *Pinus silvestris* stained with copper sulphate. Note the fine fissures containing the metal. (Photograph by Dr. S. Chou (1971)). Magnification 171 000 X.

therefore form the conceptual basis upon which experimental work may be based.

Several workers have provided experimental evidence which seems to demand the presence in cell walls of capillary spaces equivalent to capillaries about 3.8 nm diameter (see Stone, 1964). Of recent years interest has centred in this regard on density considerations and on N_2 adsorption and only these in the main will be dealt with here.

(a) *Density considerations*
The density of cell walls may be measured either by displacement of liquid or by flotation in density gradient columns. The former method is the one mostly used and it has become generally agreed

376

that the density (of the whole wall) measured by displacement of polar solvents which do not swell it is about 1.46 gm cc^{-1} (ranging from 1.42 to 1.48) and in water, which does cause swelling, about 1.53 g. cc^{-1}. The answer is the same whether measured on fine particles or on sections and the discrepancy is generally thought to be accounted for by a density of adsorbed water > 1 (e.g. 1.115 g cc^{-1} at the fibre saturation point). Equally the density of dry wood sections is 1.46 when He is used as displacement fluid and is the same in a silicone of molecular weight 280 and of molecular weight 26 000 (Weatherwax and Tarkow, 1968). This would seem to imply that in dry cell walls there are no capillary spaces available to He, to polar fluids or to the silicone, though a small void volume may remain, available only to water. Stamm had already concluded that in dry cell walls the capillaries are occluded, a view reiterated by him in more recent times (Cowling and Stamm, 1963). Accordingly (Stamm, 1964) the density of the cell wall ρ_m at any water content between zero and the fibre saturation point is given by

$$\rho_m = \rho_0/(1 + \rho_0 m/\rho_w)$$

where ρ_0 is the density of the cell wall substance (1.46), ρ_w is the average density of the adsorbed water (1.115 at F.S.P.). Hence, on drying $\rho_m \rightarrow \rho_0$.

The value of 1.46 for ρ_0 would seem to be a little on the low side, for the density of wood substance may be calculated roughly from that of its constituents. The density of crystalline cellulose is about 1.60, but the measured density varies down to 1.53 dependent upon the amount of paracrystalline material present in the sample; let us take the commonly accepted, rather low, value of 1.55. The density of lignin is given in the literature as ranging up to 1.60 or more; again we take the commonly agreed low figure of 1.40. Finally the density of hemicelluloses is about 1.49. Suppose the ratio of cellulose : lignin : hemicelluloses is 0.4 : 0.4 : 0.2 (taking a high value for lignin to give a low value for total density) the density of wood substance comes out as 1.48 as a minimum. This suggests that there must be void spaces in dry wood cell walls.

Nevertheless several workers (see Petty, 1971) have measured densities for the dry wall at much less than 1.46 (ranging from 0.7 to 1.40), suggesting a high porosity, by using a somewhat naive method whereby the bulk density of wood is divided by the fractional wall volume present measured by some microscopic method. The method assumes that the normal treatment for mounting sections (alcohol → xylol → canada balsam) results in a section with walls shrunken to the dry state. This is probably not so. Bouteljé (1962) has tried three different mounting methods and found three different shrinkages, the normal method giving the lowest values. It is to be noted that a 10 per cent deviation from full shrinkage can account for the discrepancy. Petty (1971) has recently examined the microphotometric method used by some workers in determining the wall volume and has found an error which itself fully explains the discrepancy.

It seems therefore that the value of 1.46 needs to be accepted for the dry cell wall and that the total volume for any voids is small. Stone and Scallan (1967) and Kellog and Wangaard (1969) have concluded from N_2 adsorption that the void volume is about 4 per cent. This would mean a density of wall substance at about 1.52. The voids are clearly accessible to water so that in wet cell walls the void spaces (i.e. spaces containing water only) must be very much greater though the water in only about 50 per cent of the space is mobile. This is not to say that recognisable capillary spaces are present or even tortuous fissures. The water may still be contained by a network of the matrix substances in meshes which in no sense possess a surface. Density measurements of course give no guide to the size of the fissures.

(b) *Nitrogen adsorption*

On the other hand, if the assumption is made that fissures are present with recognisable surfaces, N_2 adsorption can give a measure of the total surface and, within limits, of the size of the fissures. Since the relevant observations can be made on walls dried from water and on walls after solvent exchange drying these parameters can be determined both for the normally dried wall and on a wall which is about as porous as it was in the wet state. The method is to flow mixtures of He and N_2 over the sample at $-195°C$ and measure the changes in the gas sample by, for instance, a thermal conductivity meter. The partial pressure of nitrogen can be changed by changing either the composition of the mixture or by changing the total pressure. In this way an isotherm can be drawn relating the amount of N_2 condensed into the specimen against the partial pressure. *Assuming that capillary spaces are present*, i.e. spaces with recognisable surfaces, an estimate may be made of the total surface area of spaces, ranging in radius from those just accessible to N_2 (4 Å) up to about 30 nm. Following the analysis of Pierce (1953 and 1959), the distribution of pore radii and the distribution of pore volume between these radii may also be determined. It is to be remembered that the method takes no account of pores greater in diameter than about 30 nm.

Most workers in this field agree (Stone and Scallan, 1965) that

(a) fibres dried directly from water have an internal surface of about 1 m^2 g^{-1}.
(b) in water-swollen fibres dried by solvent exchange the area is up to 200 m^2 g^{-1}.
(c) the pores responsible are small, the most common size being in the range 1.8 − 2.2 nm.

The radius is, of course, the radius which the pores would have if they were circular cylinders, though the theory says nothing about the shape of the pores. If they were in the form of lamellae, separating the wall lamellae, then the 1.8 − 2.2 nm would represent the mean thickness of the lamellar void.

Stone and Scallan (1964 and 1965) interpret their N_2 adsorption data in this way and attempt thereby to obtain a picture of cell wall structure. The distribution of pore width in fibres of paper pulp (delignified) is as shown in Fig. 11.20. The model size is about 3.4 nm and this is more meaningful than the mean. Beginning with the dry cell wall, they assume that there are no internal voids available to N_2 and that the whole adsorption occurs at wall surfaces. The wall can then be considered as a non-porous lamella, with specific volume v, specific surface area A and thickness $t = 2v/A$. Taking $v = 0.64$ cc g^{-1} for cellulose and $A = 1$ m^2 g^{-1} by N_2 adsorption, $t = 1.28$ μm. This is said to be about the wall thickness of the fibres used by them (2 − 3 μm) so that all the adsorption could occur at the cell surfaces. This seems a peculiar argument since N_2 is adsorbed measurably only into capillaries of 30 nm radius or less. With water-swollen walls if the specific area is 100 m^2 gm^{-1}, t should be 12.8 nm and the wall would contain 100 solid lamellae each about 9.4 nm thick separated by spacings of 3.4 nm.

Stone and Scallan therefore envisage the wall as containing 100 or so lamellae consisting of microfibrils in close contact so that

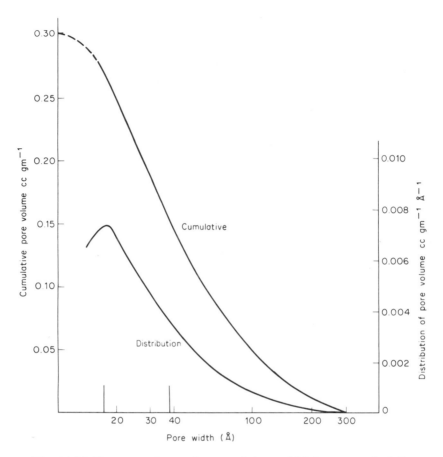

Fig. 11.20 The pore volume of pores of given width in paper pulp (after Stone and Scallan, 1965).

swelling involves a separation of these lamellae progressively to 3.4 nm apart. This would give an F.S.P. between 30 and 40 per cent, a range which includes the accepted values. It takes no account of the fact, however, that as a cell swells every lamella changes in girth, mostly increasing in girth as we have seen in dealing with the swelling wall. The microfibrils must therefore separate laterally and the microfibril lamellae themselves must take in water. This lamella concept is based upon the observation that walls are clearly lamellated both under the optical microscope and in the electron microscope. There is nothing in such observations, however, to show that matrix substances are not enclosed between the lamellae; indeed all the evidence shows that they are. When the microfibrils separate on water imbibition they must be surrounded by a swollen gel and still in this way bonded together; otherwise the tensile and swelling properties discussed in this chapter cannot be understood. It seems more likely that in swollen walls N_2 is adsorbed into meshes of the matrix surrounding the microfibrils on all sides which may well be of the order of 4.0 nm across and that, while some of these may well be distributed in lamella, some will be between microfibrils within microfibril lamellae; the postulated separation of neighbouring microfibril lamellae would then be greater than 4.0 nm as it assuredly is in some algal walls (p. 205). The fact that Stone and Scallan reach the same conclusion for lignified cell walls, for delignified cell walls, for beaten paper pulp, and even for regenerated cellulose seems to suggest again that a theory tends to support the assumptions on which it is founded.

Acccpting it that there are in the swollen wall capillary spaces of the order of 4.0 nm wide, presumably lenticular in section and tortuous, considerations of density appear to preclude any considerable number of additional voids of relatively large size, and it is recognized that such voids sometimes visible in electron microscope preparations are artefacts. If therefore plant physiologists are to continue to think of movement of water across tissues as movement along walls rather than across plasmalemmae they must concede that the flow paths must be of the order of 4.0 nm wide. It is not possible to make any precise calculations but order of magnitude estimates show that their case is then far from hopeless.

If we assume for preliminary purposes that the walls of interest are equivalent to delignified wood cell walls then we may proceed as follows. Although the discrete lamella hypothesis of Stone and Scallan is not proved the probability that there are alternating lamellae of cellulose-rich, matrix-rich lamellae is very high. Consider a cell, Fig. 11.21, b cm square, l cm long and consider any one matrix lamella a cm wide and assume that this contains only water and will allow laminar flow under a pressure gradient. This is equivalent to a lamella $4b$ high and l long since the wall is thin. If the pressure difference across the cell is Δp then the quantity of water Q passing through this lamella per second (pathway 1, Fig. 11.21) is

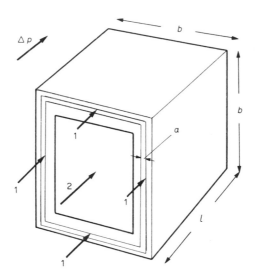

Fig. 11.21 **For explanation, see text.**

$$Q = \Delta p (4b) a^3 / 12 l \eta$$

where η is the viscosity of the flowing liquid. If there are N such lamellae in the wall, the total flow along the walls will be

$$Q = N \cdot \Delta p (4b) a^3 \, 12 l \eta$$

Inserting as typical values $b = 20 \mu m$, $l = 20 \mu m$, $a = 4.0$ nm, and taking $\Delta p = 1$ atm, $Q = 3 \times 10^{-13} N$ ml s^{-1} atm^{-1}. 3×10^{-13} is close to the amount of water flowing across the (two) plasmalemmae area $20 \mu m \times 20 \mu m$ if the permeability of a plasmalemma is 10^{-7} (pathway 2, Fig. 11.21), of the order commonly determined in higher plants. Only a few such lamellae are required therefore for wall flow to preponderate over cell flow. If the wall is about $2 \mu m$ thick, N may approach the value of 100 and since the wall will contain a much lower percentage of cellulose than does the wood wall considered above the matrix lamella may be wider. Since the lamella width appears as the cube, a small increase will have a large effect on Q. It might be that under the assumptions made the resistance through the plasmalemma would be three orders of magnitude or more greater than through the wall. In the real wall, the network of matrix polysaccharide chains will greatly enhance the resistance, but the order of magnitude difference gives considerable leeway before this would reduce the two flow paths even to the same resistance.

382

CHAPTER 12

Wall extension and cell growth

Differentiation of any plant cell from a meristem, or from an immature to a mature stage in lower plants which do not strictly have meristems, almost invariably involves an increase in size. The process may be divided somewhat loosely into two stages; an earlier stage in which the increase in volume stems largely from a synthesis of new cytoplasm, i.e. an increase in size or number, or both, of cell organelles, and a later stage in which volume increase is attributable almost entirely to an intake of water into cell vacuoles. It is with this latter stage, reflecting by far the larger increases in volume, that this chapter is concerned. Wall biosynthesis demonstrably begins during the first stage and continues throughout the second leading finally to the deposition of a secondary layer which arguably inhibits further growth. Wall extension and wall synthesis are therefore not separate processes and undoubtedly the one impinges on the other. As a consequence, although by and large this chapter will deal with growth problems and the final chapter with synthesis, there will of necessity be some reference to synthesis in this chapter and to growth in the next. Cells, and therefore the plant as a whole, reach a shape, and do so at a rate, which is entirely dependent upon the interactions between the structure of their walls and the physical and biochemical environment in which they occur. It is only a slight exaggeration, if it is an exaggeration at all, to say that plants are mostly long and thin because the macromolecules of which the walls are constituted are, as already made clear, also long and thin.

As a cell enlarges, the wall normally expands to exactly the extent necessary in such a way as to continue as a smooth unbroken envelope. ^{14}C-glucose feeding experiments demonstrate that this process involves synthesis of wall components. This was, however, deduced long before such observations were possible. If new wall material was not added during growth then increase in area of the wall, however caused, would of necessity demand a decrease in wall thickness. This can be observed microscopically mostly not to happen, for the primary wall of an extending cell has roughly the same thickness at all stages of expansion. There are of course small variations dependent presumably on the rate of synthesis relative to the rate of expansion. In parenchyma cells of coleoptiles, for instance, chemical analyses demonstrate an increase of all wall constituents during growth (Preston and Clark, 1944 — *Avena:* Blank and Frey-Wyssling, 1941) with a progressive though small *decrease* per unit wall area (Preston and Clark, 1944) and therefore a progressive decrease in wall thickness. Brown and Broadbent (1950) have reported a similar behaviour with root parenchyma cells, and Overbeck (1934) had already reached the same conclusion with respect to the setae of *Pellia epiphylla,* which can increase in length by 3100 per cent with only a 500 per cent increase in wall volume. On the other hand, Ruge (1937) has reported, and Diehl, Gorter, van Iterson and Kleinhooute (1939) have confirmed, that during the elongation of *Helianthus* hypocotyls the walls of the parenchyma cells first decrease and then increase in thickness. All these changes are, however, relatively small, and should not obscure the fundamental observation, repeated many times since these early days, that normal growth involves an equivalent wall synthesis.

There are in principle, therefore, two possibilities with regard to the mechanism by which a cell wall is induced to increase its area during cell growth. Firstly, a cell does not extend unless it is turgid, i.e. unless the cell contents are under a hydrostatic pressure — the turgor pressure — or, indeed, as shown comparatively recently (see below) unless the turgor pressure exceeds a certain critical value. It could be, therefore, that the wall becomes 'passively' extended under the stress induced by turgor pressure, the term 'passive' being qualified in the sense that metabolic activity might nevertheless be involved in bond making and breaking such as would modify the extensibility of the wall. Alternatively, if the new wall material synthesized during extension is intercalated between wall elements already deposited, this might set up 'growth pressures' analogous to swelling pressure and these might be responsible for the extension. This would be *intussusception* in the sense of von Nägeli, as against the *apposition* of von Mohl which would seem less likely to be causative in inducing such pressures.

These two processes might also act, of course, in combination. Opinion with respect to their relative importance has fluctuated down the years, and there was a long period during which it was thought that wall growth was an active process. The consensus of opinion today is that this may well in part be the case. The walls of most growing cells are extended under tensile stresses due to turgor and incorporation of new wall material is preponderately by apposition. The mechanical properties of the wall are therefore of the greatest importance. It is equally clear, however, that the bonding, and therefore the mechanical behaviour, of a wall surrounding a living protoplast is under close biochemical control and equally it is suspected that, in some cells at least, intussusception plays a modifying, if relatively minor role. The evidence comes from several sources.

12.1 Wall extension and microfibrillar orientation

The m.e.p. of the primary walls of most elongating cells lies transversely to the cell axis (Soding, 1934; Bonner, 1935 − *Avena* coleoptile parenchyma; Wergin, 1937; Anderson and Kerr, 1938 − cotton hairs; Diehl *et al.*, 1939; Ruge, 1937 − hypocotyl parenchyma; van Iterson, 1937 − staminal hairs of *Tradescantia*) or in a slow helix (Preston and Wardrop, 1949 − conifer cambium; Middlebrook and Preston, 1952 − *Phycomyces* sporangiophores; Probine and Preston, 1961 − *Nitella* internodal cells). This is the structure called by Frey-Wyssling (1930 and 1942) *tube texture*. Though at the time it was not clear that this observation had any meaning in some cases since it was made on whole cells or even on whole tissues − with the consequent complexities discussed in Chapter 4 − it was in others supported by parallel X-ray evidence and has since been confirmed by this and other methods. Even when the m.e.p. lies transversely, the walls are nevertheless optically anisotropic both in longitudinal and transverse section so that there is always some dispersion about the transverse; this is a *preferred* not a *precise* orientation. It was at first a puzzle that the orientation remained invariate though the cell was elongating sometimes to a length many times greater than the original.

Particular attention was called to the case of *Avena* coleoptile parenchyma by Bonner (1935). It is now known, and was believed in 1935, though some doubt was later expressed (see below, p. 387), that during growth the walls of these cells extend uniformly all over the surface. Wardrop (1955) proved this by noting, among other things, that the longitudinal separation between primary pit fields (Fig. 12.1) in the wall kept pace with the total cell extension. Subsequently Wilson (1957) made similar observations with the cells of *Elodea canadensis* and *Hippuris vulgaris* and Green (1954) showed that the enormous extension of the internodal cells of *Nitella* occurs through uniform extension of the wall, by observing the progressive separation of markers attached to the wall. The situation described by Bonner (1935) would certainly obtain with these and with all other cells in which wall growth is uniform. He showed that *Avena* coleoptile parenchyma could extend more than 100 per cent by growth with no net change in the m.e.p. of the wall, although a mechanical stretching of only 10 per cent sufficed to change the m.e.p. from transverse to longitudinal. Bonner, however, observed only the double walls of whole cells and Preston (1938), examining single walls only, showed that the m.e.p. was then arranged around the cell in a helix which became steeper as the cell elongated; this has since then been confirmed by Böhmer (1958) and Veen (1970). Bonner's conclusion nevertheless stands, with the logical corollary that walls are not stretched during growth, and that growth is mediated through wall synthesis. The observation of Bonner (1934) and Heyn (1935), more recently confirmed by Wardrop and Cronshaw (1958) and others, that cell extension can take place at low temperatures ($< 2°$ C) without any wall deposition might then be regarded as referring to a response to adverse circumstances. It was not until the advent of electron microscopy that this apparent paradox was resolved.

The interpretations of the first electron micrographs were clearly influenced, however, by the mistrust of simple mechanical explanations which this earlier work had implanted. The primary walls of maize coleoptiles and flax fibres (Frey-Wyssling, Mühlethaler and Wyckoff, 1948), of oat coleoptiles (Mühlethaler, 1950), of differentiating xylem elements (Bosshard, 1952; Hodge and Wardrop, 1950) and of cotton hairs (Roelofsen, 1951) were seen to contain typical cellulose microfibrils interwoven with each other but without the marked transverse orientation expected and indeed often lying in random array. It is to be recalled, however, that the polarizing microscope observations indicated that the crystallites in the wall are widely dispersed about the transverse direction and that the instrument is very sensitive to slight tendencies toward orientation (Chapter 4). Moreover, Frey-Wyssling and Mühlethaler (1950), using the growth zone of *Phycomyces* sporangiophores, and Roelofsen (1951) with cotton hairs demonstrated a tendency for the microfibrils of *innermost* wall layers to lie transversely, an observation which since then has been generalized and has proved to be important. Distributed over the wall surface, however, were localized areas in which the microfibrils lay in a very loose mesh, which Frey-Wyssling and Stecher (1951) and Stecher (1952) and later Green and Chapman (1955) regarded as centres at which microfibrils were being intercalated into the wall. They thought of these areas as transitory and hence that the wall grew as a mosaic in *mosaic growth*. Preston and Ripley (1954) were, however, unable to find such areas in cambial walls of *Pseudotsuga*, *Larix* and *Populus*; they considered such areas to be preparation artefacts and discounted the whole theory. At about the same time, Wardrop (1954) properly drew attention to the primary pit field, of whose existence there can be no question (Fig. 12.1) and modified the theory by regarding these as the centres of synthesis. Mühlethaler (1950b) was similarly misled by the appearance of the tips of parenchyma cells of maize and oat coleoptiles. Here the microfibrils were so few and so dispersed that the cell tips appeared to be open. This, coupled with his observation of longitudinal bars of longitudinally oriented microfibrils along the corners of the cells – which he, erroneously as it was to turn out, thought would inhibit longitudinal stretching – led him to propose that these cells grow by intercalation of wall material at their ends in *bipolar tip growth*. This was soon disproved by Castle (1955) for the same tissue (epidermal cells), as it was for cortical cells by Wardrop (1955) by measuring the displacement of markers affixed to the wall.

These latter observations equally disposed of mosaic growth in the cell walls concerned. Clinching evidence came with the auto-radiography of cells macerated from tissues grown in C^{14} labelled glucose (Wardrop, 1956; Setterfield and Bayley 1957), showing

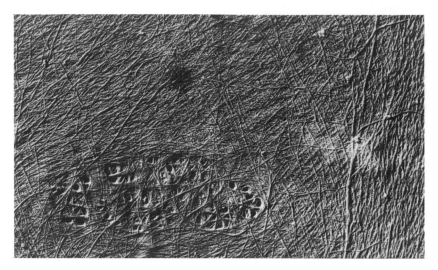

Fig. 12.1 Electron micrograph of the primary wall of a parenchyma cell of a wheat coleoptile in the zone 0 − 0.4 cm from the tip, viewed from outside the cell. Magnification 21 000 X, shadowed Pd/Au. Cell axis ↕. Note the longitudinal bars of microfibrils on either side of the primary pit field.

the radioactivity to be uniformly distributed over the wall surface with no concentration either at the two ends or at primary pit fields.

At the same time it is not to be disputed that wall growth may be mediated through a series of minute localised events and Bartnicki-Garcia (1973) has recently reopened the possibility of a form of mosaic growth which might in principle be more acceptable (p. 424).

Meanwhile, Roelofsen (1951) had already made the first of his electron microscope investigations leading to his concept of *multinet growth* which has since then, sometimes with some modification, been extended to cover many cell types. This first observation was made on cotton hairs, the general literature on the structure of which has been reviewed by Flint (1950). Roelofsen found that with the primary wall, for which the m.e.p. lies transversely, the microfibrils of the innermost layer lie dispersed about the transverse direction while in the outermost layer they are arranged almost longitudinally. This was soon extended to hairs of *Ceiba* and *Asclepias* and the growth zone of *Phycomyces* sporangiophores (Roelofsen and Houwink, 1953) and was confirmed for cotton by Tripp, Moore and Rollins (1954). Since that time observations of a similar nature have been made for *Nitella* internodal cells (Probine and Preston, 1961) for cells of *Cladophora* and *Chaetomorpha* (Frei and Preston, 1961) among others and, with cells growing in tissues, for oat coleoptile parenchyma (Wardrop, 1956; Wardrop and Cronshaw, 1958; Setterfield and Bayley, 1958), pith parenchyma of *Juncus* (Houwink and Roelofsen, 1954) and conifer cambium (Wardrop, 1958). In most of these cells it has been shown that the wall extends uniformly and, by radiotracer techniques, that new material is deposited at the innermost face of the wall.

The structural basis for the multi-net growth hypothesis, supported by these observations, is as represented diagrammatically in Fig. 12.2. The microfibrils of the innermost layer are here considered to lie transversely with some angular dispersion and are represented as a flat network. On passing outwards, through layers M_1 and M_2 to 0, the outermost lamella, the network first opens out as the microfibrils become more nearly randomly disposed and then flattened longitudinally. The argument then is that, since this gradation is seen irrespective of the stage of growth then, when any lamella in the wall, M_2 say, was the innermost lamella, it would have had the structure of the present I lamella. Consequently, beginning with the situation presented in Fig. 12.2, while a newer lamella is being deposited over I, the lamellae I and M_1 and M_2 and 0 will be stretched, I will become M_1, M_1 will become M_2 and M_2 will add to 0. At any instant of time 0 has been stretched more than M_2, and M_2 more than M_1 while I has not yet been stretched at all. Whether or not the 0 lamella has microfibrils which are actually longitudinal depends upon the degree of strain achieved. As each lamella is stretched it will of course become thinner and eventually tear and it is significant that torn outer lamella are seen on extending cells of *Cladophora* and *Chaetomorpha* (Frei and Preston, 1961). There will therefore be, beyond 0, still thinner lamella which are torn and disorganised. During this extension, the microfibrils must, of course, slip over each other, a process no doubt facilitated by the high wet-matrix/microfibril ratio in the primary walls of higher plants. Something like this process had already been suggested by Majumdar and Preston (1941) for the primary walls of collenchyma cells but they failed to realise that this was not a special case.

389

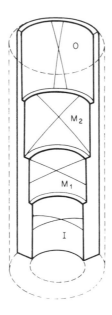

Fig. 12.2 **Diagrammatic representation of part of an elongating cell with the front wall and some layers of the back wall removed. I, innermost wall lamella; 0, outermost wall lamella; M₁, M₂, intermediate lamellae.**

As stated by Roelofsen, and as implied by the title of the hypothesis, the proposal involves a concept which is fundamental, namely that during growth the wall is passively stretched, but confines it to a situation in which the microfibrils of the innermost layer form a transversely oriented mesh. This seems a pity; for in some cells (Veen, 1970; Frei and Preston, 1961) the innermost layer is helically organised and the most notable feature is a change in the helical angle on extension rather than a change in the shape of a mesh. Such cells are nevertheless extending in precisely the same way. The term 'multi-net' must therefore be interpreted loosely.

The multi-net growth hypothesis clearly explains how it comes about that, for the cells for which it was devised, the m.e.p. of the whole wall remains transverse even though the cell may extend enormously in length (Probine and Preston, 1961). Suppose a young cell, with a wall $6\,\mu m$ thick with uniform tubular texture, extends to 10 x its length in such a way that its thickness remains constant. For simplicity, suppose the extension is stepwise and

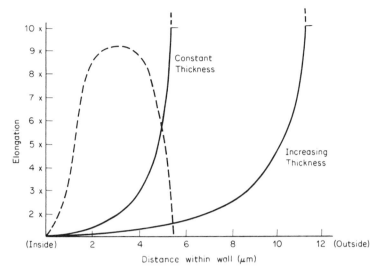

Fig. 12.3 **Distribution of strain through a wall** $6\,\mu$m **thick after an elongation of the cell by 10 x while maintaining either a constant or an increasing wall thickness. The broken curve gives the possible distribution of stress at the end of the elongation.**

imagine that first the cell, length l, increases in length to $6.0l/5.5$ and a new lamella is laid down on the inside $0.5\,\mu$m thick to retain the thickness at $6\,\mu$m. Repetition of this process tnen produces a cell which has extended 10 x and with a wall thickness of $6\,\mu$m. The extension per step can easily be calculated and therefore the strain at any depth within the wall may be determined, as presented by the smoothed curve in Fig. 12.3. Included in the figure is also the strain rate for a wall which doubles its thickness during the extension. In each case the outer step, $0.6\,\mu$m wide, represents the original wall. Since in reality this original wall would already have stretched, the real curve continues as shown by the dotted extensions. Two points emerge. Firstly, in each case the inner half of the wall has increased in length less than 2 x and, unless many of the microfibrils of the innermost layer are very widely dispersed from the transverse, this is not sufficient by far to change the net orientation from transverse to longitudinal. The m.e.p. of the outer half cannot therefore override that of the inner half and the m.e.p. of the whole wall will have remained transverse throughout. Secondly, if extension now proceeds as new inner lamellae are

391

deposited, the layers already present will move along the curve as far as strain is concerned and the distribution of strain and therefore the m.e.p. will remain constant.

12.11 Experimental tests of the multi-net growth hypothesis (M.G.H.)

Under the terms of the M.G.H. the structure of a growing wall should change if the rate of growth is varied and, conversely, the growth pattern should change if the wall structure is varied. Each of these predictions has been tested by experiment.

If *Avena* coleoptiles are placed in solutions of mannitol, longitudinal growth may be slowed or halted, dependent upon the molarity of the solution, with no or little effect on wall deposition. The wall should therefore show a smaller change than normal in the reorientation of the microfibril. This has been verified by Wardrop (1955). Conversely, coleoptiles grown at $2°$C continue to elongate slowly but wall deposition has ceased. Both inner and outer faces of the wall should then show microfibril reorientation and the again has been verified (Wardrop and Cronshaw, 1958).

Similarly, if the growth rate, or direction of growth, varies from one part of a cell to another, there should be a corresponding change in overall wall architecture. A striking case has been examined by Wardrop (1958 and 1965). Xylem fibres of angiosperms increase in length up to 5 x during differentiation and the evidence collected by Wardrop (1962) shows that this is due almost entirely to growth at the cell tips in bipolar tip growth. As a consequence the outermost layer of the primary wall of the elongating tips shows microfibrils predominantly oriented in the axial direction while in the body of the cell the reorientation has not yet reached the stage at which the microfibrils are distributed at random. An equivalent localised reorientation of wall striations was recorded long before this for cells of *Cladophora* of which only the upper half elongates during growth (Astbury and Preston, 1940).

At a more sophisticated level, a number of chemical treatments have been found which affect both the microfibril orientation in primary cell walls and the growth habit of the cell. Thus, longitudinally oriented microfibrils may be induced by kinetin (Fuchs and Lieberman, 1968), ethylene (Apelbaum and Burg, 1971) and indole acetic acid (1AA) at supraoptimal concentrations (Veen

1970b) and the cells then proceed to fatten rather than to elongate. A broadening of cells by application of supraoptimal 1AA is especially well documented (Diehl *et al.*, 1939 — *Helianthus* hypocotyl; Das, Patau and Skoog, 1956 — tobacco pith; Miller 1961 — fern gametophytes). Colchicine also has an effect (Burg, Apelbaum, Eisinger and Kang, 1971) which with *Nitella* internodal cells (Green, 1963) consists of a deposition of microfibrils in a helix 35° from the transverse (instead of the normal *c* 0°) and a fattening and twisting of the cell. An especially interesting situation was discovered by Galston, Baker and King (1953) which has since been examined in greater detail by Probine (1965). Subapical sections of pea epicotyls elongate normally in 1AA. If in addition benzimidazole (BIA) is present in a narrow range of concentration (*c*. 2×10^{-3} M), elongation is inhibited and the epicotyl broadens because the cortical cells broaden. Galston *et al.* explained this in terms of the, now discarded, mosaic growth. Probine (1965) has shown, however, that this phenomenon is consistent with the M.G.H. He has shown that while the m.e.p. of the cortical cell walls of epicotyls grown in IAA alone is uniformly transverse, the walls of the broader cells produced when BIA is present show broad ribs in which the m.e.p. is longitudinal. These ribs, which radioactivity experiments show to be newly synthesized, consist of parallel arrays of microfibrils. The changes in the ratio between cell length and cell diameter following addition of BIA at 2×10^{-3} M arc shown in Fig. 12.4. For the first 5 hours the ratio increases, for the next 7 hours or so it remains constant and thereafter it decreases. Clearly the initiation of longitudinal bars impedes longitudinal growth only slightly, but as their development continues this growth is first suppressed and in the final stage the bars are so prominent that they override the effect of the original wall and the cell begins to fatten.

In all these cases there is a clear tendency for a wall to extend in a direction at right angles to the predominant direction of the microfibrils and, in this sense, the M.G.H. is verified. Even, however, with those cells to which the M.G.H. is taken without doubt to apply, there are reasons to think that this purely mechanical explanation of growth is too crude an approximation.

(a) *Difficulties with the M.G.H.*
One of these concerns the longitudinal bars already mentioned.

393

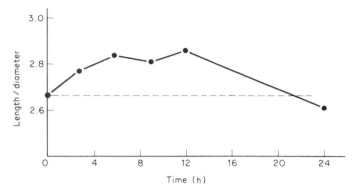

Fig. 12.4 The changing length/diameter ratio of cortical cells of pea epitcotyls grown in IAA at periods after the addition of BIA at 2×10^{-3} M (after Probine, 1965).

These occur not on the inside, but on the outside of the wall. Bars of this kind, and in this location, had already been seen on the primary walls of parenchyma cells of oat coleoptiles by Wardrop (1958) and by Setterfield and Bayley (1958). They are present throughout the growth of the wall and occur both on the inner and the outer face (Fig. 12.1). On the inner face, the microfibrils at the edges of the bars pass over to the transverse in pit field regions. Like the bars seen by Probine (1965), the outer bars are deposited *in situ* remote from the protoplasm. Roelofsen (1958) tried to meet this situation by assuming that cell turgor pressure, squeezing the wall, would displace matrix material to cell corners, carrying with it microfibrils which would then automatically be reoriented. This explanation will not do, however, since the microfibrils of the bars are *synthesized* during cell growth. On somewhat similar grounds, Sterling and Spit (1957) have concluded that with the pericyclic fibres of *Asparagus afficinale*, which elongate some 20 x during growth, M.G.H. applies only in the early stages of elongation. At later stages of growth the inner surface of the wall carries two helical sets of microfibrils each at some 60° to the cell axis while on the outer face the microfibrils lie mostly transversely. They believe that with this material both microfibrils and matrix are being deposited throughout the whole wall thickness. It is equally difficult to understand the structure of collenchyma cell walls (p. 312) in terms of M.G.H. only; as it is also of the walls of conifer cambium (which are said to show the typical multi-net structure of elongating cells) since the walls either do not extend much longitudinally (tangential walls) or expand enormously in the transverse direction due to repeated longitudinal division (radial walls).

As a final example epidermal cells do not seem to fit with the M.G.H., as first shown by Bayley, Colvin, Cooper and Martin-Smith (1957) with *Avena*. A diagrammatic representation of the wall of an epidermal cell of the root of *Sinapis alba* due to Foster (1962) is presented in Fig. 12.5; this shows more detail than in the earlier work but is in general in conformity. Epidermal cells pass through phases of elongation and widening and, of course, the outer wall finally balloons outwards to some extent. It is nevertheless difficult to harmonize this structure with the M.G.H. alone unless during growth microfibrils are pulled from the radial walls into the outer wall; and this seems unlikely. Chafe and Wardrop (1972) have recently examined epidermal cell walls in petioles of *Apium*, *Rumex*, *Abutilon* and *Avena* and have reached a structure differing from that presented in Fig. 12.5 in that the transversely oriented microfibrils of the radial walls pass over into the transverse set on the outer wall. Epidermal cells may therefore vary in this regard. As Wardrop remarks, however, this does not make these cells any easier to understand on straightforward M.G.H.

Patterns are therefore developed during growth which seem much too intricate to be explicable solely in terms of mechanical extension. The modifications required to the M.G.H. are not sufficient, however, completely to invalidate it as a basic concept. One such modification is clearly to recognize that in some cells the structure of the innermost layer may vary from time to time as in collenchyma, epidermal cells and, to a lower degree, in parenchyma cells. A second is that throughout growth, the wall may be open to interference by metabolic factors and maintained in delicate balance, perhaps in part by an interdependence of elongation and wall synthesis. At the same time, there can be no doubt but that the primary wall is extended under the stress induced by turgor pressure and that growth is to be interpreted, in part at least, in terms of the interaction between the mechanical properties of the wall, as dictated by structure, and the induced stress. It can easily be shown (Preston, 1955) that in a cylindrical cell 60μm diameter and with a wall 2μm thick the longitudinal wall stress under a turgor pressure of 10 atm should be about

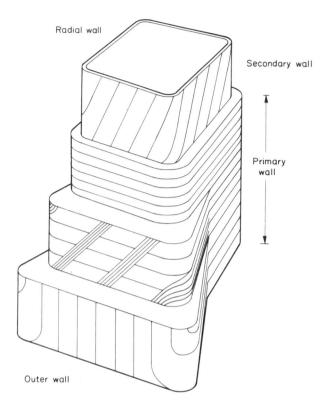

Fig. 12.5 **The microfibril orientations in the walls of epidermal cells of the root of** *Sinapis alba* **(after Foster, 1962).**

$1.6 \times 10^4 \, \mathrm{Nm^{-2}}$. Reference to Table 11.1 shows that this is sufficient to stretch the wall considerably and, of course, plasmolysis experiments show that walls are often stretched by 10 per cent or more. We now proceed to examine the mechanical properties in these terms. It is not to be anticipated that the interaction will be a simple one but we adopt a simple approach; we ask only what mechanical property may be involved, to what extent this mechanical property suffices alone to explain cell growth, and finally what metabolically-induced changes in this property are necessary if the explanation is to hold.

12.2 Wall extensibility

The assumption to be made is that the basic step in growth is a yielding of the wall under turgor-induced stress through some modulus of extensibility. It is preferable, therefore, to deal in the first instance with cells growing in isolation; for with cells in a tissue it can never be certain that the distribution of stress over the tissue is such that the turgor pressure of any one cell is resisted only by the wall of the same cell. The cell of this kind most fully investigated is the internodal cell of *Nitella opaca* (Probine and Preston, 1961) and this, as already seen, grows uniformly all over its surface. Any part of the wall is growing at the same rate as any other, and as the cell, so that any part may be used as typical of the whole. On a dry weight basis the content of cellulose is 16.7 per cent, the water soluble fraction amounting to a further 29.9 per cent (hydrolysing to give mostly glucose, galacturonic acid and galactose with traces of mannose, xylose and rhamnose) and the hemicelluloses 53.4 per cent (mostly of mannose and glucose with much smaller amounts of xylose and a trace of rhamnose). Chemically the wall is therefore a typical primary wall.

The X-ray diagram shows a tendency toward a transverse orientation of the crystallites which lie in uniplanar orientation, i.e. with the 101 planes more or less parallel to the wall surface. With *N. opaca*, however, the m.e.p. winds round the cell in a slow helix with a helical angle ranging from about 80° with young cells to about 90° with older cells (Probine and Preston, 1958) and this species was at one time thought to differ from *N. axillaris* in which the m.e.p. was said to lie transversely (Green and Chapman, 1955; Green, 1959). It is now recognized, however, that in both species the m.e.p. lies slightly off the transverse. The wall, moreover, shows typical multi-net growth. In the electron microscope the innermost and outermost lamellae have the appearances shown in Figs. 12.6a and b respectively. In each lamella there are microfibrils lying in all directions, but there is a marked tendency toward transverse orientation in the innermost lamella and toward axial orientation in the outermost lamella. The circumstance that the m.e.p. of the whole wall lies almost transversely provides a striking example of the sensitivity of the polarizing microscope to orientation and we need continually to remind ourselves that by transverse orientation we often mean orientation of the kind presented in Fig. 12.6a.

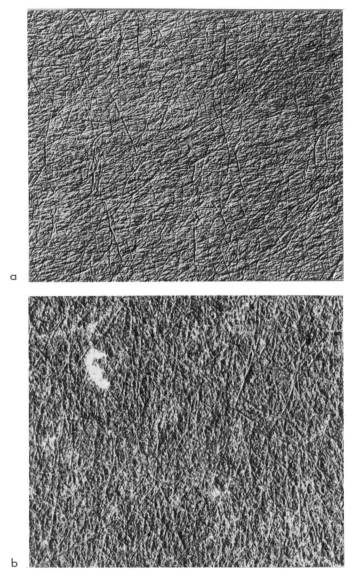

Fig. 12.6 Electron micrographs of wall lamellae from internodal cells of *Nitella opaca*; magnification 28 000 X, shadowed Pd/Au. Cell axis ↕.
(a) innermost wall lamella. Though the microfibrils show a wide spread of orientations the net tendency is toward the transverse.
(b) outermost lamella. Again the microfibrils show pronounced angular dispersion but the net tendency is axial.

An internodal cell of *Nitella* is cut off from the apical cell as a short cylinder some $50\,\mu$m long and somewhat broader laterally. This proceeds to expand into a long cylindrical cell which may reach 50 or 60 mm in length and $500\,\mu$m in diameter. This is a length increase of some 1000 x, a wall area increase of some 10 000 x and, according to Green (1958), is associated with a wall volume increase of up to 30 000 x. Since, during the period of growth of one internode several successive newer internodes are cut off from the apical cell, any one plant contains a series of inter-nodal cells at different stages of growth and therefore with different growth rates. It is, therefore, a simple matter to measure the growth rates of a number of internodal cells, to remove strips of walls for tensile testing, and so compare any desired modulus of the strips with the cell growth rate.

The specimens consist of wall strips cut either parallel or per-pendicular to the m.e.p. (and therefore sensibly longitudinally or transversely since in the specimens examined θ was rarely less than $86°$). These were used first in a determination of load/extension curves in order to determine Young's modulus. It is to be noted that these are effectively strips of wall from a plasmolyzed cell which have therefore contracted; their condition will not quite coincide with that in the wall of a turgid cell.

When a wall strip, either longitudinal or transverse, is loaded at 'zero' time and observed in a cine camera for one minute, the load removed for one minute and the cycle repeated, the extension varies with time as shown in Fig. 12.7. At the beginning of the first cycle, a deflection close to the final value is reached within about 10 s but the time constant on subsequent loading is much less — of the order of 1 s. When the load is then removed, the test strip does not regain its original length, and there is a further smaller 'set' on the second loading. On later successive cycles there is very little further change. At least a portion of the first set may be attributed to the fact that the strip is mounted air-dry and then rewetted, with a consequent swelling in length (*c.* 6 per cent) width (*c.* 4 per cent) and thickness (*c.* 100 per cent). In further work, strips were therefore conditioned by applying a load of 5 g mm^{-1} width of strip. Response to stress then consisted of an instantaneous component and a retarded elastic component with a retardation time of 1 s or less. No attempt was made to separate these and the 'instantaneous' deformation on removing the load

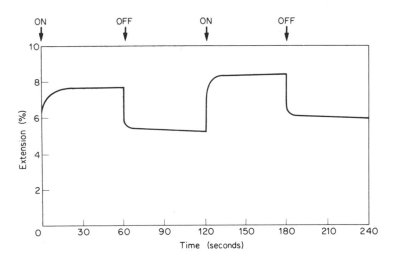

Fig. 12.7 **The variation of extension with time when a strip of** *Nitella* **cell wall is repeatedly loaded and unloaded.**

was used to compute the elastic modulus of the wall material. The strips were mostly about 0.5 mm long, 0.12 mm wide and, when dry, about 4 μm thick. Young's modulus is defined (p. 328) as

$$E = \frac{W}{bt}\frac{L}{\Delta L}$$

where W is the load, L is the length, ΔL the deformation, b the width and t the thickness of the strip. Since t cannot be measured accurately, the mass/unit area (m) of the wall was measured instead and a new modulus E' defined and used such that

$$E' = E/\rho$$

ρ being the density of the wall.

It was found that the value of E' for transverse strips showed no correlation with growth rate, but that E'_L for longitudinal strips ($< E'_T$) fell as the growth rate increased (from 0 to a value of 25 per cent per 24 h) i.e. from older to younger cells. The value of E'_T/E'_L therefore increases during growth (Fig. 12.8). The correlation

400

coefficient for E'_L against growth rate is 0.561 which, for 10 degrees of freedom, is significant at the 0.05 level, i.e. probably significant. The anisotropy of elastic properties expressed by E'_T/E'_L is, of course, precisely as expected in a wall in which the mean microfibril orientation lies transversely. It falls from about 5 in young, rapidly growing cells, to about 2 in mature cells. This latter value, it is to be recalled, is also the ratio of the transverse stress to the longitudinal stress in the wall of a cylinder which is pressurised. The higher value for growing cells implies that, in relation to the stresses imposed in the intact cell due to turgor, the wall is stronger (up to 2.5 x stronger) transversely than longitudinally.

This may satisfactorily explain why the internodal cell remains cylindrical rather than expanding like a balloon. It is, however, inconceivable that longitudinal extension is controlled by such static elastic moduli. At any instant, the wall is stretched by a fixed amount, and further elongation may be reached only if the turgor pressure increases and continues to increase during growth. This is known not to happen with root parenchyma (Burström, 1942) and there is no reason to suppose that the turgor pressure in *Nitella* varies much from 8 atm. An alternative would be to allow a steady decrease in E_L but this is better thought of in another way.

Reverting to Fig. 12.7, it will be noticed that, on each loading, the extension continues to increase slightly after the 'instantaneous equilibrium' position has been reached. This continuing extension under constant load, called 'creep', seems a far better candidate than does either the instantaneous elastic or plastic extension for the wall factor involved in growth. Following the lead, Probine and Preston (1962) have found that whereas longitudinal strips creep under constant load at rates which increase as the load increases, transverse strips also loaded parallel to their lengths do not creep even under loads which almost break them. Further, when longitudinal strips are stressed longitudinally and the stress condition expressed as the ratio of the applied stress to that to which the wall would be submitted by a turgor pressure of some standard value, say 8 atm, (to allow for variations between cells in cell diameter b and wall thickness t and therefore for variations of the stress imposed on the same strip while a part of the whole cell) then the rate of creep of wall strips varies regularly with the growth rate of the cell from which it was removed (Fig. 18.9). There is, moreover,

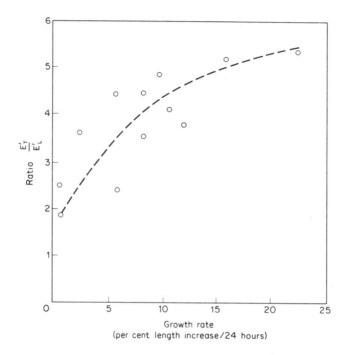

Fig. 12.8 The ratio of the transverse Young's modulus, E'_T, to the longitudinal modulus, E'_L, for wall strips from cells with different growth rates.

clearly a fairly well defined yield stress above which the rate of creep increases rapidly. In anticipation of a later discussion, it may be noted that Cleland (1959) has observed with *Avena* coleoptiles that elongation occurs only at a turgor pressure above a certain critical value (*c.* 6 atm) and therefore a critical wall stress.

There is therefore a nice parallel between cell growth and the creep rate of the wall, and this is strengthened by the further observations that K^+ ions added to the wall displace Ca^{++} ions and increase markedly both growth rate and creep. Creep is, of course, clearly a property of the matrix substance. It is to be noted in this context that growth rate is related to the proportion of water-

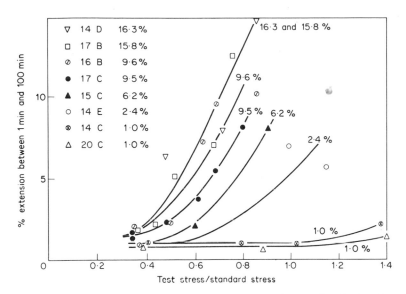

Fig. 12.9 The relation between creep rate and stress for longitudinal strips of *Nitella* wall chosen from cells of known growth rate. The number adjoined to each curve gives the percentage cell elongation during the 24 hours prior to the removal of the strip.

soluble substances in the wall ($r = 0.789$, significant at the 0.02 level for 6 degrees of freedom) (Probine and Preston, 1962).

Nevertheless, the parallel is not exact. For one thing, the creep rate as expressed in Fig. 12.9 (which takes into account both the properties of the wall and the stress to which it is subjected by the cell) is of the order of 10 times greater than the growth rate. One possible reason for this is that during creep the stress is uniaxial while in the cell the stress is multiaxial. Kamiya, Tazawa and Takata (1963) have given reason to believe that this corresponds to a correction of about 4 x, but this still leaves a discrepancy of the order of about 2.5 x. Further, though not pointed out by Probine and Preston, their results show equal amounts of creep in equal increments of log time over the period of their observations (1000 mins.) Creep rate therefore decreases rapidly with time whereas growth rate does not. This latter observation cannot entirely, if at all, be explained by the fact that in the one case the

403

wall is inert whereas in the other the original structure is being maintained by constant deposition of new lamellae from within. It seems necessary to postulate a mechanism for the continual weakening of bonds in the wall just as it seems necessary for a control mechanism in the early stages of creep to maintain the coherence of bonds. Such mechanisms can hardly be other than metabolic.

Determination of creep rates is, however, not the best way to examine the bonding involved, if only because the material is changing during the observation; indeed, little can be deduced about the bonding in this way other than that it must be between matrix substances or between the matrix and the microfibrils. Accordingly, attention has turned to the converse phenomenon, *stress relaxation*, in which the wall is strained slightly and instantaneously and the decay of stress is recorded. The appropriate investigations have been made for *Nitella* by Haughton, Sellen and Preston (1968), for *Nitella*, *Penicillus* and *Acetabularia* by Haughton and Sellen (1969) and more recently for tissues of higher plants by other workers whose conclusions will be considered later. The relevant theory will be given only in broad outline; for further details the original papers themselves must be consulted.

The specimens — longitudinal wall strips — were extended wet after a mechanical conditioning by extending and relaxing slightly a few times so that subsequent observations were repeatable. They were then extended rapidly by 1-3 per cent and the stress decay measured between 0.02 and 10^3 s after the extension. Typical curves are given in Fig. 12.10 for relaxation at a variety of temperatures. Relaxation is approximately linear with log time. The tensile modulus $E(t)$ at time t can be represented as a summation of individual moduli $E_i(t)$ each of which corresponds to a separate stress-bearing element in the wall with a relaxation time τ defined by

$$E_i(t) = E_i\, e^{-t/\tau}.$$

Therefore
$$E(t) = \sum_1^n E_i\, e^{-t/\tau}.$$

With an infinite number of elements, this may be written

$$E(t) = E_e + \int_{-\infty}^{+\infty} H_i\, e^{-t/\tau}\, d\ln\tau$$

404

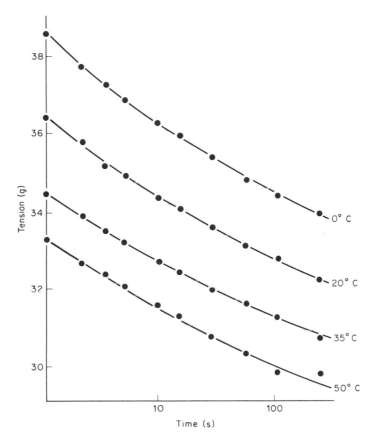

Fig. 12.10 **Decay of stress with time for longitudinal wall strips of *Nitella* following a 2 per cent increment of strain.**

where H_i is the *relaxation spectrum*, defined as the contribution to the tensile modulus of elements having relaxation times whose logarithms range from $\ln \tau$ to $\ln \tau + d \ln \tau$. The term E_e is added to allow for elements with an infinite relaxation time (i.e. E_e is the quasi-static modulus). H may be determined from the differential of stress with respect to ln time. At $0°C$ it turns out to be $10^8 \ Nm^{-2}$, falling slowly with increasing time of relaxation, two orders of magnitude less than the value of E for cellulose. A value of this order is, however, characteristic of materials whose mobility is severely restricted either by cross-linking or because they lie below the glass transition temperature. This suggested that the relaxation is occurring in the matrix substances, that these have

405

some degree of order (and there is other evidence for this, p. 189) and that the molecular rearrangements responsible for stress relaxation in *Nitella* are of a localised nature; long range conformational changes are unimportant.

Something about the nature of the bonds involved may be deduced from activation energies calculable according to the theory adopted. A procedure commonly used with crystalline polymers and with fibres is to match the curves at different temperatures by moving them vertically and horizontally until they coincide at some standard temperature, say $0°C$. The activation energy can then be calculated from the shifts. This was attempted by Haughton *et al.* (1968). Curve matching of this kind is usually, however, applied to linear systems (i.e. systems in which stress relaxation is a function only of time and not of stress and strain). Investigations of stress/strain loops, however, (Haughton and Sellen, 1969) show that the present system is non-linear. Moreover, the relaxation curves (Fig. 12.10) are featureless and this leads to uncertainty in matching and perhaps to large errors. Accordingly, Haughton and Sellen (1969) used the Eyring rate theory in terms of a non-linear model consisting of a spring in series with a non-Newtonian dash pot (i.e. a dash pot containing a liquid, the viscosity of which depends on stress). The theory shows that the horizontal movement required to superimpose curves at different temperatures yield values of the activation energy. These turn out to be 64 Kcal mole^{-1} for *Nitella* and 70 Kcal mole^{-1} for *Penicillus*. If the bonds involved are H bonds, this means that each 'flow unit' is associated with about 15 H bonds and is of the order of size of three cellobiose units. Linearity with log time means that the number of such bonds broken at one location is the same in equal periods of log time. Since growth rates do not fall logarithmically, it needs to be stressed again that biochemical processes must be involved in the living cell and we may now tentatively conclude that they must be such as to convert a log time dependence to a time dependence. Cell wall growth would then be a biochemically controlled viscoelastic flow, a concept which has been supported by Lockhart (1965) and Probine and Barber (1966), and is analogous to the 'chemical relaxation' of Ray and Ruesink (1962). Flow of precisely the kind envisaged here has received very recent support through the chemical investigations of Keegstra, Talmadge, Bauer and Albersheim (1973) though in the model of the wall which these authors present only four hydrogen bonds are involved in the flow unit. Their results are, however, equally in harmony with a larger flow unit as demanded by stress relaxation. The criticism that wall growth cannot in any sense be flow because cyanide stops growth and because temperature has a marked effect upon growth ($Q_{10} = c.$ 2 as against 1.05 for creep) (Cleland, 1967) cannot therefore be sustained.

Even with such single cells, the complex structure of the wall makes the distribution of stress across it somewhat uncertain and it is not possible at this time to be certain which lamellae in the wall carry the stress and in which therefore the bonding is important. This becomes vital when we come later to discuss the effect of growth substances. With whole tissues the problem is an order of magnitude more difficult. Even with such a comparatively simple experimental object as an oat coleoptile it is not possible to say how much of the stress induced by the turgor pressure of any one parenchyma cell is borne by, for instance, the outer-epidermal wall with its longitudinally oriented microfibrils. Nevertheless, a number of workers down the years have ventured into this precarious field, and it seems that, in general, walls of higher plants behave in much the same way as the *Nitella* wall.

The earlier work on tissues involved the bending of a tissue (used as a cantilever) under a small weight, direct tension loading of tissues, or changes in length invoked by changes in turgor pressure. This has been reviewed by Burström (1961) in a wide-ranging discussion including the effects of physiological factors on growth which are not under consideration here, and will not be discussed in detail. It has frequently been concluded that wall extensibility — particularly plastic extensibility — plays a large part in cell growth. As a parallel to the work on *Nitella* quoted above, it may be recalled that Frey-Wyssling (1948), using in part the observations of Burström (1942) claimed that in root parenchyma an 'elastic modulus' of the wall decreased as growth commenced and increased again as growth slowed down. The argument here was that if a cell under turgor pressure P (causing a wall stress T) has length L, and the length is reduced to L_o on plasmolysis, then $T/(L - L_o)$ is the elastic modulus. This tacitly assumes that the turgor pressure cell-extension curve is linear, and this is far from the case (Preston, 1955). Moreover, while the modulus does initially decrease as cell growth begins, it increases again while the cell is still growing rapidly. There is, therefore, no exact parallel between this modulus and growth rate though there is nevertheless here again a sign that mechanical properties and growth may not be separable.

Since the period covered by Burström's review, R. Cleland in particular has given attention to this problem, using the Instron testing technique. Although he adopts a somewhat surprising pre-treatment of his material, his findings are doubtless acceptable at least in the qualitative sense. A good deal of his work refers to the effects of turgor and growth substances, but this is relegated here to a later section.

Cleland (1967) uses *Avena* coleoptile sections. These he first boils in methanol, treats for 18 h in 200 μg l^{-1} pronase, dries and stores in methanol for up to 1 year. Drying undoubtedly causes changes in bonding which are not reversible and pronase treatment is guaranteed to have the same effect. The material is then rehydrated and stretched on an Instron tester, usually at 2 mm min^{-1}. On the first extension the load extension curve resembles that of OC, Fig. 11.2, except that the curve, as seems usual with primary walls, is convex to the strain axis, not the stress axis, so that the wall material is stiffer at higher strains. On removal of the load, the material contracts to L, Fig. 11.2, so that part of the extension is plastic. Repeated extension leads to the situation shown in Fig. 11.10 though the stress/strain curve is curvilinear, still convex to the strain axis. Cleland then takes the slope of this curve at a load of 25 gm as giving the elastic compliance J_E, presumably because the load on the wall at full turgor lies between 20 and 40 gm. Subtraction from the total compliance then gives J_P, the plastic 'compliance'. This is a somewhat arbitrary separation of elastic and plastic deformations. It may nevertheless be significant that J_P turns out to be stress dependent and time dependent, increasing as the growth rate increases and decreasing when it falls.

More recently Cleland (1971) has followed Probine and Preston (1962) in investigating creep properties in his material, pretreated as described above. The creep rate is again linear with log time for periods in excess of 1400 min if the coleoptile sections had been pretreated with auxin. Under these conditions the creep rate per decade of log time is nearly proportional to applied stresses between 4 and 22 x 10^6 Nm^{-2} and the creep is partly reversible (at about half the creep rate) and is non-linear. These reactions to constant stress seem indistinguishable from those of *Nitella*. With coleoptile sections not pretreated with auxin, the results are variable, the rate of creep often increasing with increase in log time and sometimes reaching rates in excess of those for auxin-pretreated sections. In comparing these reactions to constant stress with rates of growth, Cleland notes that, for sections which have been mechanically conditioned prior to observations of creep, there is a yield stress below which the rate of creep is low, recalling again the observation, to be discussed below, that cell growth does not occur if turgor pressure is below a certain value. Moreover, pretreatment of sections with KCN reduces the subsequent creep in wall material. However, Q_{10} for creep is again 1.05 so that biochemical reactions must again be implicated in growth. Cleland and Haughton (1971) have since that time examined stress relaxation in *Avena* coleoptiles; this is again non-linear. Similarly, Masuda, Yamamoto and Tanimato (1972) have examined stress relaxation in *Avena* coleoptile and pea epicotyls pretreated much as described by Cleland. They have as yet drawn no particular conclusions, but they have demonstrated that auxin has a marked effect on the relaxation spectrum. All these observations are, however, it is to be recalled, somewhat difficult to interpret in a tissue; we shall return to them again in dealing with auxin effects.

12.3 Spiral growth

Perhaps this is the moment to pause in these considerations of 'straight' growth, to consider another aspect of growth shown by some cells growing in isolation which is also clearly related to mechanical properties. As these cells elongate, one end rotates with respect to the other so that a marker placed on the wall moves helically round the cell instead of passing along a line parallel to the cell axis. This type of growth is called *spiral growth* though, strictly speaking, it should be called helical growth. Fig. 12.11 makes it clear that the rotation of one end of the cell, $\Delta \phi$, is related to the angle of spiral growth, θ, by the relation $\tan \theta = r(\Delta \phi / \Delta L)$.

Most of the early work on spiral growth was carried out on the sporangiophores of *Phycomyces*. Growth occurs only in the apical 2 mm of this organism and in this region it was demonstrated by

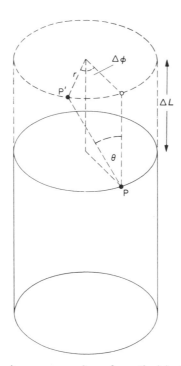

Fig. 12.11 **Diagrammatic representation of a cylindrical cell extending through a length ΔL (broken lines) in spiral growth so that a point P on the wall passes to point P′.**

Oort (1931), Oort and Roelofsen (1932) and especially by Castle (1937 and 1942) that markers placed on the wall move not only upwards but laterally. As the tip elongates, therefore, it twists. Before the appearance of the sporangium the twist is left-handed (S helix). During the swelling of the sporangium, both longitudinal growth and twist cease but after the full development of the sporangium the sporangiophore begins both to elongate and twist, this time in a right-handed (Z) direction. This continues for some time, during which the rate of twisting decreases, momentarily becomes zero, and then continues again in the left-handed twist. Several explanations of this behaviour have been attempted, all in terms of wall architecture.

The skeletal polysaccharide is here chitin, in the form of microfibrils visually identical with cellulose microfibrils. In the apical 2mm the sporangium is covered by a primary wall only, in which the m.e.p. lies in a slow left-hand helix (Middlebrook and Preston

1952). Below this, passing down the sporangiophore, the bire-fringence decreases, becomes zero and then increasingly positive, as the secondary wall is deposited with longitudinally oriented microfibrils and effectively stops growth. Castle (1936) had hazarded the guess that spiral growth might be due to an aniso-tropic response to turgor pressure but gave no details as to possible mechanisms. Heyn (1936) suggested that the wall slipped along planes of weakness but attempted no quantitative assessment and in any case argued from the structure of the secondary instead of the primary wall.

Each of these theories proposed in their different ways that the genesis of spiral growth lay in the mechanical properties of the wall. Somewhat later, a formal theory was proposed in quantitative terms (Preston, 1948) which was later tested by experiment (Middlebrook and Preston, 1958a, b). This involved a helical spring model which, like the corresponding models discussed in Chapter 10, is far too simple. Nevertheless, the theory gives sufficiently close agreement with observed results as to encourage refinement of the model.

The theory was based on an inference drawn from cellulosic walls, subsequently verified (Middlebrook and Preston, 1952a) that in the growth zone the microfibrils tend to lie in a flat helix, at an average angle α to the transverse but with some considerable angular dispersion. In the model these are replaced by a set of microfibrils lying accurately at the angle α and therefore with mechanical properties dependent in part on those of the real microfibrils and in part upon the angular dispersion. If the model microfibrils may slip past each other during wall extension, then each will behave as a spiral spring, i.e. the free end will rotate as the spring is extended. The angular rotation $\Delta\phi$ for a length increase ΔL is then given by

$$\Delta\phi/\Delta L = [\cos\alpha \sin\alpha (1 - 2n/q)]/a[\cos^2\alpha + 2n/q \sin^2\alpha]$$

where q is the Young's modulus, n the torsional rigidity of the model microfibrils and a is the radius of the helix. The model predicts correctly (1) that the growth spiral will normally have the same sign as the structural spiral (since q is normally greater than $2n$) but that the sign of the growth spiral can change (when $q < 2n$), (2) that spiralling will cease *whenever* growth ceases and should then resume with opposite sign, (3) that $\Delta\phi/\Delta L$ should be

411

inversely proportional to *a*. These predictions have all been verified (Preston and Middlebrook, 1949; Middlebrook and Preston, 1952b). Moreover, insertion of measured parameters into the equation gives a calculated value for $\Delta\phi/\Delta L$ surprisingly close to the observed value (Middlebrook and Preston, 1952b).

When electron micrographs of the wall of the growth zone became available, both Roelofsen and Frey-Wyssling (see refs. in Roelofsen, 1959) questioned the basis of this theory since the microfibrils then observed were not arranged precisely as in the model. This does not, however, detract from the theory since, by definition, this correspondence was not to be expected. Frey-Wyssling attempted an alternative explanation in terms of the mosaic theory of growth which, as we have seen, is not now accepted. The advent of the multi-net growth hypothesis, which applies to the growth zone of these sporangiophores, if anything strengthened the spring model explanation.

Subsequently, though with another organism (see below), Probine has attempted refinements of the kind described for wood in Chapter 10. Meanwhile, Frei and Preston (1961b) had used the wall structure of two other organisms to predict that each should show spiral growth and had, unexpectedly, proved that they do so and at a rate not far from that predicted (p. 215). Finally, Probine (1963) re-examined the basis of such spiral growth using the internodal cells of *Nitella opaca, Nitella axillaris* had already been shown to exhibit marked spiral growth (Green, 1954). He has adopted a stance similar to that of Barber, Meylan and Cave already discussed in Chapter 10 by considering the wall as homogeneous and with orthorhombic elastic symmetry. When a cell with such a wall is extended by pressure from within, 12 elastic moduli are necessary to explain the behaviour if the axes of elasticity (i.e. the microfibrils) lie parallel or perpendicular to the cylinder axis; three of these are shear moduli. When, however, as in *Nitella*, the microfibrillar direction lies in a slow helix round the cylinder, the axes of elastic symmetry no longer coincide with the principal axes of stress (the cylinder axis and the direction in the wall perpendicular to this) and four new moduli must be introduced which couple sheer and extension. The circumstance that the axis of maximum stress does not coincide with the axis of maximum strain causes one end of the cylinder to rotate with respect to the other. Probine verified, by observing internodal cell behaviour during changes of turgor, that twisting occurred in the direction anticipated. Lack of information on the moduli required (which would in any case be difficult if not impossible to obtain) prevented him, however, from attempting any quantitative check.

It seems, however, reasonable to conclude from these three examples that all cells growing in isolation with helically organised walls should show spiral growth and that this is dictated by the mechanical properties of the wall. When such cells, growing as part of a tissue, do not show this phenomenon, then they must be prevented from doing so by external impressed forces. We have seen in Chapter 10 the steps which need to be taken to allow for an associated phenomenon in explaining the strength properties of wood and other fibrous tissues. If, in tissues, cells are allowed to slip over each other, then it follows that the cells must tilt, and an attempt has been made to explain *spiral grain* in conifers in this way (Preston, 1949). It is to be supposed that the movement of elaters in liverworts are to be explained in similar terms. Otherwise circumstances could be envisaged in tissues, whereby an overall torque is developed in the tissue; could it be that the spiralling of tendrils and even the nutation of stem apices are similarly based?

12.4 Growth and turgor pressure

Returning now to 'straight' growth, it seems increasingly likely that the rate of deformation of a cell may be controlled by a biochemically-mediated viscoelastic 'flow' of material in the wall. If this is so, then growth rates should change as turgor pressure changes unless the resistance of the wall to stress changes with turgor pressure. Requirement for turgor pressure as the driving force of cell elongation has been repeatedly supported ever since the early statement of de Vries (1877). This requirement has recently been questioned by Burström (1964) and by Burström, Uhrstrom and Wurscher (1967) on the basis that turgor pressure in intact tissues may not be high enough. In passing, it may be noted that the latter authors have shown that E_{tissue} as measured by the method of Falk *et al.* (1958) increases some 50 x from plasmolysis to turgor, a change similar to that which may be deduced for the wall from the early plasmolysis experiments of Oppenheimer

(1930), (Preston, 1955). Cleland (1967) gives reasons, based on the notorious difficulties in measuring cell water potentials, to think, however, that this claim cannot be sustained.

Steady growth of many cells and tissues has been shown by several workers to increase as turgor pressure increases only above a certain critical turgor pressure (Cleland, 1959 and 1967; Lockhart, 1967; Ray and Ruesink 1962; Green, 1971). Cleland (1967) and Green (1971) have examined this circumstance in some detail and their findings have much in common though Cleland uses *Avena* coleoptiles and Green *Nitella*. The rate of elongation is in each case proportional to the difference between the actual turgor pressure (P) and the critical turgor pressure (Y). In *Avena* coleoptiles, Y is about 6 atm compared with the fully turgid value of c. 9 atm. Cleland again notes a stiffening in the wall as turgor pressure rises in the absence of auxin (in the sense that his J_p falls), which he tends to relate to an increased rate of wall synthesis which he demonstrates, though this may be doubted both because the time factor does not seem correct and because the newly deposited inner wall lamellae are not likely to carry much of the stress. This will be considered again when auxin affects are dealt with.

Green (1971) goes further in analysing Y. He shows that the rate of growth in *Nitella* internodal cells may be expressed as

$$r = m(P - Y)$$

where r is the relative growth rate $[(1/L)\,dL/dt]$ and turgor pressure is measured by a fine manometer inserted into the cell. m, if constant, is about $0.1\,h^{-1}\,atm^{-1}$, i.e. 10 per cent/$h^{-1}\,atm^{-1}$, for small changes in P. Experimentally induced changes in P therefore lead to changes in r. P and Y are each of the order of 5-6 atm and $P - Y$ is about 0.2 atm. The initial response to a small change in P is therefore high. After about 15 minutes, however, the cell compensates in such a way that r returns to normal with no change in P. Green gives sound reason to believe that this is not due to a change in m (e.g. because the effect is observed at $P = Y$) and must therefore be due to a change in Y. The compensatory increase in Y when P is increased appears to differ causally from the decrease in Y following a decrease in P, for only the latter is inhibited by sodium azide. Compensatory changes in Y as P is reduced, moreover, decrease in amount, as the reduction is increased, to 0 at $P = 2$ atm (where growth appears not to resume) and this is therefore

414

the lowest value of Y attainable. Green remarks that this might equate to the yield stress for creep observed by Probine and Preston (1962); a little calculation shows that this yield stress of *c.* 10^8 dynes cm^{-2} is equivalent to a turgor pressure of *c.* 1.7 atm. This turgor dependence of compensation shows that the steady state growth rate is related to turgor in part owing to time dependent changes in the wall which seem therefore to run through all studies in wall extensibility. Green (1972) has more recently extended these observations to cover *Avena* coleoptiles as well.

The requirement that $P > Y$ for growth clearly suggests that there are in the wall strong bonds which apparently need to be subjected to a minimal strain before they can be fractured and creep commence. The dependence of Y on P would then presumably involve a complex relation of P, not only to wall deposition but to metabolic factors which themselves can affect the activation energy of these bonds.

12.5 Cell growth and wall protein

Before considering auxin effects on growth it is as well to look first at the possible intermolecular bonds in the primary wall, particularly bearing in mind the presence of Hyp-protein.

Bonding between the polysaccharide moieties of the wall is likely to be of three broad kinds. (1) H bonding between $-OH$ and $-OH$ or between $-H$ and $-COOH$, (2) salt linkages between molecular chains ($-COO-Ca-OOC-v-COONa$ Na $OOC-$) and (3) van der Waal's forces ($-CH_3...H_3C-$). There may be occasional primary valences between individual molecular chains but the evidence for these is scanty though Bauer, Talmadge, Keegstra and Albersheim (1973) have very recently proposed covalent links between the xyloglucan and pectic polysaccharides in sycamore suspension cultures. One is inclined to think that H bonds will preponderate. There is, however, some evidence, in addition to the above indirect indication, that rather strong bonds do occur. As an example, Rayle, Haughton and Cleland (1970) examined the effect of pH on wall creep. The experimental material consists of 18 mm long segments of *Avena* coleoptiles cut with one end 3 mm from the tip, frozen in liquid Freon and thawed in 0.01 M PO_4 buffer and elongated under a 20 gm load at 23°C. In this material, wall synthesis does not occur. Treatment of the specimen with alkali solutions (up to pH 13) has no effect on creep rate but with solutions at pH 3.6, the creep rate increases very markedly, reaching an elongation of 30 per cent in 600 min (for comparison, this is the elongation rate of a coleoptile in optimal 1AA solutions during 8 − 10 h). This suggests that acid-labile, alkali-stable bonds are involved in extension. These cannot be H bonds since these bonds are not stable in alkali as evidenced both by the swelling of e.g. cellulose in alkali and an increased rate of deuteration of H bonds in alkali. The Hyp-arabinose link proposed by Lamport is, however, such a bond. There is strong evidence that the Hyp-protein in the wall is bonded in this way to high molecular weight polysaccharides (Boundy, Wall, Turner, Woychik and Dember, 1967; Lamport, 1970; Bauer, Talmadge, Keegstra and Albersheim, 1973) and there is the further possibility of −S=S− bonds between molecules of the protein itself. Since, over the wall, these strong bonds would be in series and parallel with H bonds between polysaccharides, then breakage would leave the H bonds as the creep-inducing bonds.

There is now much other evidence that the Hyp-protein is involved both in the mechanical properties of the wall and in rates of growth. No exact parallel has been claimed between Hyp-protein content and growth rates and perhaps such a direct relation is not to be expected. Direct evidence for an involvement of Hyp-protein with wall extensibility come from the work of Cleland (1967) showing that with oat coleoptiles removal of wall protein increases extensibility [contrary to the earlier findings of Olson, Bonner and Morré, (1965)] and from the observation (Thompson and Preston, 1968) that treatment of the walls of *Cladophora* and *Chaetomorpha* either with pronase or dithiothreitol (an S=S bond splitter) catastrophically reduces the tensile moduli. Less directly, Linskens (1964), for instance, has shown that walls with little Hyp-protein have lower tensile strength than do walls rich in this component. This seems to support the earlier suggestions by Lamport (1963) that the protein may confer rigidity upon a wall by cross linkage.

The association between the protein and cell growth is not, however, so clear cut. On the one hand pea seedlings of the Rondo variety, which is a dwarf, contain much more wall-bound Hyp than do those of the Alaska variety, which is not (0.5 per cent as against 0.25 per cent) (Winter, Meyer, Hengeveld and Wiersma,

1971). Correspondingly, application of 1AA to excised stem segments strongly inhibits the increase in Hyp-protein which would otherwise occur but stimulates elongation. This is in harmony with the conclusions of Cleland and Karlsnes (1967) that an excess of wall-bound Hyp may be responsible for the cessation of growth. Along the same lines, Ridge and Osborne (1970) and Osborne (1972) have found, again with pea seedlings, that treatment of the apex with ethylene leads to an increase in wall-bound Hyp and an inhibition of longitudinal growth, and Klug and Bayley (1965) have found an increase in wall protein as growth slows in Jerusalem artichoke tubers. All this looks like an inverse relationship between growth and protein content. On the other hand, according to Cleland and Karlsnes (1967) the Hyp content of growing cell walls increases over the growth period and Winter *et al.* (1971) found the greatest elongation in their material (after treatment with 1AA and sugar) to coincide with the greatest Hyp content. The inverse relationship does not therefore hold under those circumstances. In some instances, at least, it appears that an increase in Hyp is only one of the factors changing with extension. Pea seedlings treated with ethylene show the typical reduction of elongation and increase in diameter and this is accompanied by an increase in Hyp-protein (and peroxidase) in the wall (Osborne 1972). However, the walls of the cortical parenchyma also double in thickness, and some part of the microfibrils adopt a new, longitudinal orientation in place of the previous transverse orientation. There is therefore here a whole complex of factors at work and it is dangerous to pick on only one of them – enhanced Hyp-protein – as causative.

Perhaps, therefore, the importance of wall-bound Hyp-protein lies in a property other than its ability to bond with polysaccharides. It must not be forgotten that neither *Nitella* nor, apparently, *Valonia* wall proteins contain Hyp (p. 62) and yet each manages to grow to its own satisfaction. In this sense, suggestions by Cleland (1967) might be significant. He considers that the known blocking of wall protein synthesis by externally fed Hyp and the concomitant inhibition of growth, may be due to the blocking of the synthesis, by an 1AA-controlled process, of a pool of substances necessary for growth and used up during growth. These substances could be a Hyp protein. In that case, since exogenous 1AA and sugar (but not either separately) increases Hyp formation (Cleland,

417

1968), while sugar alone increases cellulose synthesis but 1AA and sugar increases the synthesis of non-cellulosic substances (Baker and Ray, 1965; Ray, 1962), perhaps the Hyp-protein is needed only for the intussusception of matrix polysaccharides and for their correct orientation in terms of the necessary anisotropy of wall polysaccharides. Similarly, Dashek (1970) has mooted the possibility that the incorporation of sugars into wall polysaccharides might be by glycolysation of the Hyp-protein. On the other hand, we have seen reason to believe that the protein may have enzymatic properties (p. 67) and this might open out other possibilities.

It is, however, premature to speculate upon the precise relation of wall-protein to growth, and certainly naive to discard at this stage the possibility that the function of this protein lies in wall stiffening merely because there is no fixed relation between wall protein content and growth. Until more is known about the bonding involved, and indeed about the distribution of the protein between the lamella of growing wall and over the wall area, the gross content is irrelevant. The gradually changing structure from inside the primary wall to the outside makes it quite uncertain which lamella bears the major stress and therefore which lamella could be critical as far as growth is concerned, and this information, difficult to come by as it may be, is urgently needed. It could be a simpler matter to examine the distribution of protein over a cell wall; for instance, the distribution of wall protein between the upper and lower halves of a cell of *Cladophora*, in which only the upper part is growing, and *Chaetomorpha* in which only the lower half extends.

An involvement of the bonding between wall polysaccharides in the events leading to growth is strongly indicated by the accumulating evidence of a turnover of many of these polysaccharides during growth (Margere and Lenoel, 1961; Machlachlan and Young, 1962; Matchett and Nance, 1962), in confirmation of earlier considerations (Frey-Wyssling, 1950; Preston, 1952). The observations that IAA treatment enhances turnover while Ca^{++} (which presumably is involved mostly in cross-linking) does not, form an indirect indication that turnover may in some way be involved with growth. More recently Machlachlan and Duda (1965) have shown that in pea epitcotyl sections grown either in water or sucrose solutions there is no turnover of pectic substances. The only

418

polysaccharides they found to turn over are glucose- and galactose-containing substances which are soluble in hot dilute acid and therefore constitute a hemicellulose fraction. Oddly enough, only the cellulose increased in amount per segment (tripling in sucrose solution during a 30 per cent extension). Since presumably the new lamellae responsible for this must have been associated with new hemicellulose and pectic compounds, the older lamellae must have lost a proportion of these compounds during growth. A similar situation has been found by Nelmes and Preston (1968) for apple fruit parenchyma cells through weekly analysis from pollination to ripening in storage. As with the wall protein, observations of this kind lack definition since the work is entirely with tissues.

12.6 Wall extension and auxin

The association of an increased wall plasticity with IAA-stimulated growth advanced more than 40 years ago by Heyn (1931) has since that time been supported by overwhelming evidence (see review by Setterfield and Bayley 1961) and the aim during the past 10 years or so has been to examine this relationship rather than to test its existence.

IAA has no effect on wall extensibility when turgor pressure has been reduced to zero (Cleland, 1967). Correspondingly, when a coleoptile segment extending in the presence of exogenous IAA is placed in a mannitol solution of appropriate concentration (for up to 100 minutes) growth is not enhanced above the initial rate when the segment is returned to the growth solution (Cleland, Thompson, Haughton and Rayle, 1972). Unless there exists a direct effect of pressure on the auxin-mediated events leading to extensibility changes (such as a change in conformation of a high molecular weight intermediate) this may imply that bonds broken under no-growth conditions are remade (perhaps passively) in the absence of deformations which would otherwise separate the constituent molecular groups. The well-known absence of an IAA effect on either extensibility or growth in killed tissue or excised walls further implies that wall loosening needs the proximity of living cytoplasm. It is said that protein synthesis is necessary for wall loosening (Black, Bullock, Chantler, Clark, Hanson and Jolley, 1967; Courtney, Morré and Key, 1967) and Cleland (1967), as shown above, has found it necessary to assume a pool of sub-stances, probably protein, synthesized by an auxin-induced process and used up in growth. It has further been claimed that RNA synthesis is necessary for auxin-induced effects (Key, 1969) which would put the primary effect of IAA far back in the sequence leading to the direct effect in the wall. However, the effect of IAA on growth can be detected within 1 minute, (Rayle, Evans and Hertel, 1968), a lag time far too short, as pointed out by Key (1969), to allow a primary IAA effect at the level of gene transcription. The period is equally too short to allow any explanation through a change in the bulk composition of the wall, a possibility which has, in any case, repeatedly been ruled out by wall analyses during growth. The indication seems to be that the initiation site of the auxin effect on growth cannot be far removed from the wall and cannot therefore lie far (in time) from the plasmalemma. It is interesting therefore to note that effects of IAA have been found on the electrical potential of the plasmalemma (Etherton, 1970; Tanada, 1972) and that Rayle and Cleland (1970) have raised the possibility that the similar effects of IAA and low pH on growth may derive from an alteration of membrane characteristics, perhaps by activation of a proton pump. Moreover, application of IAA causes a change in water flux as immediate as the change in growth rate (Kang and Burg, 1971); these two processes, however, have different IAA concentration optima, are induced at different IAA concentration thresholds and persist for different times after IAA removal. This is a reminder that the two processes are not necessarily connected although the similar effect of inhibition would seem to be significant. Kang and Burg (loc. cit.) go so far as to surmise that IAA might effect a change in the conformation of a molecule in the membrane, and the effect of supraoptimal concentration on microfibril orientation already mentioned (pp. 392, 393) would point in the same direction.

The ultimate effect as far as growth is concerned must presumably be the passage into the wall, or the activation within the wall, of a substance active in bond breaking and making and it could be envisaged that this should be an enzyme. There is, however, as yet no idea what this enzyme (if any) might be. It is not to be expected that it should be cellulase, even though this enzyme does increase the extensibility of isolated walls (Olson, Bonner and Morre, 1965) and the evidence is that it is not. For instance, Fan and Machlachlan (1966 and 1967) have shown that, though exegenous IAA increases the cellulase content of decapitated coleoptiles by 12 to 16 times, an increase of this order induces cessation of elongation; and indeed, that inhibition of cellulase formation by puromycin or actinomycin D greatly enhances elongation in the presence of auxin. The older idea, ably championed by Bennet-Clark (1956) that pectic substances are involved has been shelved, partly because in oat coleoptiles (for which it was introduced) the amount of pectin turns out to be much lower (0.3 per cent) than had been thought, and partly because there is no evidence for an effect of IAA either on pectinase or pectin methylesterase (Jansen, Jang, Albersheim and Bonner, 1960). The only remaining competitors appear to be hemicellulases and proteases. Although IAA increases the activity of hemicellulases (Katz and Ordin, 1967; Lee, Kivilaan and Bandurski, 1967), only proteases are known to affect wall plasticity (Thompson and Preston, 1968). In any case, the effects of all such degrading enzymes are irreversible whereas auxin effects are reversible (Cleland, 1968) so that none of them can be a candidate unless some parallel 'repairing' system is also present.

12.7 The wall growth process — a summary

The processes at work during the extension of a primary wall around the living cell — which we may term the growth process to distinguish it from the wall extensibility of isolated walls — are therefore becoming a little clearer though they can still be defined only in the vaguest terms.

We begin with the concept that the wall must grow through biochemical creep powered by the wall stresses induced by turgor. We then need to ask which lamellae in the wall of a cell growing in isolation, and which wall in a tissue, is most resistant to creep in relation to the stress upon it, because this will be rate controlling and therefore alone susceptible to IAA effects. We may hazard a guess about the distribution of stress over a single wall from the circumstances that (a) the innermost lamella in process of deposition is not strained and therefore carries no stress, (b) at any instant wall lamellae in the rest of the wall must be equally strained in the sense that each will contract by the same amount if the turgor

421

pressure is reduced to zero. The effects of orientation on Young's modulus discussed in the last chapter suggest that the reorientation through this wall, from roughly transverse to roughly axial, will involve an increasing Young's modulus from inside to outside until a lamella is reached near the outside which has been fractured by the continuing growth. Therefore it may be postulated that the distribution of stress may be of the kind represented by the (notional) broken line in Fig. 12.3. Biochemical creep of innermost lamellae will then produce little effect other than a redistribution of stress to those parts of the wall which are strongest. We may therefore suppose that creep will be effective in growth only in the regions of high stress away from the inner face of the wall and it is here that the final effects of IAA must be felt. Whatever the nature of the 'activator' of creep it must therefore diffuse through several lamellae (as defined on the scale of the electron microscope and perhaps only a few 10s of nms). Similarly, we might expect that with tissues such as oat coleoptiles, the longitudinally oriented microfibrils of the outer epidermal wall would mean a concentration of stress there and that these walls might therefore be especially important; and it might in this context be significant that Burström, Uhrstrom and Wurscher (1967) have noticed a more pronounced auxin effect with epidermal strips of pea internodes than with cortical cells.

Returning to isolated cells, in the absence of wall deposition the region of maximum stress (maintained in spite of creep by continual uptake of water into the cell) would move progressively inwards. In the normal situation, incorporation of new lamellae from within would maintain the distribution of stress static. In each case the cell grows at a rate dependent both upon the maximum stress and the rate of biochemically invoked bond control at the region of maximum stress. We now turn to the matter of the yield stress called to notice by Cleland and by Green.

The situation between the wall-deposition situation and the no-wall-deposition situation might be different. In the latter, there is a limit to the number of 'strong' bonds available so that if, as Cleland has postulated, the 'yield turgor' reflects the need either actually to break bonds or to decrease the potential barrier to breakage to a low level, then the period over which a 'yield turgor' is observable should be limited. This might be worth examining. If wall deposition is occurring, of course, this might continually add strong bonds. However, this explanation of the 'yield turgor' cannot explain the recovery from a decrease of turgor observed by Green, and the involvement of strong bonds might on these grounds be doubted. It is to be noted that when the turgor is reduced to the yield turgor ($P = Y$) the wall is still under stress, and growth is momentarily halted because there are no bonds which the reduced stress can break in sufficient number. The wall is now, however, in a somewhat different state from that in the fully turgid cell, since it has contracted. Reduction in the yield stress such as would allow growth to recommence might be induced by an effect on auxin-induced wall loosening mechanisms either through reduced turgor pressure or because in the contracted state the number of bonds per unit volume is greater and therefore a greater number may be under collective attack simultaneously in any one location. Neither of these seems highly probable and the kind of compensation following changes of turgor as observed by Green seems to have at the moment no rational explanation.

Two other phenomena, related to this one, also seem difficult of explanation in terms of the widely accepted view that the effect of IAA on the wall is to induce wall loosening. Firstly, the initial creep rate of an isolated wall is greater than the growth rate by a factor too great to be accounted for by the different distribution of stress in the two cases (p. 403). This would appear to mean that, during growth, bonds which would be broken during extension are either protected against breakage or are remade, so that the IAA-induced reaction is not to loosen bonds but either to remove the protection or to inhibit the remaking. The most recent suggestion by Keegstra, Talmadge, Bauer and Albersheim (1973) that the immediate effect of auxin is the activation of a H ion pump reducing wall pH and loosening H bonds, while attractive for its immediacy and for the rate at which the reactant (H^+) might diffuse, does not seem of itself to meet this requirement. Secondly, the lowest yield stress observed in growing cells is roughly equivalent to the yield stress to creep in isolated walls of closely similar cells and correspond to a turgor pressure of 2 atm (p. 415). This low yield stress is induced under special circumstances. The normal yield stress is much higher than this and can be equivalent to a turgor pressure of *c.* 6 atm. Under these conditions, bonds which are known to yield at 2 atm must be protected or remade after break. The biochemical events in the wall associated with growth could therefore be even more complex than has been thought. They occur in a region of the wall remote from the protoplasm through which each newly deposited lamella passes in time particularly in a growing cell which maintains its constant wall thickness. The molecular events leading to a yielding of the wall are not necessarily in process simultaneously all over the wall and in this sense the model of the growth process recently devised by Bartnicki-Garcia (1973), while not meeting the present requirements in detail, is not without merit. He considers that lytic enzymes may be passed to the plasmalemma in a vesicle and discharged into the wall where they split either inter- or intra-molecular bonds considered by him to lie in the microfibrils. The tension in the wall then causes a localised increase in area. Synthetases and precursors are then transmitted to the wall also *via* vesicles and the structure knitted together again. The thesis of this final chapter is that the bonds of interest lie in the matrix, not in the microfibrils and that the materials first secreted into the wall are not lytic enzymes but some kind of bond protecting agent, and with this modification the model seems worth pursuing.

In any event, the primary wall is clearly a complex system delicately constructed and under delicately balanced regulatory control. There can be no longer any doubt that its mechanical properties are of concern in growth only because it is part of the living system, a cell organelle rather than a dead excretory product.

CHAPTER 13

Wall biosynthesis

This assembly of polysaccharides and other chemical constituents, delicately balanced both structurally and biochemically, obviously presents a situation in which the biosynthesis must be highly complex. It is not sufficient solely to follow the biochemical pathways along which each polysaccharide constituent, for instance, has travelled; the overriding necessity must be kept in mind that these constituents must be assembled in specific spatial order and that the order may, from time to time, be changed during the development of each single wall. Ideally, therefore, the biochemical synthesis and the spatial ordering of *all* wall constituents should come under consideration since there could well be interactions at least as far as the ordering is concerned. Attention here will be confined, however, to the carbohydrate moiety only. Lignin will not be dealt with, partly because its biosynthesis involves highly complex chemistry beyond the scope of this book, and partly because the walls which become lignified are almost completely assembled before lignification begins; a lead into the literature on lignin biosynthesis will be found in the papers by Kratzl and Wardrop and others in the book edited by Côté (1965). Similarly, suberin and cutin will not be considered since these occur only in specialised cells such as epidermal and root endodermal cells, even though these cells and their walls are important. It will become clear that neither the biochemistry nor the biophysics of wall polysaccharide synthesis is as definitive as could be wished, though recent developments seem to promise something spectacular in the next few years.

On the chemical side, there seems to be general agreement that nucleotide sugar phosphates act as precursors through transglycolization and that in some way a lipid is involved, perhaps as a component of a membrane. Sugar nucleotides appear, from the thermodynamic standpoint, more promising as precursors than do other possible candidates such as sugar phosphates or oligosaccharides. The free energy of hydrolysis (ΔG^0) for UDP-D-glucose is -7600 cal mole^{-1} as against a value of -4350 cal mole^{-1} for the α-1, 4-link in glycogen and of -4850 cal mole^{-1} for α-D-glucose phosphate (Burton and Krebs, 1953).

The standard technique in testing the efficiency of precursors and enzymes has come to lie in the mixing of radioactive precursors with crude enzyme preparations obtained from a variety of plants by centrifuging homogenates in such a way as to obtain an active fraction. There is therefore no guarantee that the enzyme preparation should be active in producing only one polysaccharide or, indeed, that the extract does not contain hydrolases as well as synthetases. The product is then isolated in some appropriate way (sometimes involving the complication that carrier polysaccharides of the expected type are added) identified as closely as possible and its radioactivity monitored as a check that the supplied precursor is involved and as a measure of the efficiency of the conversion. The enzymes are commonly found to be stabilised by dithiothreitol (which is often therefore added) so that $-$SH groups may be involved at or near the active centre. In the following discussions the enzyme extract used will be classed by the force of centrifugation at which the corresponding pellet was obtained, e.g. the 10 000 g pellet.

On the biophysical, or structural, side a number of cell organelles have been identified as either certainly or possibly associated with wall synthesis. Outstanding among these are Golgi vesicles, microtubules and plasmalemma granules, but the endoplasmic reticulum, plasmalemmasones and lomasomes have also been mentioned. The cisternae of Golgi bodies bud off vesicles laterally (Figs. 13.1 and 1.1) and the electron microscopic evidence is that these migrate to the wall where their membranes may fuse with the plasmalemma (Olszewska, Gabara and Steplewski, 1966; Mühlethaler, 1967; Northcote, 1969). This movement was first described by Mollenhawer, Whaley and Leech (1961) and by Mollenhawer and Whaley (1962) in cells of the root cap, and has since that time been supported by observations on a wide variety of plants (Drawert and Mix, 1962 − *Microsterias;* Sievers, 1963 − root hairs;

Wang and Bartnicki-Garcia, 1966 — *Phytophthera;* Mühlethaler, 1967 — higher plants; Heath, Gay and Greenwood, 1971 — *Saprolegnia;* Northcote, 1969 — higher plants) in which these vesicles become associated with regions of wall deposition. Fusion of the membrane with the plasmalemma involves a deposition of the vesicle contents between the plasmalemma and the wall, contributing at least to the wall matrix substances. At an earlier stage of wall development, i.e. during anaphase and telophase of cell division, the small vesicles which accumulate along the cell equator are said to be Golgi vesicles. These fuse to form the cell plate with the same staining reactions as the vesicle contents (Whaley and Mollenhawer, 1963; Frey-Wyssling, Lopez-Saez and Mühlethaler, 1964). Since the cell plate contains lipases, esterases and acid phosphatase (Olszewska *et al.,* 1966), it may be assumed that the Golgi vesicles contain these enzymes too.

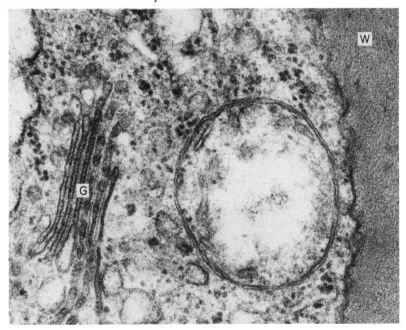

Fig. 13.1 **Electron micrograph of section of differentiating fibre from apple wood; fixed in glutaraldehyde and post-stained in osmium tetroxide, uranyl acetate and lead citrate; magnification 99 000 X. G, golgi cisternae budding off golgi vesicles; W, wall. (By courtesy of Dr. Brenda J. Nelmes)**

Microtubules, discovered by Ledbetter and Porter (1963), occur in animals and in all plants so far investigated, whether or not then

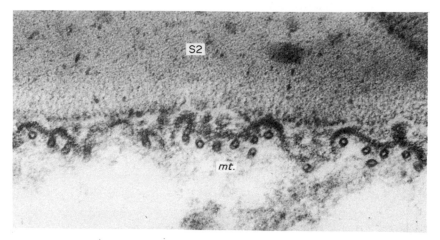

Fig. 13.2 As 13.1, showing microtubules in cross section near the plasmalemma (mt); the wall near the plasmalemma is the S2 layer. Magnification 93 600 X. (By courtesy of Dr. Brenda J. Nelmes.)

cells are producing a wall (Fig. 13.2). They are normally 23–27 nm in external diameter with a wall about 7 nm thick and of indefinite length. Both these and the plasmalemma granules have been cited as involved especially in cellulose biosynthesis and will be taken up in detail later on. The structure and possible functions of microtubules have been recently reviewed (Newcomb, 1969; Pickett-Heaps, 1973).

Lomasomes (Moore and McAlear, 1961) and plasmalemmasomes (Girbardt, 1958) are possibly different names for the same organelle, called for this reason *paramural bodies* by Marchant and Robards (1968). They are common in lower plants, particularly in fungi, but occur also in higher plants (Marchant and Robards, 1968; Manocha and Shaw, 1964). When a wall is present, they come to lie as comparatively large multivesicular bodies sited between the plasmalemma and the wall. Marchant and Robards, using a number of fungi, the alga *Anaboenopsis* and the angiosperm *Salix,* tend to the view that the two organelles may be different in origin; lomasomes may be multivesicular bodies produced in the cytoplasm (according to Pilet (1971) by evagination of either or both of the endoplasmic reticulum and the outer membrane of the nucleus) and passed through the plasmalemma, while plasmalemmasomes are produced by multiple evagination of the plasmalemma itself. On the other hand, Heath and Greenwood (1970), working with *Saprolegnia* and *Dictyuchus,* give good reason to think that these are one and the same organelle. According to them, plasmalemmasomes develop when the increase in the surface area of the plasmalemma, induced by successive fusions with Golgi vesicles or otherwise, is too rapid to be accommodated by the cell wall. The resulting evaginated multivesicular bodies may then separate from the plasmalemma to form lomasomes. These organelles are said to have pectolytic properties (Pilet, 1971) but apart from that, the only current evidence that they have anything to do with wall synthesis lies in their location, often close to sites of wall deposition. They are sometimes, however, embedded in the wall and this could mean that they hamper synthesis.

With this brief preamble we may now turn to the synthesis of the matrix substances and of cellulose, dealing successively with the possible biochemical pathways and the sites of synthesis.

13.1 Matrix substances

One of the few investigations yielding direct evidence implicating nucleotide sugars in the synthesis of pectin is that of Villemez, Lin and Hassid (1965). They have shown that a particulate fraction of mung bean (*Phaseolus aureus*) roots sedimenting between 500 g and 10 000 g will incorporate galacturonic − ^{14}C from UDP-D-GALACTOSE − ^{14}C into a polygalacturonic acid to the extent of 65 per cent, when other nucleotide galacturonic acids are inactive as precursors (Villemez, Swanson and Hassid, 1966). According to subsequent findings (Kauss and Swanson, 1969), methylation follows in a later step though the enzymes involved appear to be located together in a structural component of the cell. This immediately suggests that a membrane may be involved, of which the lipid moiety might be concerned in maintaining the structural integrity of the whole synthesising complex (Kauss, Swanson, Arnold and Odzuck, 1969). According to the scheme presented by Lamport (1970), UDP-D-galactose is derived from UDP-D-glucuronic acid via glucuronic acid from myo-inositol (Fig. 13.3) as already thought possible by Loewus in 1965 (see review by Loewus 1969). Accordingly, Loewus and his colleagues have found myo-inositol as a satisfactory precursor of the pectic compounds. Further than this, they have found that after myo-inositol-2-^{14}C has been incubated with a variety of cells of higher plants (pear pollen, sycamore cambium, maize roots and *Lemna* leaves) the ethanol-insoluble residue may be hydrolysed to give not only D-galacturonic acid but D-glucuronic acid and 4-0-methyl glucuronic acid together with pentoses, including xylose, and L-arbinose. Myo-inosital is thereby implicated in the synthesis of hemicelluloses also.

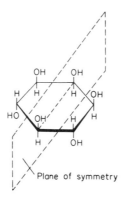

Fig. 13.3 **Myoinositol.**

At a step further back in the chain, Neish and his co-workers (Altermatt and Neish, 1956; Neish, 1958) early showed that when wheat plants are fed with uniformly ^{14}C-labelled glucose the xylan subsequently produced is also labelled. The intervention of nucleotide sugars in the biosynthesis of hemicelluloses has again been demonstrated, both by synthesis from UDP-glucose with UDP-glucuronic acid as intermediate (Streminger and Mapson, 1957) and from UDP-xylose (Pridham and Hassid, 1966; Bailey and Hassid, 1966). The xylan here is, of course, $1 \to 4$ linked. When UDP-xylose is the precursor, however, an L-araban as well as a D-xylan is produced. It has also been claimed (Barber, Elbein and Hassid, 1964) that when extracts of mung bean are incubated with a mixture of UDP-glucose and UDP-mannose, then under certain conditions a glucomannan is produced. This has recently been confirmed by Villemez (1971).

The concensus of opinion is that the matrix substances are carried to the wall in Golgi vesicles (Figs. 1.1 and 13.1), within which at least part of the synthesis may occur, though O'Brien (1973) quotes plants which deposit their walls without the intervention of these or other similar bodies. A clear demonstration was provided by Northcote and Pickett-Heaps (1966) in showing that exposure of wheat root tips to tritiated glucose leads to the appearance of the label in the Golgi vesicles of root cap cells, and that a subsequent chase with cold glucose transfers the label to the wall. Even then, however, they found that for cells behind the root cap the label appeared in the wall with little or no prior localisation in the Golgi. At about the same time Dashek and Rosen (1966) showed that wall material of pollen tubes and the contents of Golgi vesicles are both removed by pectinase. They also used the reaction of pectic compounds with hydroxylamine-ferric chloride (already reported on by Albersheim, Mühlethaler and Frey-Wyssling, 1960) to show that after treatment with this reagent both walls and Golgi vesicles become highly scattering in the electron microscope. Moreover, in pollen tubes labelled with H^3-myo-inositol, with H^3-methyl-methionine as a methyl donor, the label appears first in Golgi vesicles. Synthesis and transfer of both pectic compounds and hemicelluloses have more recently been shown to occur in pea roots in a fraction rich in Golgi bodies (Harris and Northcote, 1971).

All these observations clearly point to these vesicles as carriers of pectic compounds. Bonnett and Newcomb (1966) came to the same conclusion with the root hairs of radish. The vesicles have also been implicated in the transfer of matrix substances in algae (Hill and Machlis, 1968 — *Oedogonium;* Fowkes and Pickett-Heaps, 1971 — *Spirogyra*) and fungi (Heath, Gay and Greenwood, 1971). Heath *et al.* (1971) give reason to believe that in *Saprolegnia,* the

431

walls of which contain between 45 per cent (Novaes-Ledien, Jiménez-Martínez and Villenueva, 1967) and 85 per cent (Parker, Preston and Fogg, 1963) of hemicellulose, that hemicelluloses are also transported in the vesicles, for the periodic acid — silver hexamine reaction is similar in walls and vesicles and this reaction is due to polysaccharides (Pickett-Heaps, 1967). The reaction was not, however, identical in the two locations and this could indicate a lower DP in the vesicles. Most recently Fowkes and Pickett-Heaps (1972) have extended observations of this type to cover the bryophytes (fibre cells of *Marchantia bertercana*). The Golgi vesicles here contain fibrous material which, after peroxidation, stain with Ag hexamine just as does the wall; moreover, tritiated glucose enters Golgi vesicles first and a subsequent chase transfers the radioactivity to the wall. This function of Golgi vesicles in bryophytes had already been suggested by Schultz and Lehman (1970).

In harmony with known effects of auxins on wall synthesis, a 50 per cent increase in glucan synthetase activity has been claimed following auxin pretreatment of pea epicotyls; this again has been attributed to effects within Golgi vesicles (Ray, Shininger and Ray, 1969; Abdul-Baki and Ray, 1971). Similar effects (using 2-4-D) on the synthesis of hot water soluble glucans has also been observed by Van der Woude, Lembi and Morré (1972) using *Allium cepa,* though they claim that the synthetase here is attached to the plasmalemma.

There has been latterly, however, a tendency to involve the endoplasmic reticulum also in the transfer of material to the wall, by budding off vesicles which then pass to the plasmalemma. This role has been suggested for ER both on electron microscopic grounds alone (Robards and Kidwai, 1969; Bal and Payne, 1971) and in combination with biochemical investigation (Dashek, 1970). Dashek (loc. cit.) examined especially the transport of Hyp-protein following ^{14}C pulse experiments with cell cultures of *Acer pseudoplatanus,* and isolated three cell fractions involved. These fractions contained smooth membranes, mitochondria and Golgi vesicles, but the only organelle common to the three were the smooth membranes, probably ER membranes, and these he considers as the possible transporters of Hyp-protein. Since labelling experiments had already shown that the protein takes the form of a glycoprotein synthesized in the cytoplasm and transferred to the wall (Lamport, 1965; Olson, 1964), it follows that saccharides may also be transported in this way. Dashek points out that the particulate fractions isolated by Hassid and his school (see below), involved in incorporating nucleotide sugars, sediment at the same rate as do these smooth membranes and may therefore be the same fraction. It has been shown that glycolization of some animal proteins is mediated by smooth membranes (Cook, Laico and Eylar, 1965; Simkin and Jamieson, 1967) but the question in all such studies is whether these come from ER or the plasmalemma.

13.2 Cellulose

The biosynthesis of cellulose presents a more severe problem since this polysaccharide must be synthesized, crystallized and oriented in specific ways. Moreover, not all 1, 4-linked glucans can be classified as cellulose and in this sense it cannot as yet be claimed that *in vitro* synthesis has yet been achieved.

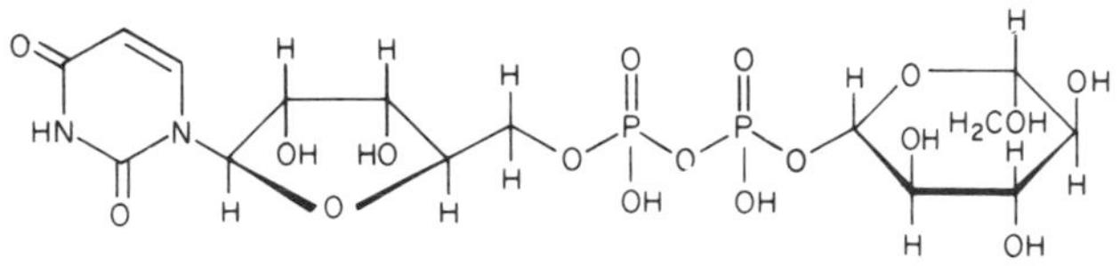

Fig. 13.4 **UDP-glucose.**

13.2.1 *Biochemical pathways*

Glaser (1958) was among the first to demonstrate, using *Aceto-bacter xylinum,* the synthesis from UDP-glucose ^{14}C (Fig. 13.4) of an alkali insoluble polysaccharide which he claimed to be cellulose. This synthesis was later supported by Brummond and Gibbons (1965) who claimed UDP-glucose as the most effective precursor of the various sugar derivatives they tried. Unfortunately they did not test GDP-glucose (Fig. 13.5) which had already been claimed by Barber, Elbein and Hassid (1964) as the only effective precursor with a 20 000 g extract from mung bean. Identification of cellulose by Barber *et al.* was claimed not only on the basis of alkali insolubility but on the presence of radio-glucose, -cellobiose, -cellotriose and -cellotetraose in an hydrolysate. Feingold, Neufeld and Hassid

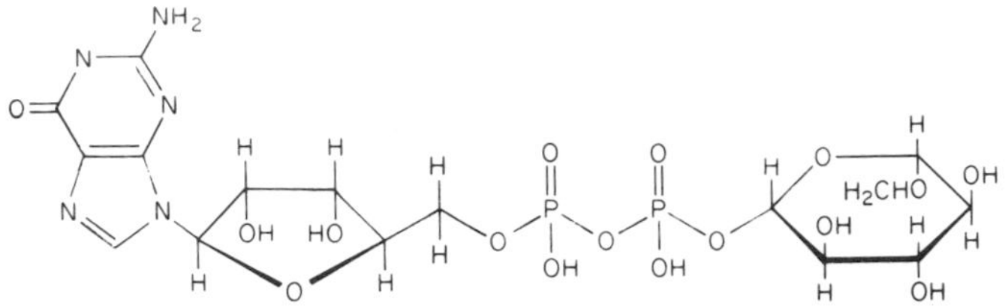

Fig. 13.5 **GDP-glucose.**

(1958) had earlier claimed that mung bean extracts synthesise from UDP-glucose a β-1-3 linked glucan similar to laminarin, callose and paramylon, and at one time it began to appear that precursors had been found specific for 1-3 and 1-4 linked polysaccharides. The situation is now, however, much less clear. Ordin and Hall (1967) have shown that extracts from the oat plant (*Avena sativa*) can use either UDP-glucose or GDP-glucose in the synthesis of an alkali insoluble polysaccharide which they regard as cellulose. With UDP-glucose, treatment of the polysaccharide with *Streptomyces* cellulase yields mostly cellobiose with a small amount of mixed β-1-4, β-1-3 linked trisaccharides. Hydrolysis of the GDP-glucose product yielded cellobiose only. They concluded (Ordin and Hall, 1968) that at least three polysaccharides are produced, one of which is cellulose and another laminarin. According to them, substrate concentration has a marked effect on the end product; with concentrations of up to 5 mM UDP-glucose, the resulting alkali-insoluble polysaccharides yield on hydrolysis equal amounts of laminaribiose and cellobiose but at concentrations greater than 5 mM only traces of cellobiose appear. On addition of dithiothreitol (UDP-glucose at 0.1 mM) the percentage of cellobiose is increased so that the enzyme involved may carry an active −SH group. A reinvestigation of the system by Flowers, Batra, Kemp and Hassid (1968), using extracts from each of *P. aureus* and *Lupinus albus*, maintains, however, the situation that UDP-glucose yields only 1-3 linked polysaccharides, though with oat extracts a β-1-4 linked glucan is also produced as Ordin and Hall (1967) had claimed.

Batra and Hassid (1969) have attempted to clarify the situation by examining the reaction of synthesised polysaccharides to a highly purified exo-β(1-3)-D-glucanase extracted from a basidiomycete. The polysaccharides synthesised from UDP-glucose were found to hydrolise to the extent of about 90 per cent in 24 hours, whereas the GDP-glucose product remained unaffected. The UDP-glucose product is therefore in the main a 1-3 linked polysaccharide. Nevertheless, Stafford and Brummond (1970) have recovered UDP-glucose as a possible cellulose precursor by showing that the addition of glucan to the mixture (using *L. albus* extracts) can cause a ten-fold increase in the presence in the product of β-1-4 links. These workers incidentally appear to be the only ones using acetolysis of the product and the claim to detect cellobiose octaacetate.

As with other wall polysaccharides, a lipid appears to be involved in the synthesis. This was first established by Colvin (1961) for *Acetobacter xylinum*, and has since then received support from Villemez and Clark (1969) (with a 50 000 g extract of *P. aureus*) and Pinsky and Ordin (1969) (with a similar extract from *A. sativum*). The lipid appears to be tightly bound to the enzyme, and the possibility has been discussed by Colvin (1972) and Lennarz and Scher (1973) that the function of the lipid might be to facilitate the transport of the glucose donor across the plasmalemma.

More recently, Robinson and Preston (1972) have re-examined the alkali-insoluble product resulting from incubation of either UDP- or GDP-glucose with a 40 000 g fraction of *P. aureus*, synthesized in sufficient quantity (and without carrier cellulose) as to allow an attempt at identification by X-ray diffraction. They found that layering the 40 000 g fraction on a Ficoll gradient column produced two fractions (A and B) active in synthesis. Fraction A, the pellet at the bottom of the gradient, consisted mainly of cytoplasmic debris with some mitrochondria and scattered smooth membranes. On the other hand, Fraction B, occurring as a white band almost at the top of the gradient, is totally homogeneous, consisting only of large segments of smooth membrane, possibly disrupted plasmalemma. This fraction may be identical with that described by Morré (1970) using onion stems.

Hydrolysates of Fraction B alone yield galactose, glucose, mannose, arabinose and xylose with glucose predominating; treatment with hot 2 per cent alkali reduces the number of detectable sugars but glucose still predominates. After incubation of the fraction with either UDPG or GPDG, both the intensity and the size of the glucose spot is greatly increased and galactose is also slightly

TABLE 13.1

X-ray spacings of products from UDPG and GDPG

G_1 D-Glucose d(Å) Intensity	G_2 Cellobiose d(Å) Intensity	G_3‡ Cellotriose d(Å) Intensity	G_4‡* Cellotretrose d(Å) Intensity	G_5‡* Cellopentose d(Å) Intensity	G_6‡*** Cellohexose d(Å) Intensity	G_7‡** Celloheptose d(Å) Intensity	G_- Cellulose II d(Å) Intensity	UDPGlc/ GDPGlc product before alkali d(Å) Intensity	UDPGlc/ GDPGlc product after alkali d(Å) Intensity
			11.29					10.78 MS	11.36 VVW
8.53 M	8.39 S	8.23 VW							8.47† VW
7.18 M		7.46 MS	7.19 MS	7.23 S	7.14 MS	7.21 MS	7.35 S		
6.09 M	6.52 W	6.63 VVW	6.59	6.49					6.56† W
		6.08 VVW					6.02 VW		
	5.60 W		5.56	5.66					
5.25 MW			5.40						5.34 MW
				5.22			5.19 M		
	5.05 W	5.08 W							
4.71 M	4.77 MS	4.60 S							
4.55 VW									
4.35 VS	4.39 VS	4.46 VVW	4.48 S	4.47 S	4.43 S	4.45 S	4.42 VS	4.56 MS	4.56 MS
							4.38 S		4.34 VVW
		4.17 VVW							
	4.04 W	4.05 MS	4.05 S	4.05 S	3.98 S	4.04 S	4.03 VS	4.08 MS	4.09 VVW
3.89 VW		3.92 VW							
	3.80 MW	3.78 MS	3.77	3.78			3.78 VW		3.85 VW
3.62 VW	3.62 W	3.60 VW	3.59	3.56				3.57 VW	
3.52 W	3.53 W						3.45 VW		
3.39 VW	3.41 MW	3.34 M	3.40	3.34 MW	3.36 W	3.38 W	3.42 M	3.37 W	3.35 VVW
3.26 VVW	3.27 W		3.30 MM						
3.15 M	3.20 W	3.16 VVW	3.23						
			3.13	3.10			3.14 S		
3.03 VVW		3.02 W							
2.90 VW	2.92 W	2.99 VW	2.98 MW	2.94 MW			2.92 VW		2.95 VW
	2.85 W	2.84 VW		2.83 MW		2.83 W			
	2.75 W	2.74 W	2.78 MW						2.75 MS
2.63 VVW	2.67 VW		2.62 M	2.64 M	2.59 MW		2.58 MS	2.63 M	2.59 MS
2.55 W	2.53 VW	2.54 VW				2.53 MW	2.56 MW		
2.50 M	2.49 W	2.51 VW	2.48	2.49				2.47 W	2.45 M
	2.44 VW								
	2.40 VW								
2.36 W	2.37 VW		2.35 MM	2.34 M	2.35 MW	2.35 W	2.39 MW		2.36 M
	2.33 VW		2.31						
2.24 MW	2.28 VW			2.25				2.28 VW	2.24 MW
2.18 VVW	2.18 VVW		2.21 W	2.22 M	2.20 M	2.22 MW	2.21 M	2.19 W	2.17 MW
	2.14 VW		2.16						
	2.10 VW		2.13						
2.08 W	2.08 VW		2.07						2.05 MW
2.03 VVW		2.04							
	2.00 W		2.01						2.00 MW
1.97 W	1.98 W		1.98	1.97 W					
1.93 W	1.88 W		1.95 W	1.92 W		1.91 W		1.92 VVW	
1.88 W									
1.84 VVW									
1.81 W	1.82 W		1.84 W	1.85 W					1.80 VW
1.76 VVW									1.75 VW
1.73 VVW	1.73 VVW								
1.68 VVW	1.67 VVW		1.69 W	1.69 W	1.68 W			1.68 VW	1.66 W
1.64 VW	1.63 VW		1.60					1.63 VVW	1.64 W
1.61 W									1.61 MW
1.58 VW									
								1.54 MW	1.54 M
1.50 VW	1.49 VW								
1.44 VVW				1.45 W					1.44 VW

Abbreviations: VS = very strong; S = strong; MS = medium strong; M = medium; MW = medium weak; W = weak; VW = very weak; VVW = very, very weak.

* Values of Trogus and Hess.
** Values of Williams.
*** Values of Wolfrom and Dacons.
† Present only in wet diagrams.

increased. Treatment with hot 2 per cent alkali leaves a fraction yielding virtually only glucose on hydrolysis. It appears that in either case only a glucan is being synthesized. The X-ray diagram (see spacing in Table 13.1) shows that the crystalline material is the same for both GDPG and UDPG. It may be argued that amorphous polysaccharides may also be produced and that these might differ between the GDPG and the UDPG product; but this may be countered since the amounts of nucleotide sugars used is barely sufficient to correspond with the crystalline product alone. The product is clearly neither cellulose I nor laminarin. The spacings in Table 13.1 make it clear that it is not cellulose II either. Indeed, the absence of a strong reflection at about 7.3 Å, the presence of arcs corresponding to longer spacings than this, and the richness of the diagram in short spacings, all suggest that perhaps the synthesized product is a mixture of low molecular weight oligosaccharides. Though these are undoubtedly 1-4 linked glucans, it cannot be claimed that they are cellulose. Since the methods adopted in this investigation did not differ in any material particular from those used by earlier workers, the conclusion that cellulose itself has not yet been synthesized seems unavoidable.

This perhaps ought not to be surprising. Crude extracts must surely contain a mixture of enzymes and some of these may be hydrolases. If this is the case, then the product may indicate some balance between the activity of enzymes, such as between synthetases and hydrolases, and the resultant product depends upon niceties of detail in the methods used.

13.2.2 Structural considerations

We now turn to the question of the organelles responsible for cellulose synthesis. It is implicit in the work described above that synthesis occurs at or near the plasmalemma and this has been stated explicitly by Villemez, McNab and Albersheim (1968). Only one group of investigators have implicated Golgi vesicles in the synthesis and only in the haptophycean alga *Pleurochrysis.* This is a group of unicellular algae of which the wall consists of a series of imbricated, highly sculptured scales which, as shown by the extensive work of I. Manton, are synthesized in Golgi vesicles. The first claim that these were in large part polysaccharide, and that the polysaccharide is cellulosic, came some three years ago

(Brown, Franke, Kleinig, Falk and Sitte, 1970) in a paper which documented the presence in Golgi vesicles of a glucan some at least of which is 1-4 linked, but failed to present conclusive proof that this is cellulose. Subsequently Herth, Franke, Stadler, Bittiger, Keilich and Brown (1972) have presented X-ray diagrams which purport to show that the native scale polysaccharide is cellulose I and that treatment with 10 per cent alkali converts this to cellulose II. These authors do not present a list of spacings but inspection of the diagrams appears still to leave such an identification uncertain. The prominent arc of cellulose I at 6.1 Å (101) is missing and the arc at c. 3.9 Å (002) does not appear precisely to match the corresponding arc of cellulose I either. Moreover, in the so-called cellulose II diagram the outer of the two rings at a little more than 4 Å seems to be missing or weak. Moreover, the DP at 3150 is low. Accepting it that the polysaccharide is microfibrillar, and admitting it that since Golgi vesicle membranes become plasmalemma membranes (by fusion with the plasmalemma) it should be no surprise that some plants should initiate the synthesis of cellulose on Golgi membranes, the case for synthesis in the Golgi vesicles of *Pleurochrysis* must be considered unproven. In face of the body of evidence tending in a different direction for a fairly wide variety of plants, it is hardly justifiable to write, as do Herth *et al.* (1972) that 'Recent findings on the formation of algal scales have questioned the general validity of such a separate pathway' (i.e. the plasmalemma-associated pathway) particularly since the authors are questioning the *general* validity by reference to one single organism. The still more recent discussion (Brown, Herth, Franke and Romanovicz, 1973), following the evidence that the structural polysaccharide is covalently linked to a protein as a glycoprotein, and following up some aspects of biosynthesis, is an important one with regard to the function of Golgi vesicles. It will not be taken up here, however, since the relevance to the present matter remains to be proved. It seems in any case doubtful that synthesis in Golgi vesicles could be expected to lead to microfibrils at least 10μm long ordered in the complex arrangements described earlier in this book without subsequent intervention of a plasmalemma-linked mechanism.

The two hypothesis for cellulose synthesis most widely considered centre in turn around plasmalemma granules and microtubules. These hypotheses each take note of the need to orient microfibrils in specific directions in the wall and each is supported both by electron microscopic and biochemical evidence. The hypotheses and the supporting evidence will be considered here separately. The earlier considerations that cellulose crystallites might be oriented either by protoplasmic streaming or by the anisotropy of stress in the wall of a turgid cell fail to explain the complexity of wall structure, are no longer accepted, and will not be considered further here; assessments of these mechanisms will be found in a number of texts including Preston (1952).

In judging the relative merits of microtubules and plasmalemma granules as possible sites of synthesis, a number of parameters of the substance to be synthesized need to be born in mind.

(a) *Wall structural factors limiting the synthetic mechanism*
1. Cellulose occurs naturally in the form cellulose I, not the form cellulose II produced when separate chains are brought together. This gives *a priori* reason to believe that the molecular chains are not synthesized individually; rather they must be synthesized collectively as a microfibril, as considered by Wardrop and Dadswell (1952). Support for this concept has been provided by Colvin, Bayley and Beer (1957) for the extracellular cellulose of *Acetobacter xylinum* and by Preston (1959) for algae and higher plants. Colvin *et al.* observed that the microfibrils grow in the medium in the absence of the amorphous precursor envisaged by earlier workers; and that they develop by extension at one end if not both. Preston drew attention to the observation of almost all electron microscopists that, however closely parallel microfibrils might lie in the wall, they are always twisted round each other. This could not happen through bundling of separately synthesized chains. Both lines of evidence showed that microfibrils must grow by end synthesis, a concept which is not now in dispute. Additional indirect evidence for end synthesis has been reviewed by Colvin (1972). The only dissientients as far as the writer is aware are Macchi, Marx-Figini and Fischer (1968) and Macchi and Palma (1969). These authors have claimed that cellulose I can be formed in yield of about 10 per cent by crystallization from dilute cellulose solutions and conclude that cellulose may be formed in the same way at the cell surface. This is an extrapolation from an *in vitro* experiment which is both naive and enormous. Moreover, aside from the low yield and low crystallinity of the product, the possibility is far from excluded that some cellulose I survived the

solution treatment; it is well known that some cellulose I may survive even the strongest mercerisation process (Rånby 1952; Manjunat and Peacock, 1969). The necessity for end synthesis (equivalent to a demand for a starter 'molecule') seems to demand an enzyme complex close to at least one end of each microfibril.

2. If the Meyer and Misch unit cell, or the necessary modifications of it discussed earlier in this book, is correct, so that adjacent cellulose chains lie antiparallel, then it seems likely both that two enzymes are necessary (Preston, 1959; Roelofsen, 1959) and that synthesis might occur at either end of a microfibril since these would be indistinguishable. The recent findings of Gardner and Blackwell (1974) (p. 140) that the chains may not lie antiparallel are of significance here. Until the difference in polarity among the chains is fully confirmed, this structural necessity for a two-enzyme complex can be resisted. There is, however, another context in which two different synthetases have been said to have become involved.

It is now almost twenty years since Frey-Wyssling (1954) postulated that the DP of primary wall cellulose was about one half of that of secondary wall cellulose and that the biosynthetic situation was in some way different between these two. This has more recently been followed and greatly amplified in a series of important papers by Marx-Figini (1963, 1964, 1966, 1967 and 1969) and by Marx-Figini and Schultz (1963 and 1966). Briefly, these workers find, by fractional precipitation and viscometry, that the primary wall cellulose of cotton has a DP_W of about 6000 and is polydisperse while the secondary wall cellulose is homodisperse with a DP of about 14 000. They find, incidentally, that with *Valonia* cellulose the DP is about 16 500 though the relevance of this determination is not evident since the *Valonia* wall cannot be classified as either primary or secondary (p. 192). From these observations the authors conclude that primary wall cellulose is produced by a time-dependent synthetic mechanism while secondary wall cellulose, in view of its constant chain length, must be synthesized with the help of a template mechanism. The template must be of the same length as the cellulose chains of the secondary wall at about 7μm and should be recognisable. Interesting though this speculation is, it goes a good way beyond the facts. There seems no good reason to reject the possibility that the synthetic mechanism is the same both for primary and secondary walls and that differences in DP and dispersity are due to other factors which differ between the two regions of wall differentiation. More-over, the evidence refers to cotton hairs only; in all other secondary walls reported upon the cellulose appears to be polydisperse. It may be significant that cellulose synthetases continue to be active after a tissue has ceased to grow but cellulose hydrolizing enzymes are confined to young meristematic regions (Datko and MacLach-lan and Perrault, 1964). This in itself could lead to a rise in DP during development. Spencer and MacLachlan (1972) have recently examined the situation in pea epicotyls by examining the DP of cellulose from the plumular hook downwards, isolating the cellu-lose by the method of Timell (1963) which does not cleave 1-4 bonds and extracts more than 95 per cent of the cellulose. In the region of the plumular hook the DP ranged from 95 to 7530 (average c. 5000) and at the basal mature end between 300 and 9230 (mean c. 9000) with a steady increase from one end to the other. The basal end must have contained secondary walls but the authors do not mention a double peak in the distribution of DP such as would be expected following Marx-Figini.

3. The microfibrils of the wall, though sometimes lying at random in the plane of the wall, mostly lie in some specific orientation, probably constant within a species, sometimes very perfect as in *Valonia* and the Cladophorales, and sometimes related to cell di-mensions, as in tracheids and fibres. It is not easy to see how this could come about unless the unique synthetase complex is situated outside the plasmalemma and unless neither the microfibril nor the synthetase complex moves during synthesis.

4. The microfibrils in a single wall lie in two directions with a third less abundant set with some species of algae, and with a fourth in some higher plants. In the algae (*Valonia* and Cladophorales) the direction changes from one lamella to another in a regular sequence and many times without mistake. The synthetase system must therefore be such as to recognise three, sometimes four, directions and no other.

5. The microfibrils of a wall lie usually in separable lamellae. This is not, however, always the case; the microfibrils of one lamella may be inextricably woven with those of the next (Frei and Preston, 1961). The synthetase system must therefore be able to deposit microfibrils in at least two directions simultaneously.

6. The shape or coherence of the cell surface is significant. When a cell of *Cladophora* or *Chaetomorpha* is plasmolyzed, the cyto-plasm still deposits a wall over the now spherical surface; but the

441

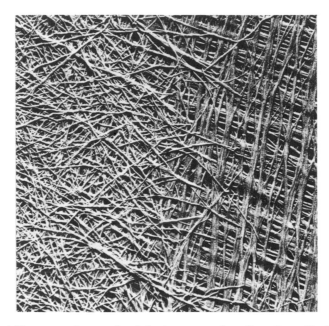

Fig. 13.6 Electron micrograph of the innermost lamellae of a wall of a cell of *Ch. melagonium* immediately prior to sporulation; magnification 30 000. The normal crossed fibrillar wall can be seen, overlayed by a lamella just deposited with random microfibrils.

microfibrils now lie at random (Frei and Preston, 1961). On recovery from plasmolysis wall deposition with microfibrils in the normal two or three directions is resumed and the random lamella becomes buried. Similarly, as the protoplast of a cell of *Cladophora* or *Chaetomorpha* prepares for sporulation and before there is any microscopically visible sign of segregation, it lays down a wall lamella in which the microfibrils lie at random rather than in the normal crossed configuration (Frei and Preston, 1961) (Fig. 13.6). In these plants, at least, the synthetase mechanism must be such that it can lose its sense of order while not losing the power of synthesis.

13.2.3 The microtubule or template hypothesis

The case for microtubules rests almost entirely on the degree to which they are found associated with sites of wall deposition and to lie parallel to the microfibrils of the adjacent walls. Close association between thickened bands on the wall has been found by

442

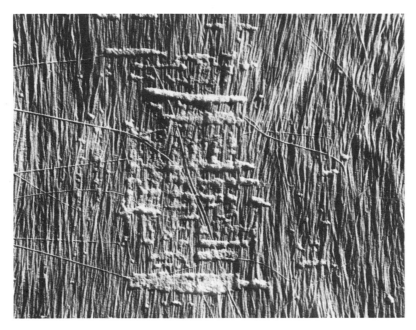

Fig. 13.7 Electron micrograph of innermost wall lamella from *Ch. melagonium*; magnification 20 400 X, shadowed Pd/Au. Note bands of granules each some 30 nm diameter, lying almost at $90°$ to the microfibrils of the lamella just completed. Microfibrils of the next lamella appear to arise from these bands.

Hepler and Newcomb (1964), Cronshaw and Bouck (1965) and Wooding and Northcote (1964) among others. Similarly, many workers, such as Chafe and Wardrop (1964) and Robards and Kidwai (1972), have shown almost precise agreement between the microfibril direction and the microtubule direction even within a single fibre which is laying down microfibrils in different directions in different locations. An example is presented in Fig. 13.2. These parallelisms do not, however, necessarily imply a cause and effect relationship; it could be that some third factor, directly or indirectly, is concerned with orienting both microtubules and microfibrils. If this is the case it might be anticipated, since microtubules and microfibrils are spatially separated, that occasions could be found in which one is oriented and not the other; and these have been found. In the root hairs of radish, for instance, the microfibrils of the wall in the zone 0-25 μm from the tip lie at random though the microtubules lie parallel (Newcomb and Bonnett, 1965). Moreover, even in fibres and tracheids,

443

microtubules and microfibrils do not lie parallel at the elongating tips (Robards and Humpherson, 1967). Again, microtubules are sometimes absent even when a cell is laying down exceedingly well-ordered microfibrils. This is true of the swarmers of the Clad-ophorales (Preston and Goodman, 1968), an interesting case since microtubules are plentiful while the swarmer is naked but dis-appear on the onset of wall deposition. A similar situation has been found in *Pediastrum* (Gawlick and Millington, 1969).

The case for microtubules cannot therefore be regarded as strong and the fact that Marx-Figini at one time considered these bodies as possible templates cannot be said to strengthen the case either. It has already been pointed out by Preston and Goodman (1968) that microtubules are quite inadequate as templates for the syn-thesis of whole microfibrils 10 or 20 nm wide, particularly remem-bering the sizes of the enzymes and precursors which have to be specifically assembled; they do not seem therefore to comply with condition (1) above and it is not easy to see how they could meet conditions (3) − (5) either.

A better choice of function for microtubules would appear to be concerned either with protoplasmic streaming, as first suggested by Ledbetter and Porter (1963) or with directing substances to the plasmalemma perhaps for wall synthesis. The streaming function has been supported by Kommick and Wohlfarth-Botterman (1965) and by McManus and Roth (1965) on the grounds that micro-tubules are numerous in myxomycete plasmodia and amoeba. The frequently observed passage of microtubules through the plasmalemma (e.g. Heyn, 1972; Nelmes, Ashworth and Preston, 1973) seems on the other hand to suggest a direction of material to the site of synthesis. The observations of Nelmes *et al.* go further in suggesting a connection with lignification. Since, finally, micro-tubules are common in the animal as well as the plant kingdom, it seems hardly supportable that their function should lie solely in terms of a structure which appears only on one of them.

13.2.4 The ordered granule hypothesis

This hypothesis stems from the observation that in *Valonia* (Preston and Kuyper, 1951; Preston, Nicolai and Kuyper, 1953), in fusi-form initials of conifer cambium (Preston and Ripley, 1954) and parenchyma cells of broad bean internodes (Williams, Preston and Ripley, 1955) granular bodies may be seen on the innermost

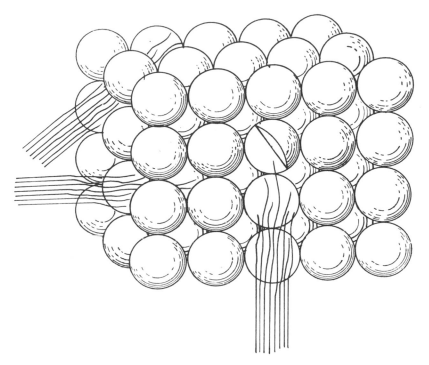

Fig. 13.8 Diagrammatic representation of an array of granules associated with the plasmelemma. Microfibrils growing through the array are drawn as bundles of threads representing (not to scale) individual cellulose chains.

lamella of the wall which are somewhat larger in diameter than are the microfibrils. Subsequently Frei and Preston (1961) observed similar granules on the inner face of the walls of plasmolysed cells of *Chaetomorpha,* lying in bands either parallel to or perpendicular to the microfibrils (Fig. 13.7) and therefore in this last case parallel to the set of microfibrils about to be produced. These observations could clearly be interpreted in terms of arrays of granules on the plasmalemma, and the observation that microfibrils pass out from these arrays (Preston and Ripley, 1951; Frei and Preston, 1961) suggested that they are synthetase granules. The hypothesis was formally stated in 1963 (Preston, 1964), in terms which seem to meet all the requirements (1) − (6) above and to take in the idea of a common orienting mechanism (p. 210), in the form of a diagram (Fig. 13.8).

445

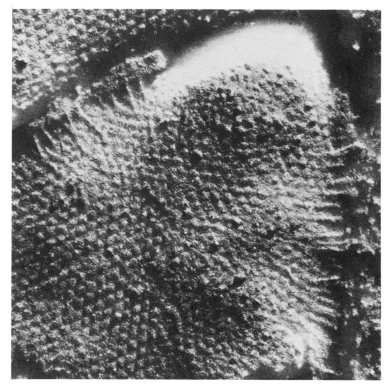

Fig. 13.9 Electron micrograph of freeze-etch replica of the plasmalemma of brewer's yeast cells seen from outside; magnification 222 000 X, shadowed Pd/Au. Note the granules, in hexagonal array, confluent with fibrils (not cellulose) of somewhat smaller diameter.

The plasmalemma is considered as covered with a layer of granules in square packing. Each granule, some 30 nm in diameter with the Cladophorales is taken to contain glucan synthetases such that, on coming into contact with a microfibril end, each chain in the microfibril is extended by progressive condensation on its end of glucose residues contributed by a nucleotide glucose donor. The chains, and the microfibril, are kept straight by the known intra-chain hydrogen bonding, and the direction of the microfibril is dictated by the square packing of the granules. Accordingly, in whatever direction a microfibril 'approaches' a granule array, it will be constrained to develop along one of the sides of the squares or along a diagonal, though then with a lower probability since a gap occurs between one granule and the next. The array will

446

therefore synthesize two abundant sets of microfibrils lying mutually at right angles and two less abundant sets lying at 45°. Only one of these diagonal sets is commonly observed and we may assume that the other is suppressed by a polarity in the granules. In the algae concerned, the orientation of the microfibrils *as laid down* is not as perfect as it later comes to be as the wall is stretched and for this reason, as well as for the need for granules formally lying along a diagonal occasionally to come into contact (to allow synthesis to proceed from one granule to the next), it must be supposed that the granules are in slight movement. In the model of Fig. 13.8, three layers of such granules are included to allow unimpeded synthesis along three directions simultaneously and geometrically to allow microfibrils to wander from one granule layer to another and therefore to interweave. Three layers are, however, clearly not a necessity.

The statement made in 1963 was closely followed by the first demonstration by the freeze-etching technique of plasmalemma granules lying in a close crystalline array (Moor and Mühlethaler, 1963) (Fig. 13.9). The granules are closely associated with wall fibrils in such a way as to suggest that the one is synthesizing the other, and, though the fibrils are not cellulosic since the organism used was yeast, they are glucan and the situation revealed in this way is suggestive. The granules are not as large as those postulated for Cladophorales but this would not seem to have significance since the fibrils are only 10 nm in width as against the 20 nm of the Cladophorales. Since that time, granules have been observed to lie in or on the plasmalemma of all cells which have been examined in this way.

Barnett and Preston (1970) have, indeed, recorded on the surface of swarmers of *Cladophora* an array of granules precisely as postulated by Preston (1964) (Fig. 13.10). Significantly, the axes of the granule lattice coincide with the microfibril directions (not shown). It has not been found possible to repeat this observation but it has to be remembered that the granules must be labile since they do not normally appear in conventional ultrathin section (though they have been seen by Robards (1969) and by Roland (1967) in differentiating xylem and in collenchyma) and that it is not self evident that the whole surface of the plasmalemma should be covered at one and the same time by granules in an array. Moreover, in these swarmers the first wall layer deposited has random microfibrils (Nicolai, 1957) and the subsequent layer with crossed

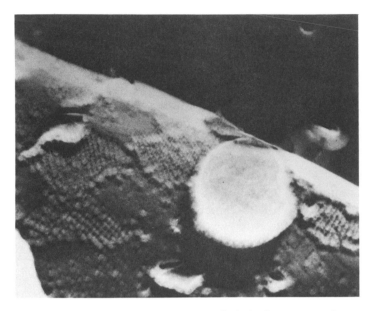

Fig. 13.10 As 13.11 but surface of swarmer of *Cladophora rupestris;* magnification 30 000 X. Note three sets of granules which clearly lie one over the other.

microfibrils tends to prevent cleavage along the plasmalemma. The possibility of observing the granular array may therefore be limited in space and is certainly limited in time. Nevertheless one small array has been seen with swarmers of *Chaetomorpha* (Robinson and Preston, 1971) and, in the same organism, plasmalemma granules have been seen attached to short fibrillar bodies (Figs. 13.11 and 13.12) which, in surface view of the plasmalemma (Fig. 13.13), all point in the same direction. Grout, Willison and Cocking (1973) have also reported that granules associated with the plasmalemma of tomato fruits sometimes have tails. The earlier conclusion of Willison and Cocking (1972) that the cellulose microfibrils of tomato fruit walls, though produced in association with the plasmalemma, are not related to granules cannot be sustained since their electron micrographs do not reveal the loci at which such a relation is to be sought.

Relevant observations have also been made on the unicellular alga *Oocystis*, referred to on p. 218. *Oocystis* shows a complex array of granules associated with the plasmalemma when viewed

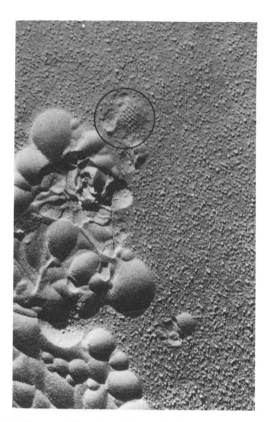

Fig. 13.11 As 13.10; magnification 55 200 X. Note the small array of granules (circled), and granules with tails.

from outside the cell. These granules, some 8.5 nm in diameter, are arranged in pairs and the pairs arranged in rows (Fig. 13.14) which·correspond in direction with that of one of the sets of microfibrils (Robinson and Preston 1972). Sometimes, as in Fig. 13.14, indeed, the granule rows occur in two sets recording both directions of microfibrils. Plasmalemma granules have also been seen on many other organisms including *Chlorella* (Staehelin, 1966) and *Cyanidium* (Staehelin, 1968) in which both they and the wall microfibrils lie at random; a particularly beautiful array has been recorded on the outer surface of the bacterium *Acineto-bacter* (Sleytr and Thornley, 1973) (Fig. 13.15) though this organism, unlike some other gram-negative bacteria, does not produce cellulose. A review has recently been presented by Staehelin and

449

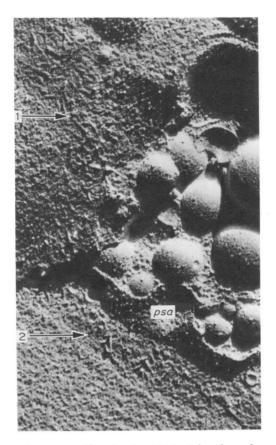

Fig. 13.12 As 13.11; magnification 55 200 X. The plasmalemma in surface view (*psa*) carries granules with tails; immediately outside this (arrow 2) is a region (fractioned obliquely) also showing granules with tails. Tangential fracture at the plasmalemma region (arrow 1) also shows granules with tails.

Probine (1970). They are digestible by proteolytic enzymes (Staehelin, 1968; Fox, 1972) and, in yeast, are mannan-protein complexes (Matile, Moor and Mühlethaler, 1967). Stockman (1972) has given reason to believe that the most plausible mechanism of cellulose synthesis thermodynamically is just such a simultaneous polymerization and crystallization on microfibril ends as these granules would induce. According to him, a new approaching molecule will then be trapped in a crystal lattice even when thermodynamics demand disorder; the new microfibril section is therefore initially 100 per cent crystalline and disordered regions are

Fig. 13.13 **As 13.12**; magnification **96 600 X**. Note granules with tails.

Fig. 13.14 As **13.10** but *Oocystis;* magnification **100 000 X**, showing a complex array of plasmalemma granules.

451

Fig. 13.15 Freeze-etched preparation showing the surface of a gram-negative bacterium *Acinetobacter* strain MJT/F5/199A. The surface is composed of a regular array of subunits, arranged in rows in two directions which are approximately at right angles. The spacing between the rows is *c*. 8 nm in one direction and 60 nm in the other. Magnification 140 000 X. (By courtesy of Sleytr and Thornley (1973)).

secondarily produced, yielding an order-disorder structure which is metastable.

The existence of granules in the required location, namely over or in the outer surface of the plasmalemma, is therefore no longer in doubt and, though complete proof is still lacking, it seems likely that these represent synthetase systems accepting precursors from Golgi vesicles, microtubules or endoplasmic reticulum, or all three, and spinning out cellulose microfibrils. The particles themselves are possibly also produced in Golgi vesicles. If this is how the microfibrils are produced, then several consequences follow, as pointed out by Preston and Goodman (1968).

The granular hypothesis gives a ready explanation of the shape of naturally occurring ends of microfibrils. These have been found by Frei and Preston (1961) and by Schnepf (1965) to be tapering, mostly to a blunt point. If we adopt the formal position that a microfibril is required to 'grow' through a linear array of synthesizing granules in contact only at a point as in Fig. 13.8, then it is clear that only a molecular chain co-linear with the line joining these points can grow uninterruptedly through the file; chains not in this position encounter a gap at the surface of each granule. If we assume that the gap does not stop synthesis permanently but induces only a delay in the continuation of the synthesis which is related to the length of the gap, then the end of a growing microfibril might be expected to be tapering, as observed. Other considerations mentioned above require that tne granules should be in small movement relative to each other, and the delay in synthesis could equally be due to the time elapsing between successive contacts of points on the surfaces of two neighbouring granules away from the centre line. Again, if within each granule chains away from the common centre line were curved or folded, each of the chains within a granule could become so long that, when straightened — as, for instance, through interchain hydrogen bonding — each would come into contact with the next granule along the file. Delay in the synthesis of lateral chains would then be involved with deiays in the formation of hydrogen bonds between chains separated by curvature.

In passing it may perhaps be mentioned that tne necessity for linear arrays of granules in contact for extensive cellulose synthesis could be responsible for the low rates of synthesis recorded *in vitro*.

Two further questions remain to be considered: (i) what factors are involved in the necessity for the presence of a microfibril end before synthesis can occur and (ii) if the chains of cellulose in the crystal lattice are antiparallel how can these be synthesized simultaneously side by side? We may examine these in turn.

In order to join two glucose molecules together one must be phosphorylated and both must be held in suitable mutual positions. It might be that the synthesizing enzyme is itself incapable of doing this so that one glucose molecule must be held in some other way. If this molecule is the terminal molecule on a chain within a microfibril end its motion will be severely restricted and it could in part at least be for this reason that the presence of a microfibril end is essential. There is, however, a further aspect to consider. Uridine or guanosine diphosphate glucose would appear to be admirably suited as immediate precursors for cellulose synthesis since the hydrogen bonding groups ($-NH_2$ and $-OH$) at the uracyl or guanine end of the molecule are situated precisely in positions to form hydrogen bonds with the successive $-OH$ groups on a cellulose chain. It could be therefore that a microfibril end is necessary in order that, by hydrogen bonding with one chain in it, a nucleosyl glucose could be held in such a position that the terminal glucose is held over another chain so that the glucose may be transferred. The attachment would then be between carbon 1 of the donor molecule and carbon 4 of the receptor molecule. This would give a simple picture of the synthesis of those chains at a microfibril end which present carbon 4. If all the cellulose chains in a microfibril were to lie pointing in the same direction then no further assumptions would need to be made; the microfibrils would then grow at one end only and whether they do so or not is not known.

With antiparallel chains, the biosynthesis of cellulose is conceptually a much more difficult process for any proposed mechanism (Preston, 1959 and 1964; Roelofsen, 1959; Colvin, 1964). On the scheme outlined above, an antiparallel arrangement would seem at first sight to demand one of two further assumptions. Either (i) a chain synthesized in any granule must subsequently fold back on itself (or be synthesized in the folded form) or (ii) each granule must contain two different enzymes. In the former situation we would presumably need to think of the forward synthesis of a chain being halted — perhaps by coming to the end of a line of granules — of the chain turning back on itself and proceeding to 'backward' synthesis until stopped by, for instance, contact with an existing fold lower down. Synthesis of the microfibril could then continue, when its end again contacted a line of granules, only from those chains which had not folded back. The difficulties with such a mechanism are manifold even excluding the known difficulties of explaining the physical properties of cellulose in terms of folded chains. The crystallization process is becoming dangerously close to a precipitation from solution, which would give cellulose II, not cellulose I.

One final point remains to be considered, namely the relation of the cellulose-synthetic mechanism to cell division. Duplication of the enzyme system after division presents in principle no problem since such a duplication is common to all biochemical systems. If, however, it is true that cellulose synthesis demands the presence of a microfibril end, then it follows automatically that so long as each cell in a tissue of dividing cells continues to synthesize cellulose the number of microfibril ends in the tissue as a whole must increase; for otherwise the number per cell would decrease at each successive division and would in some cells become zero. Though the number of microfibril ends present on the inner face of the wall is not known for any cell, its determination should not present insuperable difficulties. A rough estimate of the minimum number per cell may, in fact, be made for *Chaetomorpha aerea*. This plant is known to have between 100 and 300 nuclei per cell each of which undergoes at least one division before sporulation. We can reasonably therefore estimate the number of gametes produced per cell in the order of 300. Not all of these will develop into new plants but a large proportion of them may be observed to develop a wall. If each gameter must carry at least one microfibril end, then the number of such ends in the parent cell at the time of sporulation must have been of the order of several hundreds. Incidentally, three hundred microfibril ends per cell would mean 1 microfibril end in an area 50μm square, i.e. about 1 end per 25 electron micrographs, 20 cm square taken at $20\,000 \times$ magnification. Even if therefore this estimate is one hundred times too small, microfibril ends would still be scarce on electron micrographs, as in fact they are. It seems likely that a few hundred microfibrils per cell in *Chaetomorpha aerea* is the minimum possible. A rough maximum may be estimated in a different way. A swarmer 6μm diameter can develop a wall three microfibrils thick in two hours. Taking a microfibril as 200 Å by 100 Å in section, a simple calculation shows that the swarmer must have produced microfibrils at the rate of about 10^6 Å per minute. Since each microfibril contains some 600 cellulose chains this implies an attachment of about 10^7 glucose units per minute. In a synthesizing granule 300 Å in diameter there can be at most about 100 enzyme molecules 30 Å in diameter. The turnover mediated by enzymes is commonly of the order of several thousand moles per mole per minute so that the turnover of glucose in a single granule can be at most several hundred thousand, i.e. of the order of 10^5. This implies that attachment of 10^7 glucose units per minute requires 100 granules per swarmer, and therefore 100 microfibril ends. We might expect, therefore, a vegetative cell of *Chaetomorpha* to carry between several hundred and several tens of thousands of microfibril ends.

Between two successive divisions of a vegetative cell, therefore, a large number of new microfibril ends must be produced, possibly ranging into the thousands. Whether these are produced by enzyme degradation of the cellulose or by stresses in the wall during elongation or in some other way it is impossible to decide. Such an effect of stress would, however, be in harmony with a positive correlation between the rates of cell wall deposition and cell extension for which there is already some evidence (Chapter 12).

The conclusion is inevitable that the involvement of ordered granules in the biosynthesis and orientation of cellulose microfibrils seems, if not actually to have been established, at least to provide a system which describes satisfactorily the known geometrical requirements and to represent a hypothesis in favour of which there is now a good deal of evidence.

References

Abdul-Baki, A.A. and Ray, P.M. (1971), *Plant Physiol.* **47**, 537.

Adler, E., Björkvist, K.J. and Häggroth, S. (1948), *Acta Chem. Scand.* **2**, 93.

Adzumi, II. (1937), *Bull. Chem. Soc. Japan* **12**, 304.

Albersheim, P., Mühlethaler, K. and Frey-Wyssling, A. (1960), *J. Biochem. Biophys. Cytol.* **8**, 501.

Alexander, L.E. (1970), *X-ray diffraction methods in Polymer Science.* Wiley-Interscience, London.

Altermatt, H.A. and Neish, A.C. (1956), *Canad. J. Biochem. Physiol.* **34**, 405.

Anderson, D.B. (1927), *Sitzber. Akad. Wiss. Wien Math.-Naturw. Kl. Abt. 1,* **136**, 429.

Anderson, D.B. and Kerr, T. (1938), *Ind. Eng. Chem.* **30**, 48.

Apelbaum, A. and Burg, S.P. (1971), *Pl. Physiol.* **48**, 648.

Aspinall, G.O. (1964), in *Chemie et Biochemie de la Lignine, de la Cellulose et des Hemicelluloses.* Symposium International Grenoble, p. 421.

Aspinall, G.O. (1969), *Adv. in Carbohydrate Chem.* (M.L. Wolfrom and R.S. Tipson, eds.) **24**, 333.

Aspinall, G.O. and Kessler, G. (1957), *Chem. Ind., London.* p. 1296.

Astbury, W.T. (1945), *Nature, London.* **155**, 667.

Astbury, W.T., Marwick, T.C. and Bernal, J.D. (1932), *Proc. Roy. Soc.* **B109**, 443.

Astbury, W.T., Preston, R.D. and Norman, A.G. (1935), *Nature, London.* **136**, 391.

Astbury, W.T. and Preston, R.D. (1940), *Proc. Roy. Soc.* **B129**, 54.

Asunmaa, S. (1954), *Svensk. Papperstidn.* **58**, 308.

Asunmaa, S. and Lange, P.W. (1953a), *Svensk. Papperstidn.* **55**, 217. (1953b), *Idem.* **56**, 85.

Atkins, E.D.T., Mackie, W., Nieduszinski, I.A., Parker, K.D. and Smolko, E.E. (1973), In press.

Atkins, E.D.T., Mackie, W. and Smolko, E.E. (1970), *Nature, London.* **225**, 626.

Atkins, E.D.T. and Parker, K.D. (1969), *J. Polymer Sci.* **C28**, 69.

457

Atkins, E.D.T., Parker, K.D. and Preston, R.D. (1969), *Proc. Roy. Soc.* **B173**, 209.

Bailey, I.W. (1913), *Forestry Quart.* **11**, 12.

Bailey, I.W. (1936), *Ind. Eng. Chem. (Anal)* **8**, 52, 389.

Bailey, I.W. and Berkeley, E.E. (1942), *Amer. J. Bot.* **29**, 231.

Bailey, I.W. and Kerr, T. (1935), *J. Arnold. Arbor.* **16**, 273.

Bailey, I.W. and Vestal, M.R. (1937), *J. Arnold Arbor.* **18**, 185, 196.

Bailey, P.J. (1966), *Physical studies in relation to the permeability of the xylem of Douglas Fir.* Ph.D. Thesis, Leeds.

Bailey, P.J. and Preston, R.D. (1969), *Holzforsch.* **23**, 113.

Bailey, P.J. and Preston, R.D. (1970), *Holzforsch.* **24**, 37.

Bailey, R.W. and Hassid, W.Z. (1966), *Proc. Nat. Acad. Sci. U.S.A.* **56**, 1586.

Baker, D.P. and Ray, P.M. (1965), *Pl. Physiol.* **40**, 345.

Bal, A.K. and Payne, J.F. (1971), *Zeit. f. Pflanzenphysiol.* **66**, 265.

Balashov, V., Preston, R.D., Ripley, G.W. and Spark, L.C. (1957), *Proc. Roy. Soc.* **B146**, 460.

Bamber, R.K. (1961), *Nature, London.* **9**, 409.

Barber, G.A., Elbein, A.D. and Hassid, W.Z. (1964), *J. Biol. Chem.* **239**, 2672.

Barber, G.A., Elbein, A.D. and Hassid, W.Z. (1964), *J. Biol. Chem.* **239**, 4056.

Barber, N.F. (1968), *Holzforsch.* **22**, 97.

Barber, N.F. and Meylan, B.A. (1964), *Holzforsch.* **18**, 146.

Barer, R. and Mellor, R.C. (eds.) (1955), *Analytical Cytology.* McGraw Hill, London.

Barkas, W.W. (1949), *The swelling of wood under stress.* H.M.S.O. London.

Barnett, J.R. and Preston, R.D. (1970), *Ann. Bot.* **34**, 1011.

Bartnicki-Garcia, S. (1966), *J. Gen. Microbiol.* **42**, 57.

Bartnicki-Garcia, S. (1973), *Symp. Soc. Gen. Microbiol.* No. XXIII, p. 245.

Batra, K.K. and Hassid, W.Z. (1969), *Plant Physiol.* **44**, 755.

Bauch, J., Liese, W. and Berndt, H. (1970), *Holzforsch.* **24**, 199.

Bauch, J., Liese, W. and Schultze, R. (1972), *Wood Sci. and Technol.* **6**, 165.

Bayley, S.T., Colvin, J.R., Cooper, F.P. and Martin-Smith, C.A. (1957), *J. Biophys. Biochem. Cytol.* **3**, 171.

Beer, M. and Setterfield, G. (1958), *Amer. J. Bot.* **45**, 571.

Bauer, W.D., Talmadge, K.W., Keegstra, K. and Albersheim, P. (1973), *Plant Physiol.* **51**, 174

Belford, D.S. (1958), *Investigations into the fine structure of plant cell walls.* Ph.D. Thesis, Leeds.

Bennet-Clark, T.A. (1956), in *The Chemistry and Mode of Action of Plant Growth Substances* (R.L. Wain and F. Wightman, eds.) Butterworth, London.

Berkeley, E.E. and Kerr, T. (1946), *Ind. Eng. Chem.* **38**, 304.

Berlyn, G.P. and Mark, R.E. (1965), *For. Prod. J.* **16**, 140.

Betrabet, S.M. and Rollins, M.L. (1970), *Text. Chem. Coll.* **2** (2), 39.

Bienfait, J.L. (1926), *J. Agric. Res.* **33**, 183.

Bisset, I.J.W., Dadswell, H.E. and Wardrop, A.B. (1951), *Aust. Forestry* **15**, 17.

Bittiger, H. and Husemann, E. (1964), *Makromol. Chem.* **75**, 222. **80**, 239. (1966), *Idem.* **96**, 92.

Black, M., Bullock, C., Chantler, E.C., Clark, R.A., Hanson, A.D. and Jolley, G.M. (1967), *Nature, London.* **215**, 1289.

Black, W.A.P. (1950), *J. Marine Biol. Assoc. U.K.* **29**, 379. (1954), *Ibid.* **33**, 49.

Blank, F. and Frey-Wyssling, A. (1941), *Ber. Schweiz. bot. Ges.* **51**, 116.

Bobák, M. and Nečessaný, V. (1967), *Biologia Plantarum (Praha)* **9**(3), 195.

Böhmer, H. (1958), *Planta* **50**, 461.

Bonner, J. (1934), *Proc. Nat. Acad. Sci. U.S.A.* **20**, 393.

Bonner, J. (1935), *Jahrb. wiss. Bot.* **82**, 377.

Bonnett, H.T. and Newcomb, E.H. (1966), *Protoplasma* **62**, 59.

Bosshard, H.H. (1952), *Ber. Schweiz. bot. Ges.* **62**, 482.

Boundy, J.A., Wall, J.S., Turner, J.E., Woychik, J.H. and Demler, R.J. (1969), *J. Biol. Chem.* **242**, 2410.

Bourret, A., Chanzy, H. and Lazaro, R. (1972), *Biopolymers* **11**, 803.

Boutelje, J. (1962), *Svensk. Papperstidn.* **65**, 209.

Bouvery, H.O., Lindberg, B. and Garegg, P.J. (1960), *Acta Chem. Scand.* **14**, 742.

Bowen, T.J. (1970), *An Introduction to Ultracentrifugation.* Wiley-Interscience, London and New York.

Boyd, J.D. (1950), *Aust. J. Appl. Sci.* **1**, 294; *Aust. J. Sci. Res.* **B3**, 270, 294.

Boyd, J.D. (1972), *Wood Sci. and Technol.* **6**, 251.

Bragg, W.L. (1913), *Cambridge Phil. Soc.* **17**, 43.

Brand, J.C.D. and Speakman, J.C. (1961), *Molecular Structure. The Physical Approach.* Arnold, London.

Brown, K.C., Mann, J.C. and Peirce, F.T. (1930), *Shirley Inst. Mem.* **9**, 1.

Brown, R. (1963), in *Cell Differentiation* (G.E. Fogg, ed.) Academic Press, New York, p.1.

Brown, R.M., Franke, W.W., Kleinig, H., Falk, H. and Sitte, P. (1970), *J. Cell Biol.* **45**, 246.

Brown, R.M., Herth, W., Franke, W.W. and Romanovicz, D. (1973), in *Biogenesis of plant cell wall polysaccharides* (F. Loweus, ed.) Academic Press, U.S.A.

Browning, B.L. (ed.) (1963), *The Chemistry of Wood.* Interscience, New York.

Brummond, D.O. and Gibbons, A.P. (1965), *Biochem. Zeit.* **342**, 308.

Buer, F. (1964), *Flora (Jena)* **154**, 349.

Burg, S.P., Apelbaum, A., Eisinger, W. and Kang, B.G. (1971), *Hort. Sci.* **6**, 359.

Burström, H.G. (1942), *Ann. Agric. Coll. Sweden* **10**, 113.

Burström, H.G. (1961), *Encyclopedia of Plant Physiology* **14**, 285.

Burström, H.G. (1964), *Physiol. Plantarum* **17**, 207.

Burström, H.G., Uhrström, I. and Wurscher, R. (1967), *Physiol Plantarum* **20**, 213.

Burton, K. and Krebs, H. (1953), *Biochem. J.* **54**, 94.

Buston, H. (1935), *Biochem. J.* **29**, 196.

Carlström, D. (1957), *J. Biophys. Biochem. Cytol* **3**, 669.

Carman, P.C. (1956), *Flow of gases through porous media*. Butterworth, London.

Casperson, C. (1962), *Rep. Div. Forest Prod. C.S.I.R.O. Aust.* 1962/63.

Castle, E.S. (1936), *J. Cell, Comp. Physiol.* **7**, 445

Castle, E.S. (1937), *J. Cell. and Comp. Physiol.* **8**, 493.

Castle, E.S. (1942), *Amer. J. Bot.* **29**, 664.

Castle, E.S. (1955), *Proc. Nat. Acad. Sci. U.S.A.* **41**, 197.

Caulfield, D.F. (1971), *Textile Res. J.* **41**, 267.

Cave, I.D. (1966), *Forest Products J.* **16**, 37.

Cave, I.D. (1968), *Wood Sci. and Technol.* **2**, 268. (1969), *Idem.* **3**, 40.

Cave, I.D. (1972 a), *Wood Sci. and Technol.* **6**, 96.

Cave, I.D. (1972 b), *Wood Sci. and Technol.* **6**, 157.

Chafe, S.C. and Wardrop, A.B. (1970), *Planta* **92**, 13.

Chafe, S.C. and Wardrop, A.B. (1972), *Planta* **107**, 269.

Chou, S. (1971), *Degradation of wood as affected by preservatives*. Ph.D. Thesis, Leeds.

Chou, S. and Preston, R.D. (1973), *Holzforschung. In press.*

Chrispeels, M.J. and Sadava, D. (1969), *Science* **165**, 299.

Christensen, T. (1962), *Botanik* **2**, no. 2. (eds. Böcher, Lange and Sorensen) Copenhagen.

Clark, S.H. (1930), *Forestry* **4**, 93.

Cleland, R. (1959), *Physiol. Plantarum* **12**, 809.

Cleland, R. (1967), *Plant Physiol.* **42**, 271, 1165.

Cleland, R. (1967), *Planta* **74**, 197.

Cleland, R. (1968), *Plant Physiol.* **43**, 1625.

Cleland, R. (1968), *Science* **160**, 192.

Cleland, R. (1971), *Plant Physiol.* **47**, 805.

Cleland, R. and Haughton, P.M. (1971), *Plant Physiol.* **47**, 812.

Cleland, R. and Karlsnes, A.M. **(1967)**, *Plant Physiol.* **42**, 669.

Cleland, R., Thompson, W.F., Haughton, P.M. and Rayle, D.L. (1972), in *Plant Growth Substances 1970*, (D.J. Carr, ed.) Springer, Berlin, p. 2.

Cockrell, R.A. (1943), *Trans. Amer. Soc. Mech. Eng.* **65**, 1.

Cockrell, R.A. (1947), *Trans. Amer. Soc. Mech. Eng.* **69**, 931.

Colvin, J. Ross (1961), *Can. J. Biochem. Physiol.* **39**, 1921.

Colvin, J. Ross (1963), *J. Cell Biol.* **17**, 105.

Colvin, J. Ross (1964), in *Formation of Wood in Forest Trees* (M.H. Zimmerman, ed.) Academic Press, New York.

Colvin, J. Ross (1972), *Critical Reviews in Macromolecular Science* **1**, 47. Chemical Rubber Co.

Colvin, J.R., Bayley, S.T. and Beer, M. (1957), *Biochem. Biophys. Acta* **23**, 652.

Comstock, G.L. and Côté, W.A. (1968), *Wood Sci. and Technol.* **2**, 279.

Cook, G.M., Laico, M.T. and Eylar, E.H. (1965), *Proc. Nat. Acad. Sci. U.S.* **54**, 247.

Correns, C. (1893), *Ber. dtsch. bot. Ges.* **11**, 410.

Correns, C. (1894), *Ber. dtsch. bot. Ges.* **12**, 355.

Côté, W.A. (ed.) (1965), *Cellular Ultrastructure of Woody Plants.* Syracuse U.P., Syracuse, U.S.A.

Côté, W.A., Day, A.C. and Timell, T.E. (1968), *Wood Sci. and Technol.* **2**, 13.

Courtney, J.S., Morré, D.J. and Key, J.L. (1967), *Plant Physiol.* **42**, 434.

Courtois, J.E. and Le Dizet, P. (1970), *Bull. Soc. Chim. Biol.* **52**, 15.

Cowling, E.B. and Stamm, A.J. (1963), *J. Polymer Sci.* **c**, 243.

Cowdery, D.R. and Preston, R.D. (1966), *Proc. Roy. Soc.* **B166**, 245.

Cox, G.C. (1971), *The structure and development of cells with thickened primary walls.* Ph.D. Thesis, Oxford.

Cox, G.C. and Juniper, B. (1973), *J. Microsc.* **97**, 343.

Cronquist, A. (1960), *Bot. Revs.* **26**, 437.

Cronshaw, J. (1957), *A reinvestigation of wall structure and growth in the green alga Valonia* Ph.D. Thesis, Leeds.

Cronshaw, J. and Bouck, G.B. (1965), *J. Cell Biol.* **24**, 415.

Cronshaw, J. and Preston, R.D. (1958), *Proc. Roy. Soc.* **B146**, 37.

Cronshaw, J., Davies, G.W. and Wardrop, A.B. (1960), *Holzforsch.* **15**, 75.

Cronshaw, J., Myers, A and Preston, R.D. (1958), *Biochem. Biophys. Acta* **27**, 89.

Crook, E.M. and Johnstone, I.R. (1962), *Biochem. J.* **83**, 325.

Dadswell, H.E. and Nicholls, J.W.P. (1959), C.S.I.R.O. *Austr. Div. For. Prod. Tech. Pap.* No. 28.

Dadswell, H.E. and Wardrop, A.B. (1959), *J. Austr. Pulp and Paper Industr. Feder. Assoc.* **12**, 129.

Dadswell, H.E., Watson, A.J. and Nicholls, J.W.P. (1959), *Tappi* **42**, 521.

Darmon, S.E. and Rudall, K.M. (1950), *Disc. Faraday Soc.* **9**, 25.

Das, N.K., Patau, K. and Skoog, F. (1956), *Physiol. Plantarum* **9**, 640.

Dashek, W.V. (1970), *Plant Physiol.* **46**, 831.

Dashek, W.V. and Rosen, W.G. (1966), *Protoplasma* **61**, 192.

Datko, A.H. and Machlachlan, G.A. (1970), *Can, J. Bot.* **48**, 1165.

Dawes, C.G. (1965), *J. Phycol.* **1**, 121.

Dennis, D.T. and Preston, R.D. (1961), *Nature, London.* **191**, 667.

Diehl, G.M., Gorter, C.J., Iterson, G. van and Kleinhooute, A. (1939), *Rec. Trav. Bot. Neerl.* **6**, 709.

Dinwoodie, J.M. (1966), *Nature, London.* **212**, 515.

Dinwoodie, J.M. (1968), *J. Inst. Wood. Sci.* **4**, 37.

Dinwoodie, J.M. (1972), *Wood Sci. and Technol.*

Donnan, F.G. and Rose, R.S. (1950), *Can. J. Res.* **B28**, 105.

Dougall, D.K. and Shimbayashi, K. (1960), *Plant Physiol.* **33**, 396.

Drawert, H. and Mix, M. (1962), *Planta* **58**, 448.

Drummond, D.W., Hirst, E.L. and Percival, E. (1958), *Chem. and Indust.* 1088.

Dweltz, N.E. (1960), *Biochem. Biophys. Acta* **44**, 416.

Dweltz, N.E. (1961), *Biochem. Biophys. Acta* **51**, 283.

Echols, R.M. (1955), *Tropical Woods*, No. 102, 11.

Edelman, J. and Hall, M.A. (1964), *Nature, London.* **201**, 296.

Ellefsen, Ø., Kringstad, K. and Tønnesen, B.A. (1963), *Encyclopedia of X-rays and Gamma Rays* (Clark, G.L. ed.) Reinhold, New York.

Emerton, H. and Goldsmith, V. (1956), *Aus. J. Bot.* **10**, 108.

Ernst, A. (1903), *Beih. Bot. Zbl.* **13**, 115.

(1904), *Idem.* **16**, 199.

Escherich, W. (1956), *Protoplasma* **47**, 487.

Etherton, B. (1970), *Plant Physiol.* **45**, 527.

Falk, S., Hertz, C.H. and Virgin, H.I. (1958), *Physiol. Plantarum* **11**, 802.

Fan, D.F. and Machlachlan, G.A. (1966), *Can. J. Bot.* **44**, 1025.

(1967), *Idem.* **45**, 1837.

Plant Physiol. **12**, 35.

Feingold, D.S., Neufeld, E.F. and Hassid, W.Z. (1958), *J. Biol. Chem.* **233**, 783.

Feldmann, J. (1946), *C.R. Acad. Sci. Paris* **222**, 753.

(1954), *Huitième Congres International de Botanique, Paris,* Rapports et Communications, Section **17**, 97.

Fischer, D.G. and Mann, J. (1960), *J. Polymer Sci.* **C12**, 189.

Fischer, F.G. and Dörfel, H. (1955), *Z. Physiol. Chem.* **302**, 186.

Flint, E.A. (1950), *Biol. Rev.* **25**, 414.

Flowers, H.M., Batra, K.K., Kemp, J. and Hassid, W.Z. (1968), *Plant Physiol.* **43**, 1703.

Foster, D.H. and Wardrop, A.B. (1951), *Austr. J. Sci. Research Ser. A* **4**, 412.

Foster, R.C. (1962), *Cell wall structure and growth.* Ph.D. Thesis, Leeds.

Fowkes, L.C. and Pickett-Heaps, J.D. (1971), *J. Phycol.* **7**, 285.

Fowkes, L.C. and Pickett-Heaps, J.D. (1972), *Protoplasma* **74**, 19.

Fox, C.F. (1972), *Sci. Amer.* **226**, 30.

Françon, M. (1961), *Progress in Microscopy.* Pergamon Press, London.

Frei, E. and Preston, R.D. (1960), *Unpublished.*

Frei, E. and Preston, R.D. (1961), *Proc. Roy. Soc.*   (a) **B154**, 70; (b) **B155**, 55.

Frei, E. and Preston, R.D. (1961c), *Nature, London.* **192**, 939.

Frei, E. and Preston, R.D. (1962), *Nature, London.* **196**, 130.

Frei, E. and Preston, R.D. (1963), *Proc. Leeds Phil. Soc. Sci. Sec.* **9**, 101.

Frei, E. and Preston, R.D. (1964a), *Proc. Roy. Soc.* **B160**, 293.

Frei, E. and Preston, R.D. (1964b), *Proc. Roy. Soc.* **B160**, 314.

Frei, E. and Preston, R.D. (1968), *Proc. Roy. Soc.* **B169**, 127.

Frei, E., Preston, R.D. and Ripley, G.W. (1957), *J. Exp. Bot.* **8**, 139.

Frenzel, P. (1929), *Planta* **8**, 642.

Freudenberg, K. (1964), *Holzforsch.* **18**, 3.

Frey, R. (1950), *Ber. schweiz. bot. Ges.* **60**, 199.

Frey-Wyssling, A. (1930), *Z. wiss. Mikr.* **47**, 1.

Frey-Wyssling, A. (1936), *Protoplasma* **26**, 372.

Frey-Wyssling, A. (1940), *Holz als Roh- u. Werkstoff* **3**, 349.

Frey-Wyssling, A. (1942), *Jahrb. wiss. Bot.* **90**, 705.

Frey-Wyssling, A. (1948a), *Submicroscopic morphology of protoplasm and its derivatives.* Elsevier, Amsterdam.

Frey-Wyssling, A. (1948b), *Growth Symposium* **12**, 151.

Frey-Wyssling, A. (1950), *Ann. Rev. Plant Physiol.* **1**, 169.

Frey Wyssling, A. (1953a), *Submicroscopic Morphology of Protoplasm.* Elsevier, Amsterdam.

Frey-Wyssling, A. (1953b), *Holz als Roh- u. Werkstoff* **11**, 283.

Frey-Wyssling, A. (1954), *Science* **119**, 80.

Frey-Wyssling, A. (1955). *Biochem. Biophys. Acta* **18**, 166.

Frey-Wyssling, A. (1959), *Die Pflanzliche Zellwand.* Springer Verlag, Berlin.

Frey-Wyssling, A. and Mitrakos, K. (1959), *J. Ultrastr. Res.* **3**, 228.

Frey-Wyssling, A. and Mühlethaler, K. (1949), *Schweiz. Bauzeitung* **67**, No. 3. (1950a), *Schweiz. Landwirt. Monatshefte* **60**, 1. (1950b), *Vjschr. Naturforsch. Ges. Zürich* **95**, 45.

Frey-Wyssling, A. and Mühlethaler, K. (1962), *Die Makromol. Chemie* **62**, 25.

Frey-Wyssling, A., López-Sáez, J.F. and Mühlethaler, K. (1964), *J. Ultrastr. Res.* **10**, 422.

Frey-Wyssling, A., Mühlethaler, K. and Bosshard, H.H. (1956), *Planta* **47**, 115.

Frey-Wyssling, A., Mühlethaler, K. and Muggli, R. (1966), *Holz als Roh- u. Werkstoff* **24**, 443.

Frey-Wyssling, A., Mühlethaler, K. and Wyckoff, R.W.G. (1948), *Experientia* **6**, 12, 475.

Frey-Wyssling, A. and Stecher, H. (1952), *Experientia* **7**, 410

Frey-Wyssling, A. and Stussi, F. (1948), *Schweiz. Zeit. Forstwesen* **3**, 1.

Fritsch, F.E. (1956), *The structure and reproduction of the algae.* Vols. I & II. Cambridge University Press.

Fuchs, Y. and Lieberman, H. (1968), *Plant Physiol.* **43**, 2029.

Galston, A.N., Baker, R.S. and King, J.W. (1953), *Physiol. Plantarum* **6**, 863

Garland, H. (1939), *Ann. Mo. Bot. Gard.* **26**, 95.

Gawlick, S.R. and Millington, W.F. (1969), *Amer. J. Bot.* **56**, 1084.

Gillis, P.P. (*paper to be published*)

Girbardt, M. (1958), *Arch. Mikrobiol.* **28**, 255.

Glaser, L. (1958), *J. Biol. Chem.* **232**, 627.

Glasziou, K.T. (1959), *Physiol. Plantarum* **12**, 670.

Gorham, P.R. and Colvin, J.R. (1957), *Exp. Cell Res.* **13**, 187.

Goring, D.A.T. and Timell, T, (1962), *Tappi* **45**, 454.

Gotelli, I.B. and Cleland, R. (1968), *Amer. J. Bot.* **55**, 907.

Green, P.B. (1954), *Amer. J. Bot.* , 403.

Green, P.B. (1958), *J. Biophys. Biochem. Cytol.* **4**, 505.

Green, P.B. (1959), *Biochem. Biophys. Acta* **36**, 536.

Green, P.B. (1960), *Amer. J. Bot.* **47**, 476.

Green, P.B. (1963), in *Cytodifferentiation and Macromolecular Synthesis.* (M. Locke, ed.) Academic Press, New York, p. 203.

Green, P.B. (1970), in *Plant Growth Substances* (D.J. Carr, ed.) Springer, p. 9.

Green, P.B. (1971), *Plant Physiol.* **47**, 439.

Green, P.B. and Chapman, G.B. (1955), *Amer. J. Bot.* **42**, 685.

Gross, S.T., Clarke, G.L. and Ritter, G.J. (1939), *Paper Trade J.* **109**, 303.

Grossman, P.U.A. and Wold, M.B. (1971), *Wood Sci. and Tech.* **5**, 147.

Grout, B.W.W., Willison, J.H.M. and Cocking, E.C. (1973), *Bioenergetics* **4**, 311.

Haleem, M.A. (1971), *X-ray diffraction studies on the structure of alpha-chitin and β-D-1, 3 xylan.* Ph.D. Thesis, Leeds.

Hall, C.E. (1950), *J. Appl. Phys.* **21**, 61.

Hall, M.A. and Ordin, L. (1967), *Physiol. Plantarum* **20**, 624.

Hämmerling, J. (1944), *Arch. Protist.* **97**, 7.

Hanack, M. (1965), *Conformation Theory.* Academic Press, New York and London.

Hanic, L.A. and Craigie, J.C. (1969), *J. Phycol.* **5**, 89.

Harada, H. (1973), *Holzforschung* **27**, 12.

Harada, H. and Côté, W. (1967), *Holzforsch.* **21**, 82.

Harada, H. and Wardrop, A.B. (1960), *J. Japan Wood Res. Soc.* **6**, 34.

Harada, H., Miyazaki, Y. and Wakashima, T. (1958), *Bull. Govt. For. Exp. Stn. Tokyo* No. 104.

Harlow, W.M. (1932), *Amer. J. Bot.* **19**, 729.

Harris, J.M. and Meylan, B.A. (1965), *Holzforsch.* **19**, 144.

Harris, P.J. and Northcote, D.H. (1971), *Biochem. Biophys. Acta* **237**, 56.

Hart, C.A. and Thomas, R.J. (1967), *For. Prod. J.* **17** (11), 61.

Hartler, N. (1969), *Norsk Skogind.* **23**, 114.

Hartshorne, N.H. and Stuart, A. (1970), *Crystals and the polarizing microscope,* 4th ed. Arnold, London.

Hassel, O. and Ottar, B. (1947), *Acta Chem. Scand.* **1**, 929.

Haug, A., Larsen, B. and Smidsrød, O. (1966), *Acta Chem. Scand.* **20**, 183. (1967), *Idem.* **21**, 691, 768.

Haughton, P.M. and Sellen, D.B. (1969), *J. Exp. Bot.* **20**, 516.

Haughton, P.M., Sellen, D.B. and Preston, R.D. (1968), *J. Exp. Bot.* **19**, 1.

Haworth, W.N. (1929), *The Constitution of the Sugars,* London.

Hearle, J.W.S. (1958), *J. Text. Inst.* **49**, T389.

Hearle, J.W.S. (1963), *J. Appl. Polymer Sci.* **7**, 1207.

Heath, I.B. and Greenwood, A.D. (1970), *J. Gen. Microbiol.* **62**, 129.

Heath, I.B., Gay, J.L. and Greenwood, A.D. (1971), *J. Gen. Microbiol.* **65**, 225.

Henglein, F.A. (1958), *Encyclopedia of Plant Physiology* **6**, 452.

Hengstenberg, J. and Mark, H. (1928), *Zeit. Kristallogr.* **69**, 271.

Hepler, P.K. and Newcombe, E.H. (1964), *J. Cell. Biol.* **20**, 529.

Hermans, P.H. (1943), *Kolloid-Zeitsch.* **102**, 169.

Hermans, P.H. (1949), *The physics and chemistry of cellulose fibres.* Elsevier, New York.

Hermans, P.H. and Weidinger, A.J. (1948), *J. Appl. Phys.* **19**, 491.

Hermans, P.H. and Weidinger, A.J. (1949), *J. Polymer Sci.* **4**, 135.

Herth, W., Franke, W.W., Stadler, J., Bittiger, H., Keilich, G. and Brown, R.M. (1972), *Planta* **105**, 79.

Hess, K. (1928), *Die Chemie der Zellulose und ihrer Begleiter.* Akademische Verlagsgesellschaft M.B.H., Leipzig.

Heyn, A.N.J. (1935), *Protoplasma* **24**, 372.
Proc. Kon. Akad. Wetensch. Amst. **37**, 132.

Heyn, A.N.J. (1936), *Protoplasma* **25**, 372.

Heyn, A.N.J. (1940), *Bot. Rev.* **6**, 515.

Heyn, A.N.J. (1953), *Nature, London* **173**, 100.

Heyn, A.N.J. (1958), *J. Appl. Phys.* **26**, 519.

Heyn, A.N.J. (1966), *J. Cell Biol.* **29**, 181.

Heyn, A.N.J. (1969), *J. Ultrastruct. Res.* **26**, 52.

Hill, G.J.C. and Machlis, L. (1968), *J. Phycol.* **4**, 261.

Hirst, E.L., Jones, J.K.N. and Jones, W.O. (1939), *J. Chem. Soc.* 1880.

Hodge, A.J. and Wardrop, A.B. (1950), *Austr. J. Sci. Res.* **B3**, 265.

Honjo, G. and Watanabe, M. (1958), *Nature, London* **181**, 326.

Hough, L., Jones, J.K.N. and Wadman, W.H. (1952), *J. Chem. Soc.* 3392.

Houwink, A.C. and Roelofsen, P.A. (1954), *Acta Bot. Neerl.* **3**, 385.

Howsmon, J.A. (1949), *Textile Res. J.* **19**, 153.

Hunsley, D. and Burnett, J.H. (1968), *Nature, London* **218**, 462.

Iriki, Y. and Miwa, T. (1960), *Nature, London* **185**, 178.

Iriki, Y., Susuki, T., Nisizawa, K. and Miwa, T. (1960), *Nature, London* **187**, 82.

Israel, H.W., Salpeter, M.M. and Steward, F.C. (1968), *J. Cell Biol.* **39**, 698.

Iterson, G. van Jr. (1937), *Protoplasma* **27**, 190.

Iterson, G. van and Meeuse, A.D.J. (1941), *Kon. Ned. Akad. Wet.* Amsterdam **44**, nos. 7 & 8.

Jacobs, M.R. (1938), *Commonwealth For. Bur. Aust. Bull.* Canberra, **22**. (1945), *Idem.* **28**.

Jansen, E.F., Jang, R. and Bonner, J. (1960), *Plant Physiol.* **35**, 567.

Jansen, E.F., Jang, R., Albersheim, P. and Bonner, J. (1960), *Plant Physiol.* **35**, 87.

Jermyn, M.A. and Isherwood, F.A. (1956), *Biochem, J.* **64**, 123.

Johnson, D.T. (1968), *J. Roy. Micros. Soc.* **88**, 39.

Jones, D.W. (1968), *Biopolymers* **6**, 771.

Jones, F.W. (1938), *Proc. Roy. Soc.* **A116**, 16.

Jones, J.K.N. (1950), *J. Chem. Soc.* 3292.

Jutte, S.M. (1956), *Holzforsch.* **10**, 33.

Jutte, S.M. and Levy, J.F. (1971), *Acta Bot. Néerl.* **20**(5), 453.

Jutte, S.M. and Spit, B.J. (1963), *Holzforsch.* **17**, 168.

Kamiya, N., Tazawa, M. and Takata, T. (1963), *Protoplasma* **57**, 501.

Kang, B.G. and Burg, S.P. (1971), *Proc. Nat. Acad. Sci. U.S.A.* **68**, 1730.

Katz, M. and Ordin, L. (1967), *Biochem. Biophys. Acta* **141**, 126.

Kauman, W.G. (1966), *Holz als Roh- u. Werkstoff* **24**(11), 551.

Kauss, H. and Swanson, A.L. (1969), *Zeit. Naturforsch.* **24**, 28.

Kauss, H., Swanson, A.L., Arnold, R. and Odzuck, W. (1969), *Biochem. Biophys. Acta* **192**, 55.

Kay, D.H. (Ed.) (1965), *Techniques for Electron Microscopy.* Blackwell, Oxford.

Keegstra, K., Talmadge, K.W., Bauer, W.D. and Albersheim, P. (1973), *Plant Physiol.* **51**, 188.

Keith, C.T. and Côté, W.A. (1968), *For. Prod. J.* **18**, 67.

Kellog, R.M. and Wangaard, F.F. (1969), *Wood and Fiber* **1**, 180.

Kelsey, K.E. (1963), *C.S.I.R.O. Aust. Div. For. Prod. Technol. Pap. No.* 28.

Kelso, W.C., Gertjejansen, R.O. and Hossfeld, R.L. (1963), *Univ. Minn. Agr. Exp. Sta. Tech. Bull.* No. 242.

Kerr, T. and Bailey, I.W. (1934), *J. Arnold Arbor* **15**, 327.

Kessler, G. (1958), *Ber. schweiz. bot. Ges.* **68**, 5.

Key, J.L. (1969), *Ann. Rev. Plant Physiol.* **20**, 449.

King, N.J. and Bayley, S.T. (1965), *J. Exp. Bot.* **16**, 794.

Kisser, J. and Frentzel, H. (1950), *Schr. Reihe ost. Gez. Holsforsch.* **2**, 7.

Kisser, J. and Steininger, A. (1952), *Holz als Roh- u. Werkstoff* **10**, 415.

Kivilaan, A., Beaman, T.C. and Bandurski, R.S. (1961), *Plant Physiol.* **36**, 605.

Kobayashi, K. and Utsumi, N. (1951), *Quoted by* Harada, H. (1953), *J. Jap. For. Sci.* **35**, 393.

Koehler, A. (1931), *Trans. Amer. Soc. Mech. Engns.* **53**, 17.

Kommick, H. and Wohlfarth-Botterman, K.E. (1965), *Zellforsch. Mikroskop. Anat. Abt. Histochem.* **66**, 434.

Koshijima, T. and Timell, T.E. (1964), *Unpublished.*

Kramer, D. (1970), *Zeit. f. Naturforsch.* **25b**, 1017.

Kratzl, K. (1965), *Cellular Ultrastructure of Woody Plants* (W.H. Côté, ed.) Syracuse U.P., Syracuse, U.S.A., p. 157.

Kreger, D.R. (1957), *Nature, London* **180**, 914.

Kreger, D.R. (1960), *Koninkl. Ned. Akad. Wetenschep. Proc.* Ser. **C63**, 613. 623.

Kreger, D.R. (1962), in *Physiology & Biochemistry of the Algae* (R.A. Lewin, ed.) Academic Press. London and New York p. 315.

Krishna, Murti, G.S.R. (1965), *Phykos.* **4**, 17.

Kremers, R.E. and Reeder, B.J. (1963), in *The chemistry of Wood* (Browning, B.L. ed.) Interscience, New York.

Kübler, H. (1959), *Holz als Roh- u. Werkstoff* **17**, 1, 44, 77.

Küster, E. (1933), *Ber. dtsch. bot. Ges.* **51**, 526.

Lamport, D.T.A. (1962), *Fed. Proc.* **21**, 398.

Lamport, D.T.A. (1965), in *Advances in Botanical Research* (R.D. Preston, ed.) Academic Press, London and New York.

Lamport, D.T.A. (1967a), *Fed. Proc.* **26**, 1965.

Lamport, D.T.A. (1967b), *Nature, London* **216**, 1322.

Lamport, D.T.A. (1969), *Biochem.* **8**, 1155.

Lamport, D.T.A. (1970), *Ann. Rev. Plant Physiol.* **21**, 235.

Lamport, D.T.A. (1971), *Abstract, Meeting of Amer. Soc. Biochem.*

Lamport, D.T.A. and Northcote, D.H. (1960), *Nature, London* **188**, 665; *Biochem. J.* **76**, 52 P.

Lamport, D.T.A., Roerig, S. and Katona, L. (1972), *J. Biol. Chem.*

Lange, N.J. (1963), *Amer, J. Bot.* **50**, 280.

Lange, P. (1954), *Svensk. Papperstidn,* **57**, 501.

Lange, N.J. (1963), *Amer. J. Bot.* **50**, 280.

Lange, P. and Kjaer, A. (1957), *Norsk. Skogindustri* **11**, 425.

Larson, P.R. (1966), *For. Prod. J.* **16**, 39.

Laue, M. von (1912), *Sitz. math. phys. Klasse bayer. Akad. Wiss.* p. 303; *Ann Physik* **41**, 971 (1913).

Ledbetter, M.C. and Porter, K. (1963), *J. Cell Biol.* **19**, 239.

Lee, S., Kivilaan, A. and Bandurski, R.S. (1967), *Plant Physiol.* **42**, 968.

Leitgeb, H. (1888), *S.B. Akad. Wiss. Wien* **96**, 13.

Lennarz, W.J. and Scher, M.G. (1973), *J. Bioenergetics* **4**, 239.

Lewis, F.T. (1935), *Amer. J. Bot.* **22**, 741.

Liang, C.Y. and Marchessault, R.H. (1959), *J. Polymer Sci.* **37**, 385.

Licse, W. (1951), *Ber. dtsch. bot. Ges.* **64**, 31.

Liese, W. (1956), *Proc. Int. Conf. Electron Microscopy, London* 1954, p. 550.

Liese, W. (1957), *Naturwiss.* **7**, 240.

Liese, W. and Bauch, J. (1964), *Naturwiss.* **51**, 516.

Liese, W. and Bauch, J. (1967), *Wood Sci. and Technol.* **1**, 1.

Liese, W. and Hartmann-Fahnenbrock, M. (1953), *Biochem. Biophys. Acta* **11**, 190.

Limaye, V.J., Roelofsen, P.A. and Spit, B.J. (1962), *Acta Botan. Néerland.* **11**, 225.

Linskens, H.P. (1964), *Proc. Intnl. Symp. Pollen Physiol. and Fertilisn.* North Holand, Amsterdam.

Lochner, J.P.A. (1949), *J. Text. Inst.* **40**, T220.

Lockhart, J.A. (1965), in *Plant Biochemistry* (J. Bonner and J.E. Warne, eds.) Academic Press, New York, p. 826.

467

Lockhart, J.A. (1967), *Pl. Physiol. Lancaster* **42**, 1545.
Lotfy, M., El-Osta, M., Kellogg, R.M., and Foschi, R.O. (1974), *Wood and Fiber. In the press.*
Lotmar, W.E. and Picken, L.E.R. (1950), *Experientia* VI, 58.
Loewus, F. (1969), *Ann. N.Y. Acad. Sci.* **165**, 577.
Love, J. and Percival, E. (1964), *J. Chem. Soc.* 3346.
Lucas, H.J. and Stewart, W.T. (1940), *J. Amer. Chem. Soc.* **62**, 1792.
McManus, M.A. and Roth, L.E. (1965), *J. Cell Biol.* **25**, 305.
Macchi, E. and Palma, A. (1969), *Die Makromol. Chemie* **123**, 286.
Macchi, E., Marx-Figini, M. and Fischer, E.W. (1968), *Die Makromol. Chemie* **120**, 235.
Machlachlan, G.A. and Duda, C.T. (1965), *Biochem. Biophys. Acta* **97**, 288.
Machlachlan, G.A. and Perrault, J. (1964), *Nature, London* **204**, 81.
Machlachlan, G.A. and Young, M. (1962), *Nature, London* **195**, 1319.
Mackie, I.M. and Percival, E. (1959), *J. Chem. Soc.* 1151.
Mackie, I.M. and Percival, E. (1961), *J. Chem. Soc.* 3010.
Mackie, W. (1969), *Carbohyd. Res.* **9**, 247.
Mackie, W. and Preston, R.D. (1968), *Planta* **79**, 249.
Mackie, W. and Sellen, D.B. (1969), *Polymer* **10**, 621.
Mackie, W. and Sellen, D.B. (1971), *Biopolymers* **10**, 1.
Majumdar, G.P. and Preston, R.D. (1941), *Proc. Roy. Soc.* **B130**, 20.
Manjunat, B.R. and Peacock, N. (1969), *Text. Res. J.* **39**, 70.
Manley, R. St. John (1964), *Nature, London* **204**, 1155.
Manley, R. St. John (1964), *J. Polymer Sci.* **1**, 1875.
Manley, R. St. John (1971), *J. Polymer Sci.* Part A-2, **9**, 1025.
Manocha, M.S. and Shaw, M. (1964), *Nature, London* **203**, 1402.
Marchant, R. and Robards, A.W. (1968), *Ann. Bot.* **32**, 457.
Marchessault, R.H. and Liang, C.Y. (1962), *J. Polymer Sci.* **59**, 357.
Margere, C. and Lenoel, P.D. (1961), *Biochem. Biophys. Acta* **4**, 275.
Mark, R.E. (1965), in *Cellular Ultrastructure of Woody Plants* (W.A. Côté, ed.) p. 493, Syracuse, U.P., Syracuse, U.S.A.
Mark, R.E. (1967), *Cell Wall Mechanics of Tracheids.* Yale U.P., New Haven and London.
Mark, R.E., Kaloni, P.N., Tang, R.C. and Gillis, P.P. (1969), *Science* **164**, 72.
Marx-Figini, M. (1963), *Die Makromol. Chemie* **68**, 227.
  (1964), *Das Papier* **18**, 546.
  (1966), *Nature, London* **210**, 754.
  (1967), *J. Polymer Sci.* **C16**, 1947.
  (1969), *Ibid. C. Polymer Symp.* No. 28.
Marx-Figini, M. and Schultz, G.V. (1963), *Die Makromol. Chemie* **62**, 49.
  (1966), *Biochem. Biophys. Acta* **112**, 81.
Masuda, Y., Yamamoto, R. and Tanimoto, E. (1972), in *Plant Growth Substances 1970* (D.J. Carr, ed.) Springer, Berlin. p. 17.
Matchett, N.H. and Nance, J.F. (1962), *Amer. J. Bot.* **49**, 311.
Matile, P., Moor, H. and Mühlethaler, K. (1967), *Arch. f. Mikrobiol.* **58**, 201.

Matzke, E.B. (1956), *Proc. Nat. Acad. Sci. U.S.A.* **42**, 26.

Matzke, E.B. and Duffy, R.M. (1955), *Amer. J. Bot.* **42**, 937.

Matzke, E.B. and Duffy, R.M. (1956), *Amer. J. Bot.* **43**, 205.

Meier, H. (1955), *Holz als Roh- u. Werkstoff* **9**, 323.

Meier, H. (1956), *Proc. Stockholm Conf. Electron Microsc.* p. 298.

Meier, H. (1957), *Holzforsch.* **11**, 41.

Meier, H. (1958), *Biochem. Biophys. Acta* **28**, 229.

Meier, H. (1961), *J. Polymer Sci.* **51**, 11;
   in *Formation of Wood in Forest Trees* (1964), (ed. M.H. Zimmerman) p. 137, Academic Press, New York.

Meier, H. (1962), *Acta Chem. Scand.* **16**, 2275.

Meredith, R. (1946), *J. Text. Ind.* **27**, T 205.

Meretz, W. (1962), *Beitrage z. Biol. d. Pflanzen* **37**, 147.

Meyer, K.H. (1942), *High polymers. Vol. IV. Natural and Synthetic High Polymers.* Interscience, New York.

Meyer, K.H. and Mark, H. (1928), *Ber. dtsch. chem. Ges.* **61B**, 593.

Meyer, K.H. and Mark, H. (1929), *Z. physik. Chem.* **B2**, 115.

Meyer, K.H. and Mark, H. (1930), *Der Aufbau der hochpolymeren organischen Naturstoffe.* Akad. Verlagsgesellschaft, Leipzig.

Meyer, K.H. and Misch, H. (1936), *Helv. Chem. Acta* **20**, 232.

Meyer, K.H. and Pankow, G.N. (1935), *Helv. Chem. Acta* **18**, 589.

Meylan, B.A. (1966), *For. Products J.* **17**, 51.

Meylan, B.A. (1968), *For. Prod. J.* **18**, 75.

Meylan, B.A. and Probine, M.C. (1969), *For. Products J.* **19**, 30.

Middlebrook, M.J. and Preston, R.D. (1952a), *Biochem. Biophys. Acta* **9**, 32.
   (1952b), *Ibid.* **9**, 115.

Miller, J.H. (1961), *Amer. J. Bot.* **48**, 816.

Mirande, R. (1913), *Ann. Sci. Nat.* (9 sér.) **18**, 147.
   C.R. Hebd. Séanc. Acad. Sci. Paris **156**, 475.

Miwa, T., Iriki, Y. and Susaki, T. (1961), *Collegues Internationaux du centre de la Recherche Scientifique* No. 103, Paris

Mix, M. (1966), *Arch. Mikrobiol.* **55**, 116.
   (1968), *Ber. dtsch. bot. Ges.* **80**, 715.
   (1972), *Arch. Mikrobiol.* **81**, 197.

Mollenhawer, H.H. and Whaley, H.G. (1962), *5th Intl. Congr. Electron Microscopy Philadelphia* Academic Press, New York.

Mollenhawer, Whaley, W.G. and Leech, J.H. (1961), *J. Ultrastr. Res.* **5**, 193.

Moor, H. (1964), *Z. Zellforsch.* **62**, 546.

Moor, H. and Mühlethaler, K. (1963), *J. Cell Biol.* **17**, 609.

Moore, R.T. and McAlear, J.H. (1961), *Mycologia* **53**, 144.

Morré, D.J. (1970), *Plant Physiol.* **45**, 791.

Moscaleva, W.E. (1957), *Acad. Sci. U.S.S.R. Inst. For.* **6**, 42.

Moss, B.L. (1948), *Ann. Bot.* **12**, 267.

Muggli, R., Elias, H.G. and Mühlethaler, K. (1969), *Makromol. Chem.* **121**, 290.

Muggli, R., Mühlethaler, K. and Elias, H.G. (1969), *Angew. Chem. Inst. Ed.* **8**, 384.

Mühlethaler, K. (1949), *Biochem. Biophys. Acta* **3**, 527.

Mühlethaler, K. (1950a), *Biochem. Biophys. Acta* **5**, 1; (1950b), *Ber. schweiz. bot. Ges.* **59**, 614.

Mühlethaler, K. (1960), *Beih. Z. Schweiz. Forstres.* **30**, 55.

Mühlethaler, K. (1967), *Ann. Rev. Plant Physiol.* **18**, 1.

Murmanis, L. and Sachs, J.B. (1969), *Amer. Wood Pres. Assoc. Proc.* **64**, 70.

Murphey, W.N. (1963), *For. Products J.* **13**, 151.

Myers, A. and Preston, R.D. (1959), *Proc. Roy. Soc.* **150**, 447, 456.

Myers, A., Preston, R.D. and Ripley, G.W. (1955), *Proc. Roy. Soc.* **144**, 450.

Nägeli, C. von (1844), *Z. wiss. Bot. Schleiden Nägeli* **1**, 134.

Nägeli, C. von and Schwendener, S. (1877), *Das Mikroskop* 2.Auflage. Leipzig.

Naum, Y.R. (1955), *Bull. Torrey Bot. Club* **82**, 480.

Nečessaný, V. (1957), *Svensk. Papperstidn.* **60**, 10.

Nečessaný, V., Jurášek, L., Sopko, R. and Bobák, M. (1965), *Nature, London* **206**, 639.

Neish, A.C. (1958), *Canad. J. Biochem. Physiol.* **36**, 187.

Nelmes, B.J. and Preston, R.D. (1968), *J. Exp. Bot.* **19**, 496.

Nelmes, B.J., Ashworth, D. and Preston, R.D. (1973), *J. Cell Biol.*

Newcomb, E.H. (1951), *Proc. Soc. Exp. Biol. N.Y.* **76**, 504.

Newcomb, E.H. (1969), *Ann. Rev. Plant Physiol.* **20**, 253.

Newcomb, E.H. and Bonnett, H.T. (1965), *J. Cell Biol.* **27**, 575.

Nicolai, M.F.E. (1957), *Nature, London* **180**, 491.

Nicolai, M.F.E. and Frey-Wyssling, A. (1938), *Protoplasma* **30**, 401.

Nicolai, M.F.E. and Preston, R.D. (1952), *Proc. Roy. Soc.* **B140**, 244.

Nicolai, M.F.E. and Preston, R.D. (1959), *Proc. Roy. Soc.* **B151**, 244.

Nieduszynski, I.A. (1969), *Physical studies of cellulose microfibrils.* Ph.D. Thesis, Leeds.

Nieduszynski, I.A. and Atkins E.D.T. (1970), *Biochem. Biophys. Acta* **222**, 109.

Nieduszynski, I.A. and Marchessault, R.H. (1972), *Can. J . Chem.* **50**, 2130.

Nieduszynski, I.A. and Preston, R.D. (1970), *Nature, London* **25**, 273.

Nilsson, S.B., Hertz, C.H. and Falk, S. (1958), *Physiol. Plantarum* **11**, 818.

Northcote, D.H. (1969), *Proc. Roy. Soc. Lond.* **B173**, 21.

Northcote, D.H. and Pickett-Heaps, J.D. (1966), *Biochem. J.* **93**, 159.

Northcote, D.H., Goulding, K.J. and Horne, R.W. (1958), *Biochem. J.* **70**, 391.

Noveas-Ledieu, M., Jiménez-Martínez, A. and Villenueva, J.R. (1967), *J. Gen. Microbiol.* **47**, 237.

Nunn, J.R. and von Holdt, M.M. (1957), *J. Chem. Soc.* 1590.

Nyburg, S.C. (1961), *X-ray analysis of organic structures.* Academic Press, New York and London.

O'Brien, T.P. (1973), *Bot. Revs. In press.*

Olson, A.C. (1964), *Plant Physiol.* **39**, 543.

Olson, A.C., Bonner, J. and Morre, J. (1962), *Abstracts of Contributed Papers to Amer. Chem. Soc. Meeting, Atlantic City,* p. 16c.

Olson, A.C., Bonner, J. and Morré, J. (1965), *Planta* **66**, 126.

Olszewska, M.J., Gabara, B. and Steplewski, Z. (1966), *Protoplasma* **61**, 60.

Oort, A.J.P. (1931), *Proc. Kon. Akad. Wet. Amsterdam* **34**, 564.

Oort, A.J.P. and Roelofsen, P.A. (1932), *Proc. Kon. Akad, Wet. Amsterdam* **38**, 898.

Ordin, L. and Hall. M.A. (1967), *Plant Physiol.* **42**, 473.

Ordin, L. and Hall, M.A. (1968), *Idem.* **43**, 1703.

Osborne, D.J. (1972), in *Plant Growth Substances 1970* (D.J. Carr, ed.) Springer, Berlin, p. 534.

Oster, G. and Pallister, A.W. (eds.) (1956), *Physical Techniques in Biological Research.* Vol. III. Academic Press, London and New York.

Ott, E. and Spurlin, H.M. (1954-55), *Cellulose and Cellulose Derivatives.* Vols. I-III. Interscience, New York.

Overbeck, F. (1934), *Z. Bot.* **27**, 129.

Page, D.H. (1966), *Pulp and Paper Mag. of Canada* **67**, T2.

Page, D.H. (1969), *J. Microsc.* **90**, 137.

Parker, B.C. (1964), *Phycologia* **4**(2), 63.

Parker, B.C., Preston, R.D. and Fogg, G.E. (1963), *Proc. Roy. Soc. Lond.* **B158**, 435.

Parker, K.D. (1969), *Nature, London* **220**, 784.

Pease, D.C. (1964), *Histological Techniques for Electron Microscopy.* Academic Press, New York and London.

Peat, S., Turvey, J.R. and Rees, D.A. (1961), *J. Chem. Soc.* 1590.

Percival, E.G.V. and Chanda, S.K. (1950), *Nature, London* **166**, 787.

Percival, E. and McDowell, R.H. (1967), *Chemistry and Enzymology of Marine Algal Polysaccharides.* Academic Press, London and New York.

Percival, E. and Ross, R. (1948), *Nature, London* **162**, 895.

Petty, J.A. (1970), *Proc. Roy. Soc.* **B175**, 149.

Petty, J.A. (1971), *Holsforsch.* **21**, 24.

Petty, J.A. (1972), *Proc. Roy. Soc.* **B181**, 395.

Petty, J.A. and Preston, R.D. (1969), *Proc. Roy. Soc.* **B172**, 137.

Petty, J.A. and Puritch, G.S. (1970), *Wood Sci and Technol.* **4**, 40.

Phillips, E.W.J. (1933), *Forestry* **7**, 109.

Phillips, E.W.J. (1941), *Emp. For. J.* **20**, 74.

Pickett-Heaps, J.D. (1967), *J. Histochem. and Cytochem.* **15**, 442.

Pickett-Heaps, J.D. (1973), in *Dynamic Aspects of Cell Growth,* (A.W. Robards, ed.) McGraw Hill, London.

Pierce, C. (1953), *J. Phys. Chem.* **57**, 149.

Pierce, C. (1959), *Idem.* **63**, 1076.

Pilet, P.E., *Les Parois Cellulaires.* Doin, Paris.

Pimentel, G.C. and McClellan, A.L. (1960), *The Hydrogen Bond*. Reinhold, New York.

Pinsky, A. and Ordin, L. (1969), *Plant and Cell Physiol.* **10**, 771.

Pockels, H. (1906), *Lehrbuch der Kristalloptik*. Leipzig.

Poincaré, H. (1889-92), Théorie mathematique de la lumiére, Carré, Paris. II, 282.

Pollard, J.K. and Steward, F.C. (1959), *J. Exp. Bot.* **10**, 17.

Preston, R.D. (1931a), *The organisation of the cell wall applied to the elucidation of problems associated with growth* Ph.D. Thesis, Leeds.

Preston, R.D. (1931b), *Proc. Leeds. Phil. Soc. Sci. Sec.* **2**, 185.

Preston, R.D. (1934), *Phil. Trans.* **B224**, 131.

Preston, R.D. (1938), *Proc. Roy. Soc.* **B125**, 372.

Preston, R.D. (1939), *Ann. Bot.* **3**, 507.

Preston, R.D. (1942), *Forestry* **16**, 32.

Preston, R.D. (1946), *Proc. Roy. Soc.* **B133**, 327.

Preston, R.D. (1947), *Proc. Roy. Soc.* **B134**, 202.

Preston, R.D. (1948a), *Biochem. Biophys. Acta* **2**, 115.

Preston, R.D. (1948b), *Biochem. Biophys. Acta* **2**, 370.

Preston, R.D. (1949), *Forestry* **23**, 48.

Preston, R.D. (1951), *The size and shape factor in colloidal systems. Disc. Faraday Soc.* No. 11, 165.

Preston, R.D. (1952), *The Molecular Architecture of Plant Cell Walls*, Chapman and Hall, London.

Preston, R.D. (1955a), in *Encyclopedia of Plant Physiology* (W. Ruhland, ed.) Springer-Verlag, Berlin **I**, 731.

Preston, R.D. (1955b), in *Encyclopedia of Plant Physiology* (W. Ruhland, ed.) **I**, 750.

Preston, R.D. (1959), in *Intnl. Rev. Cytol* (G.H. Bourne and J.F. Danielli, eds.) **VIII**, 33.

Preston, R.D. (1962), in *Interpretation of Ultrastructure. Symp. Intnl. Soc. Cell Biol.* **1**, 325.

Preston, R.D. (1963), in *Fibre Structure* (J.W.S. Hearle and R.H. Peters, eds.) Butterworth's, London, p. 258.

Preston, R.D. (1964a), *Endeavour* **XXIII**, 158.

Preston, R.D. (1964b), in *Formation of Wood in Forest Trees* (M. Zimmerman, ed.– Academic Press, New York, p. 169.

Preston, R.D. (1971), *J. Microsc.* **93**, 7-13.

Preston, R.D. and Allsopp, A. (1939), *Biodynamica* No. 53.

Preston, R.D. and Astbury, W.T. (1937), *Proc. Roy. Soc.* **B122**, 76.

Preston, R.D. and Clark, C.S. (1944), *Proc. Leeds Phil. Lit. Soc.* **4**, 201.

Preston, R.D. and Cronshaw, J. (1958), *Nature, London* **181**, 248.

Preston, R.D. and Duckworth, R.B. (1946), *Proc. Leeds. Phil. Lit. Soc.* **4**, 343.

Preston, R.D. and Goodman, R.N. (1968), *J. Roy. Microscop. Soc.* **88**, 513.

Preston, R.D. and Kuyper, B. (1951), *J. Exp. Bot.* **2**, 247.

Preston, R.D. and Middlebrook, Mavis J. (1949), *J. Text. Inst.* **40**, T715.

Preston, R.D. and Middlebrrok, Mavis J. (1949), *Nature, London* **164**, 217.

Preston, R.D., Nicolai, E., Reed, R. and Millard, A. (1948), *Nature, London* **162**, 665.

Preston, R.D., Nicolai, E. and Kuyper, B. (1953), *J. Exp. Bot.* **4**, 40.

Preston, R.D. and Ripley, G.W. (1954), *Nature, London* **174**, 76.

Preston, R.D. and Ripley, G.W. (1954), *J. Exp. Bot.* **5**, 410.

Preston, R.D. and Singh, K. (1950), *J. Exp. Bot.* **1**, 214.

Preston, R.D. and Singh, K. (1952), *J. Exp. Bot.* **3**, 162.

Preston, R.D. and Wardrop, A.B. (1949), *Biochem. Biophys. Acta* **3**, 549.

Preston, R.D. and Wardrop, A.B. (1949), *Biochem. Biophys. Acta* **3**, 585.

Preston, R.D. Wardrop, A.B. and Nicolai, E. (1948), *Nature, London* **162**, 957.

Preston, R.D., White, R.K. and Robinson, D.G. (1972),

Pridham, J.B. and Hassid, W.Z. (1966), *Biochem. J.* **100**, 21.

Probine, M.C. (1963), *J. Exp. Bot.* **14**, 101.

Probine, M.C. (1965), *Proc. Roy. Soc.* **B161**, 526.

Probine, M.C. and Barber, M.F. (1966), *Austr. J. Biol. Sci.* **19**, 439.

Probine, M.C. and Preston, R.D. (1958), *Nature, London* **182**, 1657.

Probine, M.C. and Preston, R.D. (1961), *J. Exp. Bot.* **12**, 261.

Probine, M.C. and Preston, R.D. (1962), *J. Exp. Bot.* **13**, 111.

Punnett, T. and Derrenbacker, E.C. (1966), *J. Gen. Microbiol.* **44**, 105.

Ramachandran, G.N. (1968), in *Structural Chemistry and Molecular Biology* (A. Rich and N. Davidson, eds.) Freeman, San Francisco.

Ramachandran, G.N. and Ramaseshan, S. (1961), *Encyclopedia of Physics* (S. Flüggi, ed.) Reinhold, New York.

Ramachandran, G.N., Ramakrishnan, C. and Sasisekharen, L. (1963), in *Aspects of Protein Structure* (Ramachandran, G.N. ed.) Academic Press, London.

Rånby, B.G. (1951), *The size and shape factor in colloidal systems. Disc. Faraday Soc.* No. 11, 158.

Rånby, B.G. (1952), *Acta Chem. Scand.* **6**, 101.

Rånby, B.G. (1958), *Encyclopedia of Plant Physiology* (W. Ruhland, ed.) **6**, 768. Springer-Verlag, Berlin.

Rånby, B.G. and Ribi, G. (1950), *Experientia* **6**, 12.

Rao, V.S.R., Sundararajan, P.R., Ramakrishnan, C. and Ramachandran, G.N. (1967), *Conformation of Biopolymers* Vol. 2 (G.N. Ramachandran, ed.) Academic Press, New York.

Ray, P.M. (1962), *Amer. J. Bot.* **49**, 928.

Ray, P.M. and Ruesink, A.W. (1962), *Develop. Biol.* **4**, 377.

Ray, P.M., Shininger, T.L. and Ray, M.M. (1969), *Proc. Nat. Acad. Sci. U.S.* **64**, 605.

Rayle, D.L. and Cleland, R. (1970), *Plant Physiol.* **46**, 250.

Rayle, D.L., Evans, M.L. and Hertel, R. (1968), *Proc. Nat. Acad. Sci. U.S.A.* **65**, 184.

Rayle, D.L., Haughton, P.M. and Cleland, R. (1970), *Proc. Nat. Acad. Sci. U.S.A.* **67**, 1814.

Rayne, A.E. (1945), *Royal Aircraft Establishment Rep.* no. CH 421.

Rees, D.A. (1969), in *Advances in Carbohydrate Chemistry* (M.L. Wolfrom and R.S. Tipson, eds.) Academic Press, London and New York, p. 267.

Rees, D.A. and Skerrett, R.J. (1968), *Carbohydrate Res.* **7**, 334.

Reeves, R.E. (1958), *Ann. Rev. Biochem.* **27**, 15.

Ribi, G. (1950), *Arkiv. Kemi* **2**, No. 40.

Ridge, I. and Osborne, D.J. (1970), *J. Exp. Bot.* **21**, 843.

Ritter, G.J. (1925), *Industr. Engng. Chem. (Anal.)* **17**, 1194.
    (1928), *Idem.* **20**, 941.
    (1929), *Idem.* **21**, 289.

Ritter, G.J. and Mitchell, G.L. (1939), *Paper Trade J.* **108**, 33.

Robards, A.W. (1967), *J. Roy. Microsc. Soc.* **87**, 329.

Robards, A.W. (1969), *Planta* **88**, 376.

Robards, A.W. and Humpherson, P.G. (1967), *Planta* **77**, 233.

Robards, A.W. and Kidwai, P. (1969), *New Phytol.* **68**, 343.

Robards, A.W. and Kidwai, P. (1972), *Cytobiologie* **6**, 1.

Robinson, D.G. and Preston, R.D. (1971a), *J. Cell Sci.* **9**, 591.

Robinson, D.G. and Preston, R.D. (1971b), *J. Exp. Bot.* **22**, 635.

Robinson, D.G. and Preston, R.D. (1971c), *Br. phycol. J.* **6**(2), 113.

Robinson, D.G. and Preston, R.D. (1972), *Biochem. Biophys. Acta* **273**, 336.

Robinson, D.G. and Preston, R.D. (1972), *Planta* **104**, 234.

Robinson, D.G. and White, R.K. (1972), *Br. phycol. J.* **7**, 109.

Robinson, W. (1920), *Phil. Trans. Roy. Soc. Lond.* **210**, 49.

Roelofsen, P.A. (1951), *Biochem. Biophys. Acta* **7**, 43.

Roelofsen, P.A. (1958), *Acta Bot. Néerl.* **7**, 77.

Roelofsen, P.A. (1959), *The Plant Cell Wall.* Borntraeger, Berlin.

Roelofsen, P.A. and Houwink, A.L. (1953), *Acta Bot. Néerl.* **2**, 218.

Roelofsen, P.A. and Kreger, D.R. (1951), *J. Exp. Bot.* **2**, 332.

Roelofsen, P.A., Dalitz, V.Ch. and Wijman, C.F. (1953), *Biochem. Biophys. Acta* **11**, 344.

Rogers, H.J. and Perkins, H.R. (1968), *Cell Walls and Membranes* Spon, London.

Roland, J.C. (1966), *J. Microscopie* **5**, 323.

Roland, J.C. (1967), *J. Microscopie* **6**, 399.

Ross, K.F.A. (1967), *Phase contrast and interference microscopy for cell biologists.* Edward Arnold, London.

Ruch, F. and Hentgartner, J. (1960), *Beih. z. Schweiz. Forstver.* **30**, 75.

Rudall, K.M. (1955), *Symp. Exptl. Biol.* **9**, 149.

Rudall, K.M. (1962), in *The scientific basis of medicine annual review*. Athlone Press, University of London.

Ruge, U. (1937), *Z. Bot.* **31**, 1.

Sachs, J. (1882), *Textbook of Botany*. O.U.P. London and New York.

Sachsse, H. (1965), *Schriftenreihe der Forstlichen Facultät d. Universität Göttingen u. Mitteil. d. Niedersächsischen Forst. Versuchsanstalt*, **35**, 1. Sauerlände V. Frankfurt.

Sager, R. and Palade, G.E. (1957), *J. Biochem. Biophys. Cytol.* **3**, 463.

Saiki, H., Furokawa, I. and Harada, H. (1972), *Bull. Tokyo Univ. Forest.* No. 43.

Schachman, H.K. (1959), *Ultracentrifugation in Biochemistry*. Academic Press, New York and London.

Scherrer, H. (1918), *Nachr. Göttingengesell.* **98**.

Schnepf, E. (1965), *Planta* **67**, 213.

Schnepf, E. and Koch, W. (1966), *Z. Pfl. Physiol.* **55**, 97.

Schnepf, E., Koch, W. and Deichgräber, G. (1966), *Arch. Mikrobiol.* **55**, 149.

Schultz, D. and Lehman, H. (1970), *Cytobiologie* **1**, 343.

Schurz, J. (1953), *Naturwiss.* **40**, 438.

Scurfield, G. (1972), *Aust. J. Bot.* **20**, 9.

Scurfield, G., Silva, S.R. and Wold, M.B. (1972), *Micron* **3**, 160.

Scurfield, G. and Wardrop, A.B. (1962), *Aust. J. Bot.* **10**, 93.

Sebastian, L.P., Côté, W.A. and Skaar, C. (1965), *For. Prod. J.* **15**, 394.

Setlow, R.B. and Pollard, E.C. (1962), *Molecular Biophysics*. Pergamon, London-Paris.

Setterfield, G. and Bayley, S.T. (1957), *Can. J. Bot.* **35**, 435.

Setterfield, G. and Bayley, S.T. (1958), *Biophys. Biochem. Cytol.* **4**, 377.

Setterfield, G. and Bayley, S.T. (1961), *Ann. Rev. Plant Physiol.* **12**, 35.

Sievers, A. (1963), *Protoplasma* **16**, 88.

Simkin, J.L. and Jamieson, J.C. (1967), *Biochem. J.* **103**, 38P.

Sisson, W.A. (1938), *Science* **87**, 358; *Contrib. Boyce Thompson Inst.* **9**, 381.

Sisson, W.A. (1941), *Contrib. Boyce Thompson Inst.* **12**, 31.

Sisson, W.A. and Clark, G.L. (1933), *Ind. Engng. Chem. Anal. Ed.* **5**, 296.

Sleytr, Uwe B. and Thornley, Margaret J. (1973), Submitted to *J. of Bact.*

Smith, D.N.R. and Banks, W.B. (1971), *Proc. Roy. Soc.* **B177**, 197.

Smith, F. and Montgomery, R. (1959), *The chemistry of plant gums and mucilages*. Reinhold, New York.

Smith, W.J. (1959), *Queensland For. Service Res. Note* no. 8.

Spark, L.C., Darnborough, G. and Preston, R.D. (1958), *J. Text. Inst.* **49**, T309.

Spencer, F.S. and Machlachlan, G.A. (1972), *Plant Physiol.* **49**, 58.

Sponsler, O.L. and Dore, W.H. (1926), *Colloid Symp. Monogr.* **41**, 174.

Spurr, A.R. (1957), *Amer. J. Bot.* **44**, 637.

Stacey, K.A. (1956), *Light scattering in physical chemistry*. Butterworth's, London.

Staehelin, L.A. (1966), *Z. Zellforsch.* **74**, 325.

Staehelin, L.A. (1968), *Proc. Roy. Soc.* **B171**, 249.

Staehelin, L.A. and Probine, M.C. (1970), in *Advances in Botanical Research* **III**, 1.

Stafford, L.E. and Brummond, D.O. (1970), *Phytochem.* **4**, 253.

Stamm. A.J. (1929), *J. Agric. Res.* **38**, 23.

Stamm, A.J. (1932), *J. Phys. Chem.* **36**, 312.

Stamm, A.J. (1935), *Physics* **6**, 334.

Stamm, A.J. (1964), *Wood and Cellulose Science.* Ronald, New York.

Stamm, A.J. and Wagner, E. (1961), *For. Prod. J.* **11**, 141.

Stark, G.R., Dawson, C.R. (1962), in *The Enzymes* (Boyer, D.D., Lardy, H. and Myrbäck, K. Eds.) Academic Press, N.Y.

Stecher, H. (1952), *Microscopie* **7**, 30.

Steere, R.L. (1957), *J. Biophys. Biochem. Cytol.* **3**, 45.

Sterling, C.S. and Spit, B.J. (1957), *Amer. J. Bot.* **44**, 851.

Stirling, C. (1970), *Amer. J. Bot.* **57**, 172.

Stern, F. and Stout, H.P. (1954), *J. Text. Inst.* **45**, T1896.

Steward, F.C. and Chang, L.O. (1963), *J. Exp. Bot.* **14**, 379.

Steward, F.C. and Mühlethaler, K. (1953), *Ann. Bot.* **N.S.17**, 295.

Steward, F.C., Israel, H.W. and Salpeter, M.N. (1967), *Proc. Nat. Acad. Sci. U.S.* **58**, 541.

Steward, F.C., Mott, R.L., Israel, H.W. and Ludford, P.M. (1970), *Nature, London* **225**, 760.

Steward, F.C., Thompson, J.F., Millar, F.K., Thomas, M.D. and Hendricks, R.H. (1951), *Plant Physiol.* **26**, 123.

Stewart, M.C. (1966), *The chemistry of secondary growth in trees.* C.S.I.R.O. Paper No. 43.

Stewart, M.C., Dawes, C.J., Dickens, B.M. and Nicholls, J.W.P. (1969), *Austr. J. Marine Freshwater Res.* **20**, 143.

Stöckman, V.E. (1972), *Biopolymers* **11**, 251.

Stokes, A.R. (1948), *Proc. Phys. Soc.* **61**, 115.

Stone, J.E. (1964), *Pulp and Paper Mag. Can.* **65**, T3.

Stone, J.E. and Scallan, A.M. (1964), *Pulp and Paper Res. Inst. Can. Tech. Rep.* No. 382.

(1965), *Idem.* No. 392, No. 407, No. 418.

Stone, J.E. and Scallan, A.M. (1967), *Tappi* **50**, 496.

Stone, J.E., Scallan, A.M. and Aberson, G.M.A. (1966), *Can. Pulp Pap. Mag.* **5**, T263.

Stout, H.P. and Jenkins, A. (1955), *Ann. Sci. Textiles Belges* 231.

Strasburger, E. (1962), *Lehrbuch der Botanik. 28 Auflage neu bearbeitet.* Stuttgart: Gustav Fischer Verlag.

Strauss, J. and Campbell, W.A. (1963), *Life Sci.* **1**, 50,

Streminger, J.L. and Mapson, L.W. (1957), *Biochem. J.* **66**, 567.

Sullivan, J.D. (1968), *Tappi* **51**, 501.

Talmadge, K.W., Keegstra, K., Bauer, W.D. and Albersheim, P. (1973), *Plant Physiol.* **51**, 158.

Tanada, T. (1972), *Nature, London* **236**, 460.

Thiele, H. and Anderson, G. (1955), *Kolloid Zeit.* **143**, 21.

Thomas, R.J. and Kringstad, K.P. (1971), *Holzforsch.* **25**, 143.

Thompson, E.W. and Preston, R.D. (1967), *Nature, London* **213**, 684.

Thompson, E.W. and Preston, R.D. (1968), *J. Exp. Bot.* **19**, 690.

Thornber, J.P. and Northcote, D.H. (1961), *Biochem. J.* **81**, 449.

Timell, T.E. (1963), *J. Polymer Sci.* **C2**, 109.

Timell, T.E. (1965), in *Ultracellular Structure of Woody Plants* (W.A. Côté, ed.) Syracuse U.P., Syracuse, U.S.A., p. 127.

Tracey, M.V. (1957), *Revs. Pure and Appl. Chem.* **7**, 1.

Traynard, Ph. and Ayroud, A.M. (1952), *Rev. gen. Bot.* **59**, 561.

Traynard, Ph., Ayroud, A.M., Eymery, A., Robert, A. and Coligny, S. (1954), *Holzforsch.* **8**, 42.

Tripp, V.W., Moore, A.T. and Rollins, M.L. (1951), *Text. Res. J.* **21**, 227.

Tripp, V.W., Moore, A.T. and Rollins, M.L. (1954), *Text. Res. J.* **24**, 956.

Tschammler, H., Kratzl, K., Leutner, R., Steininger, A. and Kisser, J. (1953), *Mikroskopie* **8**, 238.

Tsoumis, G. (1965), in *Cellular Ultrastructure of Woody Plants* (W.A. Côté, ed.) Syracuse U.P., Syracuse, U.S.A.

Turnbull, J.M. (1940), *J. South African For. Assoc.* **5**, 62.

Turvey, J.R. and Rees, D.A. (1958), *Third International Seaweed Symposium*. Abstracts Ed. C.O. Leocha Galway: O'Gorman.

Tuszon, J. (1903), *Ber. dtsch. bot. Ges.* **21**, 276.

Van Oordt Hulsof, B. van den Houven (1957), *Acta botan. Néerl.* **6**, 420.

Veen, B.W.(1970a), *Proc. Kon. Ned. Akad. Wet. Amsterdam* **C73**, 57, 113. (1970b), *Ibid.* 118.

Villemez, C.L. (1971), *Biochem. J.* **121**, 151.

Villemez, C.L. and Clark, A.F. (1969), *Biochem. Biophys. Res. Comm.* **36**, 57.

Villemez, C.L., Lin, T.-Y. and Hassid, W.Z. (1965), *Proc. Nat. Acad. Sci. U.S.A.* **54**, 1626.

Villemez, C.L., McNab, J.M. and Albersheim, P. (1968), *Nature, London* **218**, 878.

Villemez, C.L., Swanson, A.L. and Hassid, W.Z. (1966), *Arch. Biochem. Biophys.* **116**, 446.

Vintila, E. (1939), *Holz als Roh- u. Werkstoff* **2**, 345.

Virgin, H. (1955), *Physiol. Plantarum* **8**, 954.

Vogel, A. (1953), *Diss. E.T.H. Zürich.*

Vries, H. de (1877), *Untersuchungen uber die mechanischen Ursachen der Zellstreckung.* W. Engelmann, Leipzig.

Wang, M.C. and Bartnicki-Garcia, S. (1966), *Biochem. Biophys. Res. Commun.* 24, 832.

Wardrop, A.B. (1948), *Proc. 9th Ann. Pulp and Paper Res. Conf.* Div. For. Products Australia.

Wardrop, A.B. (1949), *Nature, London* 164, 366.

Wardrop, A.B. (1950), *Nature, London* 165, 272.

Wardrop, A.B. (1951), *Aust. J. Sci. Res.* B4, 391.

Wardrop, A.B. (1954a), *Aust. J. Bot.* 2, 165.

Wardrop, A.B. (1954b), *Holzforsch.* 8, 12.

Wardrop, A.B. (1955), *Aust. J. Bot.* 3, 137.

Wardrop, A.B. (1956a), *Biochem. Biophys. Acta* 21, 200.

Wardrop, A.B. (1956b), *Aust. J. Bot.* 4, 193.

Wardrop, A.B. (1957), *Tappi* 40, 225.

Wardrop, A.B. (1958), *Aust. J. Bot.* 6, 299.

Wardrop, A.B. (1962a), *Nature, London* 194, 497.

Wardrop, A.B. (1962b), *Bot. Rev.* 28, 241.

Wardrop, A.B. (1963), *Svensk. Papperstidn.* 66, 231.

Wardrop, A.B. (1964), in *Formation of Wood in Forest Trees* (ed. W. Zimmerman). Academic Press, New York, p. 95.

Wardrop, A.B. (1965), in *Cellular Ultrastructure of Woody Plants* (ed. W. Côté) Syracuse U.P., Syracuse, U.S.A.

Wardrop, A.B. (1969), *Aust. J. Bot.* 17, 229.

Wardrop, A.B. and Addo-Ashong, F.W. (1965), *Fracture. 1st Tewkesbury Symp. 1963.* Melbourne U.P.

Wardrop, A.B. and Bland, D.E. (1959), *Proc. Int. Congr. Biochem.* 4th Vienna. p. 93. Pergamon, London.

Wardrop, A.B. and Cronshaw, J. (1958), *Aust. J. Biol.* 6, 89.

Wardrop, A.B. and Dadswell, H.E. (1950), *Aust. J. Sci. Res.* B3, 1.

Wardrop, A.B. and Dadswell, H.E. (1952), *Aust. J. Sci. Res.* B5, 223, 385.

Wardrop, A.B. and Dadswell, H.E. (1953), *Holzforsch.* 7, 33.

Wardrop, A.B. and Dadswell, H.E. (1955), *Aust. J. Bot.* 3, 177.

Wardrop, A.B. and Dadswell, H.E. (1957), *Holzforsch.* 11, 33.

Wardrop, A.B. and Davies, G.W. (1961), *Holzforsch.* 15, 129.

Wardrop, A.B. and Foster, R,C. (1964), *Aust. J. Bot.* 11-12, 135.

Wardrop, A.B. and Harada, H. (1965), *J. Exp. Bot.* 16, 356.

Wardrop, A.B. and Preston, R.D. (1947), *Nature, London* 160, 911.

Wardrop, A.B. and Preston, R.D. (1950), *Biochem. Biophys. Acta* 6, 36.

Wardrop, A.B. and Preston, R.D. (1951), *J. Exp. Bot.* 2, 20.

Wardrop, A.B., Dadswell, H.E. and Davies, G.W. (1961), *Appita* 14(6), 185.

Wardrop, A.B., Liese, W. and Davies, G.W. (1959), *Holzforsch.* 13, 115.

Warwicker, J.O. and Wright, A.C. (1967), *J. Appl. Polymer Sci.* 2, 659.

Weatherwax, R.C. and Tarkow, H. (1968), *For. Prod. J.* 18, 83.

Welch, M.B. (1933), *J. and Proc. Roy. Soc. New South Wales* 66, 492.

Welch, M.B. (1935), *Idem.* **68**, 249.

Welland, II.J. (1954), *J. Polymer Sci.* **13**, 471.

Wergin, W. (1937), *Naturwiss.* **25**, 830.

Werz, G. (1957), *Z. Naturf.* **12 B**, 739.

Whaley, W.G. and Mollenhawer, H.H. (1963), *J. Cell Biol.* **17**, 216.

Wiener, O. (1912), *Abstr. Sachs. Ges. Akad. Wiss.* **32,**

Williams, W.T., Preston, R.D. and Ripley, G.W. (1955), *J. Exp. Bot.* **6**, 451.

Willison, J.H.M. and Cocking, E.C. (1972), *Protoplasma* **75**, 397.

Wilson, K. (1955), *Ann. Bot.* **N.S.19**, 289.

    (1957), *Ibid.* **21**, 1.

Winter, H., Meyer, L., Hengeveld, E. and Wiersma, P.K. (1971), *Acta. Bot. Néerl.* **20**, 489.

Wirth, P. (1946), *Ber. schweiz. bot. Ges.* **56**, 175.

Wooding, F.B.P. and Northcote, D.H. (1964), *J. Cell Biol.* **23**, 327.

Woude, van der, W.J., Lembi, C.A. and Morré, D.J. (1972), *Biochem. Biophys. Res. Comm.* **46**, 245.

Yao, J. and Stamm, A.J. (1967), *For. Prod. J.* **17**, 33.

Note on reference (Gardner and Blackwell, (1974)) *from page 140.*
Using the two-chain unit cell of Meyer and Misch (as a close approximation to the eight-chain cell) Gardner and Blackwell have attempted a least squares refinement of the structure. The comparison of observed intensity data with calculated data indicates a strong preference for a parallel-chain model rather than an anti-chain model. This is a development of the highest importance.

Author Index

Abdul-Baki, A.A. 432, 457
Aberson, G.M.A. 280, 476
Addo-Ashong, F.W. 347, 478, 351
Adler, E. 58, 457
Adzumi, H. 369, 457
Albersheim, P. 406, 416, 421, 423, 431, 437, 457, 458, 465, 466, 477
Alexander, L.E. 137, 457
Allsopp, A. 170, 290, 472, 473
Altermatt, H.A. 430, 457
Anderson, D.B. 313, 314, 385, 457
Anderson, G. 236, 477
Apelbaum, A. 392, 393, 457, 459
Arnold, R. 430, 466
Ashworth, D. 444, 470
Aspinall, G.O. 51, 53, 54, 457
Astbury, W.T. 99, 165, 170, 195−197, 205, 206, 231, 290, 392, 457, 472
Asunmaa, S. 289, 290, 457
Atkins, E.D.T. 137−139, 232, 234−236, 267, 268, 457, 470
Ayroud, A.M. 289, 477

Bailey, I.W. 277, 279, 289, 290, 294, 300, 306, 309, 315, 317, 369, 458
Bailey, P.J. 162, 304, 311, 375, 458
Bailey, R.W. 431, 458
Baker, D.P. 418, 458
Baker, R.S. 393, 463
Bal, A.K. 432, 458
Balashov, V. 332, 458
Bamber, R.K. 310, 458
Bandurski, R.S. 66, 421, 466, 467
Banks, W.B. 372, 475
Barber, G.A. 431, 433, 458
Barber, N.F. 342, 361, 362, 363, 365, 406, 412, 458, 473
Barer, R. 108, 458
Barkas, W.W. 355, 458
Barnett, J.R. 447, 458

Bartnicki-Garcia, S. 60, 388, 424, 427, 458, 478
Batra, K.K. 434, 458, 462
Bauch, J. 309, 310, 458
Bauer, W.D. 406, 415, 416, 423, 466
Bayley, S.T. 41, 60, 316, 387, 389, 394, 395, 419, 439, 458, 466, 475
Beamen, T.C. 66, 466
Beer, M. 314, 439, 461, 458
Belford, D.S. 42, 458
Bennett-Clark, T.A. 421, 458
Berkeley, E.E. 173, 294, 458
Berlyn, G.P. 259, 458
Bernal, J.D. 195, 457
Berndt, H. 310, 458
Betrabet, S.M. 167, 459
Bienfait, J.L. 351, 459
Bisset, I.J.W. 323, 459
Bittiger, H. 181, 438, 459, 465
Björkvist, K.J. 58, 457
Black, M. 419, 459
Black, W.A.P. 228, 459
Blackwell, G. 140, 183, 463
Bland, D.E. 290, 478, 302
Blank, F. 384, 459
Bobák, M. 289, 459, 470
Böhmer, H. 386, 459
Bonner, J. 60, 66, 385, 416, 421, 459, 465, 471
Bonnett, H.T. 431, 443, 459
Bosshard, H.H. 300, 308, 386, 459, 463
Bouck, G.B. 443, 461
Boundy, J.A. 416, 459
Bourret, A. 179, 459
Boutelje, J. 361, 378, 459
Bowen, T.J. 44, 459
Boyd, J.D. 352, 459
Bragg, W.L. 127, 459
Brand, J.C.D. 15, 459
Brown, K.C. 328, 459, 331

Brown, R. 60, 384, 459
Brown, R.M. 38, 438, 459, 465
Browning, B.L. 54, 459
Brummond, D.O. 433, 435, 459, 476
Buer, F. 224, 459
Bullock, 419, 459
Burg, S.P. 392, 393, 420, 459, 457, 466
Burnett, 167, 465
Burström, H. 407, 413, 459, 460
Burton, K. 426, 460
Buston, G. 54, 460

Campbell, W.A. 66, 476
Carlström, D. 141, 142, 460
Carman, P.C. 373, 460
Casperson, C. 302, 460
Castle, E.S. 387, 411, 460
Caulfield, D.F. 146, 173, 179, 460
Cave, I.D. 282, 287, 323, 334, 340, 341, 342, 348, 365, 412, 460
Chafe, S.C. 395, 443, 460
Chandra, S.K. 54, 471
Chang, L.O. 64, 476
Chantler, E.C. 419, 459
Chanzy, H. 179, 459
Chapman, G.B. 387, 397, 464
Chou, S. 162, 376, 460
Chrispeels, M.J. 65, 460
Christensen, T. 275, 460
Clark, A.F. 435, 477
Clark, C.S. 384, 472
Clark, R.A. 419, 459
Clark, S.H. 355, 460
Clarke, G.L. 282, 285, 316, 464, 475
Cleland, R. 60, 64, 402, 407−409, 414−417, 419, 420, 460, 464, 474
Cocking, E.C. 448, 464, 479
Cockrell, R.A. 359, 361, 363, 460
Colvin, J. Ross 2, 167, 171, 174, 181, 395, 435, 439, 454, 458, 460, 461
Comstock, G.L. 311, 461
Cook, G.M. 433, 461
Cooper, F.P. 395, 458
Correns, C. 50, 195, 205, 257, 461
Côté, W.A. 289, 290, 301, 307, 311, 349, 369, 425, 461, 464, 466, 475
Courtney, J.S. 419, 461
Courtois, J.E. 54, 461
Cowdrey, D.R. 282, 285, 323, 334, 336, 337, 340, 341, 343, 461
Cowling, E.B. 377, 461
Cox, G.C. 314, 461
Craigie, J.C. 186, 205, 464
Cronquist, A. 271, 461
Cronshaw, J. 40, 41, 43, 171, 195, 197, 199, 201, 202, 204, 207, 299, 237, 271, 312, 386, 389, 392, 443, 461, 478

Crook, E.M. 60, 461

Dadswell, H.E. 277, 289, 299, 302, 317, 323, 324, 349, 361, 439, 461, 478
Dalitz, V. Ch. 227, 474
Darmon, S.E. 141, 461
Darnborough, G. 332, 475
Das, N.K. 393, 461
Dashek, W.V. 418, 431, 432, 433, 461
Datko, A.H. 441, 461
Davies, G.W. 289, 307, 312, 461, 478
Dawes, C.G. 60, 203, 224, 461, 476
Dawson, C.R. 66, 476
Day, A.C. 289, 290, 301, 461
Dember, R.J. 416, 459
Dennis, D.T. 171, 225, 461
Derrenbacker, E.C. 60, 473
Dickens, B.M. 60, 203, 476
Diehl, G.M. 384, 385, 393, 461
Dinwoodie, J.M. 349, 350, 353, 461
Donnan, F.G. 44, 462
Dore, W.H. 45, 134, 138, 475
Dörfel, H. 231, 462
Dougall, D.K. 59, 462
Drawert, H. 426, 462
Drummond, D.W. 231, 462
Duckworth, R.E. 313, 472
Duda, C.T. 418, 468
Duffy, R.M. 7, 469
Dweltz, N.E. 141, 142, 462

Echols, R.M. 323, 462
Edelman, J. 60, 66, 462
Eisinger, W. 393, 459
Elbein, A.D. 431, 433, 458
Ellefsen, Ø 171, 462
Emerton, H. 300, 462
Ernst, A. 50, 462
Escherich, W. 51, 462
Etherton, B, 420, 462
Evans, M.I. 420, 462
Eylar, E.H. 433, 461

Falk, S. 344, 345, 413, 438, 459, 462, 470
Falke, H. 38, 459
Fan, D.F. 421, 462
Feingold, D.S. 433, 462
Feldmann, J. 274, 462
Fischer, D.G. 136, 462
Fischer, E.W. 439, 468
Fischer, F.G. 231, 462
Flint, E.A. 388, 462
Flowers, H.M. 434, 462
Fogg, G.E. 51, 256, 432, 471
Foster, D.H. 306, 478
Foster, R.C. 395, 396, 462
Fowkes, L.C. 431, 432, 462

SCIENCE

Fox, C.F. 450, 462
Françon, M. 108, 462
Franke, W.W. 38, 438, 459, 465
Frei, E. 43, 51, 52, 53, 94, 95, 136, 140,
    156, 186, 189, 204, 206, 207, 210,
    215, 217, 223, 225, 227, 229, 232,
    233, 236, 237, 239, 243, 245, 247,
    248, 252, 254, 255, 256, 267, 272,
    300, 389, 390, 412, 441, 442, 445,
    453, 462, 463
Frenzel, H. 349, 466
Frenzel, P. 309, 463
Freudenberg, K. 58, 463
Frey, R. 224, 463
Frey-Wyssling, A. 66, 164, 165, 166, 168,
    169, 174, 175, 186, 205, 224, 252,
    306, 308, 348, 349, 351, 356, 360,
    375, 384, 385, 386, 387, 407, 412,
    318, 427, 431, 457, 459, 463, 470
Fritsch, F.E. 193, 238, 239, 469
Fuchs, Y. 392, 463
Furokawa, I. 348, 475

Gabara, B. 426, 471
Galston, A.N. 393, 463
Gardner, K.H. 140, 183, 463
Garland, H. 347, 351, 463
Gawlick, S.R. 444, 463
Gay, J.L. 427, 431, 465
Gertjejansen, R.O. 373, 466
Gibbons, A.P. 433, 459
Gillis, P.P. 183, 353, 463, 468
Girbardt, M. 428, 463
Glaser, L. 433, 463
Glasziou, K.T. 66, 463
Goldsmith, V. 300, 462
Goodman, R.N. 444, 453, 472
Gorham, P.R. 174, 464
Goring, D.A.T. 48, 464
Gorter, C.J. 384, 461
Gotelli, I.B. 60, 64, 464
Goulding, K.J. 225, 470
Green, P.B. 225, 271, 387, 393, 397,
    399, 412, 414, 415, 464
Greenwood, A.D. 427, 429, 431, 465
Grew, N. 4
Gross, S.T. 316, 464
Grossman, P.U.A., 351, 464
Grout, B.W.W. 448, 464

Häggroth, S. 58, 457
Haleem, M.A. 143, 464
Hall, C.E. 64, 471
Hall, M.A. 60, 66, 434, 462, 463, 471
Hämmerling, J. 254, 464
Hanack, M. 35, 000
Hanic, L.A. 186, 205, 464
Hanson, A.D. 419, 459

Harada, H. 299, 300, 306, 307, 317,
    348, 464, 475
Harlow, W.M. 288, 464
Harris, J.M. 363, 364, 464
Harris, P.J. 431, 464
Hart, C.A. 311, 464
Hartler, N. 349, 464
Hartmann-Fahnenbrock, M. 307, 311,
    467
Hartshorne, N.H. 101, 464
Hassel, O. 34, 464
Hassid, W.Z. 429, 431, 433, 458, 462,
    473, 477
Haug, A. 232, 237, 464
Haughton, P.M. 404, 406, 409, 415, 419,
    460, 464, 470
Haworth, W.N. 34, 38, 45, 464
Hearle, J.W.S. 332, 334, 465
Heath, I.B. 427, 429, 431, 465
Hendricks, R.H. 64, 476
Hengeveld, E. 416, 479
Henglein, F.A. 190, 465
Hengstenberg, J. 145, 465
Hepler, P.K. 443, 465
Hermans, P.H. 144, 465
Hertel, R. 420, 476
Herth, W. 438, 465, 459
Hertz, C.H. 344, 462, 470
Hess, K. 2, 465
Heyn, A.N.J. 66, 147, 175, 386, 411,
    419, 444, 465
Hill, G.J.C. 431, 465
Hirst, E.L. 231, 462, 465
Hodge, A.J. 167, 311, 386, 465
Honjo, G. 136, 465
Hook, R. 2
Horne, R.W. 225, 470
Hossfield, R.L. 373, 466
Hough, L. 225, 465
Houwink, A.C. 388, 389, 465, 474
Howsmon, J.A. 353, 465
Humpherson, P.G. 444, 474
Hunsley, D. 167, 465
Husemann, E. 181, 459

Iriki, Y. 51, 52, 465
Isherwood, F.A. 40, 171, 466
Israel, H.W. 64, 65, 465, 474
Iterson, G. van Jr. 9, 226, 384, 385, 465

Jacobs, M.R. 352, 465
Jamieson, J.C. 433, 475
Jang, R. 421, 465
Jansen, E.F. 421, 465
Jenkins, A. 332, 476
Jermyn, M.A. 40, 171, 466
Jiménez-Martínez, A. 432, 470
Johnson, D.T. 167, 466

Johnstone, I.R. 60, 461
Jolley, G.M. 419, 459
Jones, D.W. 146, 466
Jones, F.W. 144, 466
Jones, J.K.N. 225, 231, 465
Jones, W.O. 231, 465
Jurásek, L. 289, 470
Jutte, S.M. 302, 307, 309, 310, 466

Kaloni, P.N. 183, 468
Kang, B.G. 93, 420, 466, 459
Karlsnes, A.M. 60, 417, 460
Katona, L. 64, 467
Katz, M. 421, 466
Kauman, W.G. 190, 466
Kauss, H. 430, 466
Kay, D.H. 160, 466
Keegstra, K. 406, 415, 416, 423, 466
Keilich, G. 438, 465
Keith, C.T. 349, 466
Kellog, R.M. 378, 466
Kelsey, K.E. 357, 361, 466
Kelso, W.C. 373, 466
Kemp, J. 434, 462
Kerr, T. 173, 289, 290, 294, 300, 385, 457, 458, 466
Kessler, G. 51, 457
Key, J.L. 419, 420, 461, 466
Kidwai, P. 432, 443, 474
King, J.W. 393, 463
King, N.J. 60, 466
Kisser, J. 57, 349, 466, 477
Kivilaan, A. 66, 421, 466, 467
Kjaer, A. 289, 467
Kleinig, H. 38, 438, 459
Kleinhooute, A. 384, 461
Kobayashi, K. 312, 466
Koehler, A. 355, 359, 360, 466
Kommick, H. 444, 466
Koshijima, T. 54, 466
Kramer, D. 179, 290, 466, 467
Kratzl, K. 57, 58, 425, 470, 477
Krebs, H. 426, 460
Kreger, D.R. 52, 189, 224, 225, 226, 264, 466, 474
Kringstad, K. 171, 311, 462, 477
Krishnamurti, G.S.R. 225, 467
Kübler, H. 353, 467
Kundu, B.C. 293
Kuyper, B. 444, 472

Laico, M.T. 433, 461
Lamport, D.T.A. 59, 60, 63, 64, 66, 67, 416, 433, 467
Lange, P. 289, 467
Lange, P.W. 290, 457
Lange, N.J. 228, 467
Larson, B. 232, 464

Larson, P.R. 324, 467
Laué, M. von, 127, 467
Lazaro, R. 179, 459
Ledbetter, M.C. 427, 444, 467
Ledizet, P. 56, 461
Lee, S. 421, 467
Leech, J.H. 426, 469
Lehman, H. 432, 475
Leitgeb, H. 50, 467
Lembi, C.A. 432, 479
Lennarz, W.J. 435, 467
Leneol, P.D. 418, 468
Leutner, R. 57, 477
Levy, F.T. 10, 11, 467
Liang, C.Y. 189, 468
Lieberman, H. 392, 463
Liese, W. 307, 309, 310, 311, 312, 458, 467
Limaye, V.J. 224, 225, 467
Lin, T.-Y. 429, 477
Linskens, H.P. 60, 416, 467
Lochner, J.P.A. 344, 467
Lockhart, J.A. 406, 414, 467
Loewus, F. 430, 468
Lopez-Saez, J.F. 427, 463
Lotmar, W.E. 142, 468
Love, J. 52, 240, 468
Lucas, H.J. 231, 468
Ludford, 64, 476

McAlear, J.H. 428, 469
McClellan, A.L. 15, 472
McDowell, R.H. 42, 43, 53, 59, 471
McManus, M.A. 444, 468
McNab, J.M. 437, 477
Macchic, E. 439, 468
Machlachlan, G.A. 418, 421, 441, 461, 462, 468, 475
Machlis, L. 431, 465
Mackie, I.M. 44, 50, 51, 264, 468
Mackie, W. 44, 52, 232, 236, 252, 457, 468
Majumdar, G.P. 312, 313, 389, 468
Malpighi, M. 4
Manjunat, B.R. 440, 468
Manley, R. St. John, 181, 182, 468
Mann, J. 136, 462
Mann, J.C. 438, 459
Manocha, M.S. 419, 468
Manton, I. 437
Mapson, L.W. 431, 476
Marchant, R. 428, 429, 468
Marchessault, R.H. 189, 468
Margere, C. 418, 468
Mark, H. 45, 134, 138, 145, 465, 469
Mark, R.E. 183, 289, 335, 340, 347, 351, 458, 468
Martin-Smith, C.A. 395, 458

Marwicke, T.C. 195, 457
Marx-Figini, M. 44, 48, 438, 440, 468
Masuda, Y. 409, 468
Matchett, N.H. 418, 468
Matile, P. 450, 468
Matzke, E.B. 7, 469
Meeuse, A.D.J. 9, 465
Meier, H. 52, 56, 240, 243, 251, 289, 290,
    297, 298, 300, 469
Mellor, R.C. 108, 458
Meredith, R. 331, 332, 469
Meretz, W. 7, 469
Meyer, K.H. 45, 134, 136, 138, 141, 165,
    469
Meyer, L. 416, 479
Meylan, B.A. 279, 282, 287, 323, 340,
    342, 361, 362, 363, 364, 365, 412,
    458, 464, 469
Middlebrook, M.J. 293, 297, 344, 385,
    411, 412, 469, 473
Millar, F.K. 60, 476
Millard, A. 164, 473
Millington, W.F. 444, 463
Mirande, R. 50, 51, 469
Misch, H. 136, 469
Mitchell, G.L. 316, 474
Mitrakos, K. 375, 463
Miwa, T. 51, 52, 256, 465, 469
Miyasaki, Y. 299, 464
Mix, M. 226, 228, 426, 462, 469
Mohl, H. von 5
Mollenhawer, H.H. 426, 427, 469, 479
Montgomery, R. 56, 475
Moor, H. 157, 447, 450, 468, 469
Moore, A.T. 59, 389, 477
Moore, R.T. 44, 428, 469
Morey, D.R. 281
Morré, D.J. 60, 416, 419, 432, 435, 461,
    469, 471, 479
Moscaleva, W.E. 348, 469
Moss, B.L. 236, 469
Mott, R.L. 64, 476
Muggli, R. 175, 183, 463
Mühlethaler, K. 164, 166, 174, 175, 197,
    199, 201, 308, 386, 387, 426
Murmanis, L. 310, 427, 431, 447, 457,
    463, 469, 470, 476
Myers, A. 40, 171, 204, 207, 237, 461,
    460

Nägeli, C. von 6, 50, 257, 470
Nance, J.F. 418, 468
Naum, Y.R. 7, 470
Nečesany, V. 277, 289, 459, 470
Neish, A.C. 58, 430, 457, 470
Nelmes, B.J. 161, 419, 427, 428, 444, 470
Neufeld, E.F. 433, 462
Newcomb, E.H. 66, 428, 431, 443, 459,

465, 470
Nicholls, J.W.P. 60, 203, 324, 361, 476
Nicolai, M.F.E. 50, 164, 170, 173, 186,
    205, 207, 217, 223, 224, 225, 226,
    444, 447, 470, 473
Nieduszinski, I.A. 137, 138, 139, 145,
    146, 147, 179, 236, 457, 470
Nilsson, S.B. 344, 345, 470
Nisizawa, K. 52, 465
Norman, A.G. 170, 290, 457
Northcote, D.H. 41, 59, 66, 225, 226,
    426, 427, 431, 464, 467, 470, 477
Novaes-Ledieu, M. 432, 470
Nyburg, S.C. 137, 470

O'Brien, T.P. 431, 471
Odzuck, W. 430, 466
Olson, A.C. 160, 416, 433, 471
Olszewska, M.J. 426, 427, 471
Oort, A.J.P. 410, 471
Ordin, L. 421, 434, 435, 466, 471, 472
Osborne, D.J. 60, 417, 471, 474
Oster, G. 108, 471
Ott, E. 49, 471
Ottar, B. 34, 464
Overbeck, F. 384, 471

Page, D.H. 280, 349, 471
Palade, G.E. 228, 475
Pallister, A.W. 108, 471
Palma, A. 439, 468
Pankow, G.N. 141, 469
Parker, B.C. 51, 226, 228, 256, 432, 471
Parker, K.D. 143, 236, 267, 268, 270,
    457, 471
Patau, K. 393, 461
Payne, J.F. 432, 458
Peacock, N. 440, 468
Pease, D.C. 162, 471
Peat, S. 272, 471
Peirce, F.T. 328, 459
Percival, E.G.V. 54, 471
Percival, E. 42, 43, 44, 50, 51, 52, 53,
    59, 229, 231, 240, 264, 462, 468,
    471
Perkins, H.R. 40, 54, 474
Perrault, P.F. 441, 461
Petty, J.A. 309, 311, 370, 371, 373, 377,
    378, 471
Phillips, E.W.J. 304, 319, 471
Picken, L.E.R. 142, 468
Pickett-Heaps, J.D. 428, 431, 432, 462,
    470, 471
Pierce, C. 379, 471
Pilet, P.-E. 66, 429, 471
Pimentel, G.C. 15, 472
Pinsky, A. 435, 472
Pockels, H. 101, 472

Poincaré, H. 101, 472
Pollard, E.C. 15, 475
Pollard, J.K. 64, 472
Porter, K. 427, 444, 467
Preston, R.D. 40, 43, 47, 50, 51, 52, 53,
    60, 64, 65, 66, 94, 95, 99, 136, 140,
    145, 146, 147, 162, 164, 165, 167,
    168, 169, 170, 171, 173, 176, 179,
    180, 186, 189, 195, 196, 197, 199,
    201, 202, 204, 205, 206, 207, 210,
    215, 217, 219, 221, 223, 224, 225,
    226, 227, 229, 232, 236, 237, 239,
    243, 245, 247, 248, 252, 254, 255,
    256, 257, 267, 272, 279, 280, 282,
    285, 290, 293, 294, 296, 297, 300,
    304, 311, 312, 313, 314, 315, 317,
    319, 320, 323, 324, 325, 332, 334,
    336, 337, 340, 341, 343, 344, 356,
    361, 364, 370, 373, 375, 384, 385,
    387, 389, 390, 395, 397, 401, 403,
    404, 407, 408, 411, 412, 414, 416,
    418, 419, 421, 432, 435, 439, 440,
    441, 442, 444, 445, 447, 448, 449,
    453, 454, 457, 458, 460, 461, 462,
    463, 464, 468, 469, 470, 471, 472,
    473, 474, 475, 477, 478, 479
Pridham, J.B. 431, 473
Priestley, J.H. 59
Probine, M.C. 225, 323, 389, 390, 393,
    394, 397, 401, 403, 406, 408, 412,
    450, 469, 473, 476
Punnett, T. 60, 473
Puritch, G.S. 371, 373, 471

Ramachandran, G.N. 35, 46, 101, 268,
    473
Ramakrishnan, C. 46, 268, 473
Ramameshan, S. 101, 473
Ranby, B.G. 167, 169, 171, 440, 473
Rao, V.S.P. 46, 473
Ray, M.M. 432, 473
Ray, P.M. 406, 414, 418, 432, 457, 473
Rayle, D.L. 415, 419, 460, 474
Rayne, A.E. 349, 474
Reed, R. 164, 473
Reeder, B.J. 290, 467
Rees, D.A. 271, 471, 477
Reeves, R.E. 34, 474
Ribi, G. 167, 169, 473
Ridge, I. 60, 417, 474
Ripley, G.W. 165, 223, 300, 332, 387,
    444, 445, 458, 470, 473, 479
Ritter, G.J. 288, 316, 464, 474
Robards, A.W. 302, 428, 429, 432, 443,
    444, 447, 468, 474
Robinson, D.G. 178, 180, 219, 221, 435,
    448, 449, 473, 474
Robinson, W. 349, 474

Roelofsen, P.A. 189, 224, 227, 252, 386,
    387, 388, 389, 390, 400, 412, 440,
    454, 465, 467, 471, 474
Rogers, H.J. 40, 54, 474
Roerig, S. 64, 467
Roland, J.C. 314, 315, 447, 474
Rollins, M.L. 59, 167, 389, 459, 477
Romanovicz, D. 438, 459
Rose, R.S. 44, 462
Rosen, W.G. 431, 461
Ross, K.F.A. 108, 474
Ross, R. 229, 471
Roth, L.E. 444, 468
Rudall, K.M. 141, 142, 143, 461
Ruesink, A.W. 406, 414, 473
Ruge, U. 384, 385, 475

Sachs, J. 306, 475
Sachs, J.B. 310, 470
Sachsse, H. 302, 475
Sadava, D. 65, 460
Sager, R. 228, 475
Saiki, H. 348, 475
Salpeter, M.M. 64, 465
Sasisekharen, L. 268, 473
Scallan, A.M. 280, 378, 379, 380, 381,
    476
Schachman, H.K. 44, 475
Scher, M.G. 435, 467
Scherrer, H. 144, 475
Schnepf, E. 179, 453, 475
Schultz, D. 432, 475
Schultz, G.V. 44, 48, 432, 440, 468,
    475
Schultz, R. 309, 458
Schurz, J. 229, 475
Schwendener, S. 257, 470
Scurfield, G. 302, 475
Sebastian, L.P. 369, 475
Sellen, D.B. 44, 52, 404, 406, 464, 468
Setlow, R.B. 15, 475
Setterfield, G. 41, 314, 316, 387, 389,
    394, 419, 458, 475
Shaw, M. 429, 468
Shimbayashi, K. 59, 462
Shininger, T.L. 432, 473
Sievers, A. 426, 475
Simkin, J.L. 433, 475
Singh, K. 99, 293, 296, 325, 472
Sisson, W.A. 197, 227, 282, 475
Sitte, P. 38, 438, 459
Skaar, C. 369, 475
Skoog, F. 393, 461
Sleytr, U.B. 449, 452, 475
Smidsrød, O. 232, 464
Smith, D.N.R. 372, 475
Smith, F. 56, 475
Smith, W.J. 361, 475

Smolko, E. 232, 236, 457
Sopko, R. 289, 470
Spark, L.C. 332, 475
Speakman, J.C. 15, 459
Spencer, F.S. 441, 475
Spit, B.J. 224, 307, 309, 394, 467
Sponsler, O.L. 45, 134, 138, 195, 475
Spurlin, H.M. 49, 471
Spurr, A.R. 314, 475
Stacey, K.A. 44, 475
Stadler, J. 438, 465
Staehelin, L.A. 449, 450, 476
Stafford, L.E. 435, 476
Stamm, A.J. 369, 377, 461, 476, 479
Stark, G.R. 66, 476
Steere, R.L. 157, 476
Stecher, H. 387, 463, 474
Steininger, A. 57, 349, 466, 477
Steplewski, Z. 426, 471
Sterling, C. 314, 394, 476
Stern, F. 326, 476
Steward, F.C. 64, 65, 197, 199, 465, 472, 476
Stewart, M.C. 60, 203, 231, 468, 476
Stöckman, V.E. 170, 450, 476
Stokes, A.R. 144, 476
Stone, J.E. 280, 378, 379, 381, 476
Stout, H.P. 326, 332, 476
Strasburger, E. 50, 476
Strauss, J. 66, 476
Streminger, J.L. 431, 476
Stuart, A. 101, 464
Stussi, F. 348, 463
Sullivan, J.D. 174, 477
Sundararajan, P.R. 46, 473
Susuki, T. 52, 465
Swanson, A.L. 429, 430, 466, 477

Talmadge, K.W. 406, 415, 416, 423, 466
Tanada, T. 420, 477
Tang, R.C. 183, 468
Tanimato, E. 409, 468
Tarkow, H. 377, 478
Thiele, H. 236, 477
Thomas, M.D. 60, 476
Thomas, R.J. 311, 464, 477
Thompson, E.W. 60, 64, 65, 416, 421, 477
Thompson, J.F. 64, 476
Thompson, W.F. 419, 460
Thornber, J.P. 41, 477
Thornley, M.J. 449, 452, 475
Timell, T.E. 48, 53, 54, 288, 290, 301, 441, 461, 464, 477
Tönnesen, B.A. 171, 462
Tracey, M.V. 2, 477
Traynard, Ph. 289, 477
Tripp, V.W. 59, 389, 477
Tschamler, H. 57, 477
Tsoumis, G. 308, 477

Tupper-Carey, R.M. 59
Turnbull, J.M. 360, 477
Turner, J.E. 416, 459
Turvey, J.R. 271, 272, 471, 477
Tuszon, J. 316, 477

Uhrstrom, I. 425, 460
Utsumi, N. 312, 466

Van Oordt-Hulsof, B. 315, 477
Veen, B.W. 386, 390, 477
Vestal, M.R. 279, 294, 306, 315, 458
Villemez, C.L. 429, 431, 435, 437, 477
Villenueva, J.R. 432, 470
Vintila, E. 360, 477
Virgin, H. 344, 462, 477
Vries, H. de 413, 477
Wadman, W.H. 225, 465
Wagner, 369, 476
Wakashima, T. 299, 464
Wall, J.S. 416, 459
Wang, M.C. 427, 478
Wangaard, F.F. 378, 466
Wardrop, A.B. 167, 170, 173, 203, 277, 289, 290, 296, 298, 299, 300, 302, 303, 306, 307, 308, 311, 312, 314, 315, 316, 317, 323, 324, 347, 349, 381, 385, 386, 387, 389, 392, 394, 395, 425, 439. 443. 460, 461, 465, 473, 475, 478
Warwicker, J.O. 173, 478
Watanabe, M. 136, 465
Watson, A.J. 324, 461
Weatherwax, R.C. 377, 478
Weidinger, A.J. 140, 479
Welch, M.B. 359, 360, 479
Welland, H.J. 140, 479
Wergin, W. 385, 479
Werz, G. 50, 479
Whaley, W.G. 426, 427, 469, 479
White, R.K. 155, 180, 221, 473, 474
Wiener, O. 106, 479
Wiersma, P.K. 416, 479
Wijman, C.F. 227, 474
Williams, W.T. 444, 479
Willison, J.H.M. 448, 464, 479
Wilson, K. 197, 201, 479
Winter, H. 416, 417, 478
Wohlfarth-Botterman, K.E. 444, 466
Wold, M.E. 351, 464
Wooding, F.B.P. 443, 479
Woude, W.J. van der 432, 479
Woychik, J.H. 416, 459
Wright, A.C. 173, 478
Wurscher, R. 413, 460
Wyckoff, R.W.G. 164, 386, 463

Yamamota, R. 409, 468
Yao, J. 369, 479
Young, M. 418, 468

Subject Index

Abies 293, 319, 323
  alba 373
  balsamea 308
  grandis 373
Acetabularia 50, 239, 244, 254
  structure 240 *et seq*
  crenulata 252
Acetobacter xylinum
  cellulose, crystallite size 145
    synthesis 439
    microfibrils 166, 174
Alaria esculenta 230
Alginic acids 230 *et. seq.*
Allium cepa 174
Alnus 301
Apjohnia 192, 202
  chemical constitution 203
  laetevirens 202
Arabinose 37
Ascophyllum alginic acid 232
Avena coleoptile
  growth 384, 386, 416, 419
    and turgor pressure 414
  wall composition 41

Bamboo fibre 98
  microfibril angle 293, 297, 325
Batophora 239, 248, 254,
  structure 240 *et. seq.*
  oerstedi 2, 41, 244, 249
Benzene 20
Betula verrucosa 290
Biosynthesis 425 *et. seq.*
  and auxin 432
  and cell organelles 426, 429
  cellulose 433 *et. seq.*
    biochemistry 433–437
    Golgi vesicles 437, 438
    microtubules 438
    plasmalemma granules 444–456

matrix substances 429
  biochemistry
  Golgi vesicles 431–433
Birefringence
  determination 86 *et. seq.*
  form- 106
  interpretation 105
Boehmeria nivea 131
Bonds
  covalent 17, 19
  energies 16, 17, 23, 24, 26
  hydrogen 25
  ionic 17, 21
Botrydium 224
Breaking strain 329
Breaking stress 329, 347
Brown algae 228 *et. seq.*
  cellulose spacings 230
Bryopis 50, 52, 256, 258, 259
  structure 255 *et. seq.*
  plumosa 261
Bulk modulus 329

Cambium 317, 318
  wall composition 41
Cannabis sativa 293
Caulerpa 50, 257, 262
  structure of 255 *et. seq.*
  prolifera 262, 263
Cedrus 293, 320
  libani 319
Cell shape 7 *et seq.*
Cellobiose 36, 37
Cellulose I
  birefringence 105, 107
  chain length 48
  chemical structure 44 *et seq.*
  conformation 46, 47
  conversion to cellulose II 140
  crystallite size

fresh material 173
line broadening 144 *et seq.*
low angle scattering 146, 147
elementary fibril 168, 174 *et seq.*, 221
folded chains 181–183
micelles 144
microfibrils
    fresh walls 172
    internal structure 169 *et seq.*
    mechanical properties 333
    mutual disposition 186–191
    uniplanar orientation 168, 184,
    185, 197, 230
synthesis 212, 433 *et seq.*
unit cell
    2-chain 134 *et seq.*
    8-chain 139
    variability 229
X-ray derived structure 133 *et seq.*
Cellulose II 143
occurrence in nature 223, 226, 227
unit cell 141
Chaetomorpha 130, 136, 173, 180, 186,
203, 426
    melagonium 41, 105, 209, 212, 442,
    443,
        cellulose crystallite size 145, 147
        chemical constitution 204
        habit 203, 204
        lamellation 205, 208, 209
        swarmers 218
        composition 42
        growth 214–216
    princeps 208, 209, 210
Chitin 1
unit cell 142, 143
Chlamydomonas 228
Chlorella 7
composition 225
Chorda filum 229–231
Chordaria flagelliformis 230
Cladophora 136, 180, 186, 203, 426
    cellulose crystallite size 147
    chemical constitution 204
    habit 203, 204
    lamellation 205
    swarmers 218, 447, 448
    wall composition 42
    gracilis 205, 209, 210
    prolifera 205, 206, 208, 209, 210
    rupestris 205, 209, 210
Cladostephus spongiosus 230
Clay minerals 223
Codium 50, 239, 254
    structure 240 *et seq.*
    fragile 252
Collenchyma cells 312–315
Compensators

de Sénarmont 88
elliptical 93
Coniophora 289
Corchorus capsularis 293
Cosmarium 228
Cotton hair 173
    cellulose crystallite size 145
    mechanical properties 328, 331, 332
    microfibrils 166, 167
Creep 401
Crossed crystal plates 98 *et seq.*
Crossed fibrillar structure 195, 196, 206
Cutleria multifida 230
Cymopolia 239, 248, 254
    structure 240 *et seq.*
    barbata 250

Dasycladus 242, 248, 251
    vermicularis 241, 252, 253
    structure 240 *et seq.*
Dichotomosiphon 51, 255, 256
Derbesia 228, 238, 239
Dictyosphaeria 192
Dictyota dichotoma 230
Dumontia incressata 230

Elachistea 230
Electron 16 *et seq.*
    associated wavelength 148
Elementary fibril — *see* cellulose I
Eremosiphaerea 226
Enteromorpha
    wall composition 42, 225
Enzymes, wall 65
Exclusion principle 18

Fibre
    diagram 131, 133
        genesis 131
    microfibril angle 278 *et seq.*
        and mechanical properties 331
        et seq.
        variation 317–326
    X-ray diffraction 128
Fractionation 40
Fraxinus 301
Freeze-etching 157–159
Fucus serratus
    wall composition 42
    X-ray spacings 230

Galactose 36
Galacturonic acid 37
Gelidium 230
Gigartina stellata 230
Glaucocystis 42, 180, 192, 218
    microfibril size 147, 178, 179
    nostochinearum 219

form 2, 22
  microfibrils 221
  structure 223
Glucose 33 *et seq.*
Glucuronic acid 37
Gonatozygon monotaenium 228
GDP Glucose 434
Gonium 228
Green algae 192 *et seq.*
Griffithsia flosculosa
  wall composition 42
  X-ray spacings 230
Growth
  and auxin 419—421
  and microfibril orientation 385 *et seq.*
  and creep 401—404
  and stress relaxation 404—406
  and turgor pressure 413—415
  and wall proteins 415—419
  and Young's Modulus 400, 401
  bonds involved 406, 416, 418
Guluronic acid 37

Halicoryne 52, 239
  structure 240 *et seq.*
Halicystis 226, 227, 238
  wall composition 227
Halimeda 257
  structure 256 *et seq.*
Helianthus 384
Hemicelluloses 41
  biosynthesis 429—433
  chemistry 53 *et seq.*
  relation to cellulose 170, 171
Himanthalia 230
Holocellulose 41
Hydrodictyon 226, 238
  africanum 226
  composition 226
  reticulatum 226
Hydroxyproline
  in peroxidase 67
  in wall protein 60 *et seq.*

Larix 320, 387
  laricina 290
  leptolepis 293, 319, 371
Laminaria
  alginic acid 232
  cellulose composition 171
  saccharina 230
Lignin 41, 57, 58, 170
  association with cellulose 170, 171

Major extinction direction
  determination 79
  interpretation 95 *et seq.*

Mannan
  Chemistry 52 *et seq.*
  seaweed 340 *et seq.*
  microfibrils 251 *et seq.*
Mannan II 247, 248
Mannose 34
Mannuronic acid 37
Microfibril *see* Cellulose
  angle
    and mechanical properties 331
    et seq.
    optical determination
      in surface view 85
      on sections 95
      in superposed lamellae 99
    values 293
Microscope, electron 148 *et seq.*
  contrast 153, 161
  electron diffraction in 152
  resolving power 149, 150
  specimen preparation 154 *et seq.*
Microscope, Interference 115 *et. seq.*,
  determination of wall composition
  117
  determination of wall thickness 118
Microscope, optical 68 *et seq.*
  resolving power 69, 71 *et seq.*
Microscope, phase 108 *et seq.*
Microscope, polarising 75 *et seq.*
  compensators
    de Senarmont 88
    elliptical 93
  extinction directions 77
  index ellipsoid 97
  major extinction position 79
  path difference 80 *et seq.*
    interpretation 94
  Poincaré sphere 102
  test plate 84
Microsterias rotata 228
  denticulatum 228
Middle lamella 5, 9
Molecular models 27 *et seq.*
Mosaic growth 387
Mougeotia 224
Mutinet growth 388 *et seq.*
  and wall strain 391
  difficulties 393—6
  structural basis 380, 381
Myoinositol 430

Nitella internodal cells 385, 417
  chemical composition 397
  growth 386
    and turgor pressure 414
  microfibril size 225
  spiral growth 412
  structure 397

wall extensibility 397 *et seq.*
Young's modulus 400
Nitella mucronata 225
opaca 225
Neomeris 239, 254
structure 240 *et seq.*
Netrium digitus
Newton's colour scale 83

Ochromonas malhamensis 179
Oedogonium 228
Oocystis 192, 218
form 220
microfibril size 167, 180, 221
plasmallemma granules 451
structure 223
Optics
Airey disc 74
Becke line 87
interference 70, 73
path difference 70, 71

Parenchyma cells 316
mechanical properties 344
Pea seedlings
wall protein 416, 417
Pectic acid 56
Pediastrum 226
Pellia epiphylla 384
Pelvetia canaliculata 42
Penicillus 257
dumetosus 257, 260, 261, 264
structure of 256 *et seq.*
Phloem fibres 315
Phycomyces sporangiophore 385
spiral growth 409—412
structure 387, 380, 389
Picea 301, 309, 322, 323
Abies 290, 373
excelsa 289
sitchensis 277, 334, 371
Pilaiyella littoralis 230
Pine cambium 41
Pinus 309
caribea 323
radiata 300, 323, 340, 363
strobus 310
sylvestris 290, 293, 304, 320, 373
Pits, bordered 303—311
aspiration 310, 311
pores, 369—375
Plasmalemma 7, 429, 444 *et seq.*
Pleurotenium trabecula 228
Polyglucuronic acid 231
structure 232 *et seq.*
polyguluronic acid 231
structure 232 *et seq.*
Polymannuronic acid 231

structure 232 *et seq.*
Polyporus 288
Polysiphonia fastigiata 230
Populus niger 289, 387
Pores, pit membrane 369—375
capillary 375—382
Porosity 267 *et seq.*
Porphyra 271—275
umbilicalis 272
Potato, mechanical properties 344—346
Primary wall 5, 9, 11
extensibility 397—409
growth 383 *et seq.*
microfibril orientation 385 *et seq.*
proteins 64 *et seq.*
Young's modulus 400
Protein 59
structural 59, 415
amino acid sequences 63
and growth 385 *et seq.*
association with oligosaccharides 63, 64
Pseudotsuga 304, 310, 386
menziesii 308, 371
taxifolia 305, 323, 324

Ramie fibres 131
crystallite size 144
X-ray diagram 144
Red algae 230, 237 *et seq.*, 271 *et seq.*
Refractive indices, determination 86 *et seq.*
Rhizoclonium 203, 225
Rhodymenis palmata
cellulose composition 171, 237
wall composition 42
X-ray spacings 230

Scytosiphon lomentarius 230
Secondary wall 5, 11, *see* separate genera
Shear modulus 329, 340
Siphonocladus 192
Sisal 293
structure and mechanical properties 332—335
Slip planes 349—351
Spiral growth 215, 409 *et seq.*
Spirogyra 224
Spongomorpha arcta 225
Stress relaxation 404—406
Sucrose 38
Swarmers 217 *et seq.*
Swelling 353 *et seq.*
single cells 356—359
wood 359 *et seq.*
Sycamore wall composition
cambium 41

phloem 41
xylem 41

Tetramonas laevis 228
Thuja plicata 371
Tracheid
  birefringence 277
  composition 276, 288–291, 292
  determination of m.e.p. 85
  helical thickenings 311
  lamellation 277, 278, 288 *et seq.*
  length variation 317–326
  microfibril angle 278 *et seq.*
    and mechanical properties
      331 *et seq.*
    variation 317–326
  microfibril size 167 *et seq.*
  model structure 299, 301
  reaction wood 301, 302
  warty layer 312
Trentepohlia 224
Tribonema
Tsuga heterophylla 308

UDP glucose 433
Ulva lactuca
  cellulose composition 171
  wall composition 225
Uncertainty principle 17, 18
Unit cell
  determination 126 *et seq.*
  reciprocal 125
  three dimensional 123
  two dimensional 120

Valonia 192, 193 *et seq.*, 302, 417, 444
  cellulose, content 195
    crystallite size 143, 173
    microfibrils 165–167, 175, 179, 197

unit cell 136
lamellation 196
vesicle structure 199–203
wall protein 65
Valonia aplanospores 193, 201, 217
  utricularis 193, 194
  ventricosa 193, 194, 196, 198, 209, 210
  macrophysa 195
van der Waals radii 16
Vaucheria 224
Vessel elements 315–317

Wood *see* individual species, Tracheid,
Fibre
  breaking stress 347, 348
  density, 376, 378
  failure
    tension 347–348
    compression 348–352
  growth stresses 352
  mechanical properties 334 *et seq.*
  porosity 367 *et seq.*
  swelling 353 *et seq.*

Xylan-β-1, 3-linked
  birefringence 258
  chain length 52
  chemical structure 50 *et seq.*
  helical structure 266 *et seq.*
  microfibrils 258 *et seq.*
  occurrence 227, 255 *et seq.*
  X-ray diagram 263–267
    spacings 269
Xylan β-1, 4-linked 54
Xylose 36

Youngs modulus 377, 328, 340, 344,
345, 346